REPRODUCTIVE PHYSIOLOGY IN PLANTS

Research Progress in Botany

REPRODUCTIVE PHYSIOLOGY IN PLANTS

Philip Stewart, PhD

*Head, Multinational Plant Breeding Program; Author;
Member, US Rosaceae Genomics, Genetics and
Breeding Executive Committee; North Central Regional Association
of State Agricultural Experiment Station Directors, U.S.A.*

Sabine Globig

*Associate Professor of Biology, Hazard Community
and Technical College, Kentucky, U.S.A.*

Apple Academic Press

TORONTO   NEW JERSEY

Research Progress in Botany Series

Reproductive Physiology in Plants

© Copyright 2011*
Apple Academic Press Inc.

First Published in the Canada, 2011
Apple Academic Press Inc.
3333 Mistwell Crescent
Oakville, ON L6L 0A2
Tel. : (888) 241-2035
Fax: (866) 222-9549
E-mail: info@appleacademicpress.com
www.appleacademicpress.com

First issued in paperback 2021

ISBN 13: 978-1-77463-242-0 (pbk)
ISBN 13: 978-1-926692-64-7 (hbk)

Philip Stewart, PhD
Sabine Globig

Cover Design: Psqua

Library and Archives Canada Cataloguing in Publication Data
CIP Data on file with the Library and Archives Canada

CONTENTS

ACKNOWLEDGMENTS AND HOW TO CITE

The chapters in this book were previously published in various places and in various formats. By bringing these chapters together in one place, we offer the reader a comprehensive perspective on recent investigations into this important field.

We wish to thank the authors who made their research available for this book, whether by granting permission individually or by releasing their research as open source articles or under a license that permits free use provided that attribution is made. When citing information contained within this book, please do the authors the courtesy of attributing them by name, referring back to theiroriginal articles, using the citations provided at the end of each chapter.

INTRODUCTION

Plant reproduction can be accomplished by sexual or asexual means. Sexual reproduction produces offspring by the fusion of gametes, resulting in offspring genetically different from the parent or parents. Asexual reproduction produces new individuals without the fusion of gametes, genetically identical to the parent plants and each other, except when mutations occur. In seed plants, the offspring can be packaged in a protective seed, which is used as an agent of dispersal.

Understanding plants' reproductive physiology is of utmost importance to human beings since we depend so heavily on plants for food and other aspects of our lives. The most common form of plant reproduction utilized by people is seeds, but a number of asexual methods are utilized which are usually enhancements of natural processes, including cutting, grafting, budding, layering, division, sectioning of rhizomes or roots, stolons, tillers (suckers), and artificial propagation by laboratory tissue cloning. Asexual methods are most often used to propagate cultivars with individual desirable characteristics that do not come true from seed. Fruit tree propagation is frequently performed by budding or grafting desirable cultivars (clones) onto rootstocks that are also clones, propagated by layering.

Since vegetatively propagated plants are clones, they are important tools in plant research. When a clone is grown in various conditions, differences in growth can be ascribes to environmental effects instead of genetic differences. The chapters in this volume cast further light onto this vital field of research.

— **Philip Stewart, PhD**

A Plant Germline-Specific Integrator of Sperm Specification and Cell Cycle Progression

Lynette Brownfield, Said Hafidh, Michael Borg, Anna Sidorova, Toshiyuki Mori and David Twell

ABSTRACT

The unique double fertilisation mechanism in flowering plants depends upon a pair of functional sperm cells. During male gametogenesis, each haploid microspore undergoes an asymmetric division to produce a large, non-germline vegetative cell and a single germ cell that divides once to produce the sperm cell pair. Despite the importance of sperm cells in plant reproduction, relatively little is known about the molecular mechanisms controlling germ cell proliferation and specification. Here, we investigate the role of the Arabidopsis male germline-specific Myb protein DUO POLLEN1, DUO1, as a positive regulator of male germline development. We show that DUO1 is required for correct male germ cell differentiation including the expression of key genes required for fertilisation. DUO1 is also necessary for male germ cell division, and we show that DUO1 is required for the germline expression of the G2/M regulator AtCycB1;1 and that AtCycB1:1 can partially rescue defective germ cell division in duo1. We further show that the male germline-restricted expression of DUO1 depends upon positive promoter elements and not upon a proposed repressor binding site. Thus, DUO1 is a key regulator in the production of functional sperm cells in flowering plants that has a novel integrative role linking gametic cell specification and cell cycle progression.

Introduction

The gametes of flowering plants are formed by discrete haploid gametophyte structures consisting of only a few cells that develop within the diploid reproductive floral organs. During spermatogenesis, each single haploid microspore divides asymmetrically to produce a larger vegetative cell that eventually gives rise to the pollen tube and a smaller germ, or generative, cell (reviewed in [1],[2]). In contrast to germline cells in metozoans [3], angiosperm male germ cells do not undergo regenerative stem cell divisions, but divide once to form a pair of sperm cells. These sperm cells are delivered to the embryo sac via the pollen tube, where they fuse with egg and central cells to produce embryo and endosperm respectively. This process of double fertilization depends upon two functional sperm cells and is considered one of the major advances in the evolutionary success of flowering plants. Despite this importance, the molecular mechanisms underlying many component processes, including the production of both male and female gametes, remain largely unknown.

Recent transcriptomic analysis of isolated Arabidopsis sperm cells shows that sperm cells express a distinct and diverse set of genes [4] and there is evidence for extensive male germ cell gene expression in maize and lily [5],[6]. Several male germline-specific genes have been characterized in Arabidopsis including AtMGH3, encoding a histone H3.3 variant [7],[8], AtGEX2, encoding a putative membrane associated protein [9], and AtGCS1 (HAP2), encoding a sperm cell surface protein required for fertilisation [10],[11]. Homologues of AtGCS1 are found in many genera [5],[12],[13] that include the green alga Chlamydomonas and the rat malarial parasite Plasmodium berghei, where they are required for gamete interactions and membrane fusion [13]. Although gene expression in angiosperm sperm cells is extensive and essential for gamete functions little is known about its regulation. A transcriptional derepression mechanism, in which expression of male germline expressed genes is repressed in all non-germline cells by a protein called Germline Restrictive Silencing Factor (GRSF), has recently been proposed [14]. A binding site for the GRSF protein was identified in the promoter region of the Lily male germline gene LGC1, and mutations in this sequence led to the ectopic activation of the LGC1 promoter in non-germline cells in lily and Arabidopsis. Although similar binding sites have been found in the promoter regions of several germline genes in Arabidopsis, including the germline-specific transcription factor gene DUO1 [14], a functional role for these sites or of GRSF activity in regulating gene expression in Arabidopsis pollen has not been shown.

Germ cell division resulting in the sperm cell pair in each pollen grain, is essential for double fertilization and recent data supports the capacity of both sperm cells to fertilize the egg cell in Arabidopsis [15]. Several mutants have been described in which germ cell division is disrupted [16]–[18]. Mutations in the conserved cell cycle regulator CDKA1 [16],[17] and in the F-BOX protein FBL17 [18] prevent germ cell division and result in mature pollen with a single germ cell. Defects in Chromosome Assembly Factor 1 (CAF1) can also disrupt germ cell division [19]. Interestingly, the single germ cells in these mutants are capable of fertilization, with cdka1 and fbl17 mutant germ cells fertilizing the egg cell to produce an embryo that aborts early in development due to the lack of endosperm production. These mutations clearly demonstrate that germ cell division and specification can be uncoupled, but do not identify how these processes may be coordinated to produce twin sperm cells competent for double fertilization.

DUO POLLEN1 (DUO1) is a unique male germ cell-specific R2R3 Myb protein that is also required for germ cell division in Arabidopsis [20]. Unlike cdka1 and fbl17 single germ cells, duo1 germ cells do not lead to successful fertilization, suggesting that in addition to germ cell cycle defects, key features of gamete differentiation and function are impaired in duo1. Here we further characterize DUO1 as an essential, positive regulator of sperm cell production in plants. We use various molecular markers and ectopic expression assays to show that DUO1 is both necessary and sufficient for the expression of male germline genes. We show that DUO1 is required for the expression of the Arabidopsis G2/M regulator CyclinB1;1 (AtCycB1;1) in the male germline and that AtCycB1:1 can partially rescue defective germ cell division in duo1. Our findings reveal a novel integrative role for the germline-specific DUO1 protein, in cell specification and cell cycle progression necessary for twin sperm cell production. Furthermore, we show that restriction of DUO1 expression to the male germline is not dependent on a putative GRSF binding site but involves positive elements in the promoter.

Results/Discussion

DUO1 Is a Key Regulator of Sperm Cell Specification

To investigate the potential role of DUO1 in regulating sperm specification we examined the expression of three male germline markers, AtMGH3, AtGEX2 and AtGCS1, in mutant duo1 pollen. We exploited marker lines with promoter regions of these germline genes linked to GFP. First we characterised the expression of these markers in a coordinated manner using confocal laser scanning microscopy (CLSM) throughout development of wild-type pollen (Figure 1A–C), and compared their profiles with the expression of a DUO1:mRFP fusion protein under control of the DUO1 promoter (DUO1-DUO1::mRFP; Figure 1D). The expression of all three germ cell markers is undetectable in free microspores when DUO1 is not expressed (Figure 1, Panel 1). Fluorescence is first detected in the germ cell during or soon after engulfment by the vegetative cell, appearing at a similar time to the expression of DUO1 (Figure 1, Panel 2). As the pollen matures the level of GFP accumulates in germ cells before mitosis and remains high in mature sperm cells (Figure 1A–C, Panels 3–5). The accumulation of GFP in progressive stages is illustrated by the reduced autofluorescence signal arising from the pollen wall, reflecting the reduced exposure needed to capture a relatively unsaturated germ cell GFP signal. DUO1 expression persists during pollen development, although its abundance does not obviously increase in tricellular and mature pollen (Figure 1D). Our analysis shows that in common with AtMGH3 and AtGEX2, the expression of AtGCS1, previously thought to be sperm cell-specific in Arabidopsis [11], is detected in germ cells soon after asymmetric division (Figure 1C). The expression of all male germ cell markers shortly after the asymmetric division shows that sperm cell specification begins early after inception of the germline prior to passage of germ cells through mitosis.

The three male germline markers were introduced into heterozygous duo1 plants that produce 50% wild type pollen and 50% mutant pollen, and GFP expression was scored. Virtually all the wild type pollen showed GFP fluorescence in twin sperm cells while there was no fluorescence, or rarely a weak GFP signal, in the single germ cell in duo1 pollen (Figure 1E–G, I–K). When these markers were introduced into the cdka;1 mutant in which the arrested germ cell is able to fertilize the egg cell, fluorescence was observed in the single germ cells in mutant pollen (Figure 1E–G). This result confirms that germ cell division and cell fate specification are uncoupled in cdka;1 mutant pollen, similar to the observed expression of germ cell markers in arrested but functional germ cells in CAF1 mutants [19]. The absence of GFP in mutant duo1 germ cells demonstrates that DUO1 is necessary for the expression of several germline-expressed genes, and explains why duo1 pollen is infertile (it lacks proteins including AtGCS1 that are essential for fertilization). In contrast, when the DUO1 promoter was used to express a nuclear-targeted histone H2B::mRFP marker protein, fluorescence was detected in mutant duo1 germ cells, similar to its expression in wild type sperm cells and in cdka;1 germ cells (Figure 1H,L), indicating that DUO1 promoter activation does not depend upon DUO1 itself.

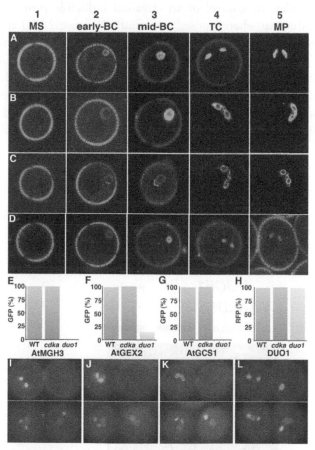

Figure 1. Expression of male germline-specific genes in wild type and duo1 pollen.
Expression of AtMGH3-H2B::GFP (A), AtGEX2-GFP (B), AtGCS1-AtGCS1::GFP (C) and DUO1-DUO1::mRFP (D) during wild type pollen development, observed with CLSM. Panels are numbered 1 (left) to 5 (right). For all markers, fluorescence is not detected in microspores (MS; Panel 1), a weak signal is detected in the germ cell during or soon after engulfment (early-BC; Panel 2), fluorescence increases in mid-bicellular pollen (mid-BC; Panel 3) and remains in tricellular (TC; Panel 4) and mature pollen (MP; Panel 5). (E–L) Expression of germline expressed genes in heterozygous duo1 plants. The percentage pollen showing GFP or RFP in sperm cells of wild type (WT) pollen or the single germ cell in cdka;1 and duo1 mutant pollen in plants homozygous for AtMGH3-H2B::GFP (AtMGH3, E), AtGEX2-GFP (AtGEX2, F), AtGCS1-AtGCS1::GFP (AtGCS1, G) and DUO1-H2B::mRFP (DUO1, H). Individual examples viewed by fluorescence microscopy in I to L. AtMGH3-H2B::GFP (I), AtGEX2-GFP (J) and AtGCS1-AtGCS1::GFP (K) are not expressed, or have reduced expression in duo1 pollen while DUO1-H2B::RFP (L) is expressed. Each image has a wild type pollen grain to the left and a duo1 mutant grain to the right (see lower DAPI images).

To independently confirm the regulation of germline genes by DUO1 we ectopically expressed DUO1 in seedlings, and in pollen vegetative cells, where At-MGH3, AtGEX2 and AtGCS1 are not normally expressed. As DUO1 contains a recognition site for microRNA159 we used a resistant DUO1 cDNA (mDUO1) with an altered nucleotide sequence at the miR159 binding site, but encoding the native amino acid sequence [21]. Transgenic seedlings in which the mDUO1 cDNA was placed under the control of an estradiol inducible promoter [22] showed mDUO1 induction when exposed to estradiol (Figure 2A). Expression of the male germline genes, AtMGH3, AtGEX2 and AtGCS1, was also induced, with high levels of transcripts present only in plants exposed to estradiol and containing mDUO1 (Figure 2A). Similarly, when a DUO1::mRFP fusion was ectopically expressed in pollen vegetative cells using the LAT52 promoter [23], we observed ectopic expression of the AtMGH3 marker in vegetative cell nuclei (Figure 2B,C). Thus ectopic expression of DUO1 is sufficient for activation of germ cell-specific gene expression in a range of non-germline cells.

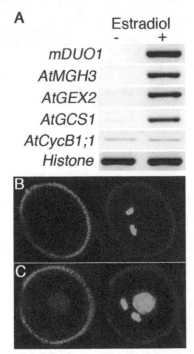

Figure 2. Ectopic expression of DUO1 results in expression of male germline specific genes.
(A) RT-PCR analysis of mDUO1, AtMGH3, AtGEX2, AtGCS1 and AtCycB1;1 expression in whole seedlings transformed with the mDUO1 cDNA (see methods) under the control of an estradiol inducible promoter grown on media without estradiol (–) or with estradiol (+). Histone H3 was used as a control. (B, C) Mature pollen grains showing AtMGH3-H2B::GFP expression specifically in sperm cells in the absence of LAT52-DUO1::mRFP (B), or in both the vegetative cell nucleus and sperm cells in the presence of LAT52-DUO1::mRFP (C). Left and right panels correspond to RFP and GFP signals viewed by CLSM.

DUO1 Is Required for AtCycB1;1 Expression in the Male Germline

The phenotype of duo1 shows that in addition to the activation of male germline genes, DUO1 is required for germ cell division. Mutant duo1 germ cells complete DNA synthesis (S) phase but fail to enter mitosis (M) [20],[24], suggesting that DUO1 may regulate the expression of essential G2/M factors. As the Arabidopsis CDK regulatory subunit AtCycB1;1 shows enhanced expression at G2/M [25],[26] and is expressed in developing pollen, we investigated AtCycB1;1 as a potential downstream target of DUO1. To monitor the expression of AtCycB1;1 we used the pCDG marker which contains the AtCycB1;1 promoter region and mitotic destruction box fused to the β-glucuronidase (GUS) reporter [25]. First we analysed the marker in wild type pollen (Figure 3A–F). Individual pollen grains at different stages of development (as determined by DAPI staining) were analysed for GUS activity, which results in the formation of indigo microcrystals. Microspores and bicellular pollen shortly after mitosis contain numerous indigo crystals, with the number peaking close to mitosis (Figure 3A–C), indicating that expression of AtCycB1;1 is linked to asymmetric division. Expression is then abolished in bicellular pollen (Figure 3D). Close to germ cell mitosis, single indigo crystals are present specifically in germ cells (located by DAPI staining; Figure 3E) indicating expression of AtCycB1;1 in the germ cell before division. The protein is degraded after mitosis and is absent in tricellular pollen (Figure 3F).

We then counted the number of pollen grains with GUS staining at different stages of development in wild type and heterozygous duo1 plants. In both wild type and heterozygous duo1 plants, polarized microspores and vegetative cells shortly after asymmetric division showed almost 100% staining, indicating expression of AtCycB1:1 (Figure 3G). Thereafter vegetative cell staining declined and was absent from late-bicellular stage pollen (Figure 3G). Germ cell staining was subsequently observed in ~100% of pollen from wild type plants close to mitosis, but was reduced by approximately half in heterozygous duo1 plants at this stage (Figure 3H). As half of the pollen population is mutant in heterozygous duo1 plants, and wild type pollen show GUS staining, this reduction in staining is consistent with a lack of AtCycB1;1 expression in mutant duo1 pollen. This indicates that DUO1 is required for the expression of AtCycB1;1 in male germ cells.

We then analysed the expression of AtCycB1;1 transcripts in seedlings after steroid induction of mDUO1. In contrast to the germline markers, AtCycB1;1 was expressed at a low level in seedlings not exposed to estradiol and the presence of DUO1 did not affect the level of AtCycB1;1 transcripts (Figure 2A). Thus, although DUO1 is required for germline expression of AtCycB1;1 the presence of DUO1 is not sufficient to induce AtCycB1;1 mRNA in seedlings. Transcription of the AtCycB1;1 gene is known to be regulated by a number of factors, including activators such as three repeat [27] or other Myb proteins [28] and TCP20 [29] and repressors such as TOUSLED [30]. Thus, DUO1 may be unable to overcome these controls in seedlings, and may affect AtCycB1;1 transcription in the male germline through an indirect mechanism or through effects on AtCycB1;1 protein stability.

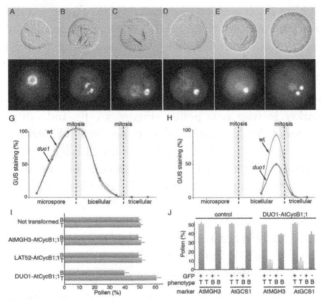

Figure 3. AtCycB1;1 expression in developing pollen.
(A–F), pCDG-dependent GUS staining (upper panel) and DAPI staining (lower panel) in isolated spores: (A, B), unicellular microspores, (C, D, E), early, mid-and late bicellular pollen and (F), tricellular pollen. (G, H) The frequency of pCDG-dependent GUS staining in microspores and vegetative cells close to mitosis is similar in duo1 heterozygotes and wild type plants (G), whereas GUS staining in germ cells, is reduced by approximately half in duo1 heterozygotes, where 50% of the pollen is WT and the other 50% mutant (H). The stage of pollen development is indicated below each graph and the approximate time of mitosis is indicated by grey squares with a dashed line. (I) DUO1-AtCycB1;1 is able to partially complement the bicellular phenotype of duo1 pollen. The amount of tricellular pollen (T) increases and the amount of bicellular pollen (B) decreases when heterozygous duo1 plants are transformed with DUO1-AtCycB1;1 (n = 31 T1 lines) compared with plants either not transformed (n = 3 individuals) or transformed with control constructs AtMGH3-AtCycB1;1::GFP (n = 17 T1 lines) or LAT52-AtCycB1;1 (n = 17 T1 lines). Bars represent the average percentage of pollen with error bars showing standard deviation. (J) Germline markers are not activated in the complemented tricellular pollen. In non-complemented plants ~50% of the pollen is tricellular (T) with marker expression and ~50% is bicellular (B) without marker expression. When the bicellular phenotype is partially complemented by DUO1-AtCycB1;1, ~10% of pollen is tricellular without marker expression, while there is a decrease in the amount of bicellular pollen. Bars represent the average percentage of pollen from 3–6 individual plants with the error bars showing standard deviation.

To investigate the role of AtCycB1;1 in the failure of duo1 male germ cells to enter mitosis we determined whether AtCycB1;1 is sufficient to rescue the germ cell mitosis defect in duo1 pollen. We used the DUO1 promoter to drive AtCycB1;1 expression in the male germline. The proportion of bicellular or tricellular pollen grains from heterozygous duo1 plants either not transformed or transformed with either of two control constructs (MGH3-AtCycB1;1::GFP, which is not expressed in mutant pollen, and LAT52-AtCycB1;1, which is expressed only in the vegetative cell) did not vary significantly from 50% (Chi2 p<0.05) (Figure 3I). In contrast, in heterozygous duo1 plants transformed with DUO1-AtCycB1:1 the majority of lines (31/49) showed a significantly reduced frequency of bicellular pollen and a corresponding increase in tricellular pollen (Figure 3I). This suggests that restoring AtCycB1;1 in duo1 mutant germ cells is sufficient to promote mitosis in a proportion of the population. Complementation was however incompletely penetrant, which may result from the use of the DUO1 promoter that may not produce native amounts of AtCycB1;1. It is also possible that other factors with a role in G2/M transition, such as other AtCycB family members that are also expressed during pollen development [31], may also be absent in duo1 pollen.

To determine if the presence of DUO1-AtCycB1;1 in duo1 pollen restored only the ability to proceed through mitosis or germline specification as well, we analysed expression of the AtMGH3 and AtGCS1 markers in duo1 plants showing partial complementation (Figure 3J). In contrast to plants without DUO1-AtCycB1;1 where almost all tricellular pollen expresses GFP, plants displaying partial complementation produce ~10% of pollen that is tricellular but does not express the markers. As there is also a ~10% decrease in bicellular pollen, this new class of tricellular pollen is most likely duo1 pollen in which the division defect has been complemented by the DUO1-AtCycB1;1 construct, but in which the markers have not been activated. Consistent with this, DUO1-AtCycB1;1 complemented duo1 pollen showed no male transmission. Thus, complementation of the bicellular phenotype by AtCycB1;1 only affects cell division and does not restore expression of germline gene expression and sperm cell function.

DUO1 Expression Is Restricted to the Male Germline Independent of a Putative GRSF Binding Site

Closer examination of mature pollen grains ectopically expressing DUO1 in the vegetative cell revealed a distinctive morphology with reduced cytoplasmic density, larger vacuoles and numerous large cytoplasmic inclusions. This phenotype was only found in pollen containing vegetative nucleus GFP and analysis of pollen viability revealed up to 50% non-viable pollen with the aberrant pollen not being viable. Similar phenotypes are not seen in pollen of plants transformed with LAT52-H2B::GFP where the transgene is transmitted normally (data not shown). Furthermore, Arabidopsis plants constitutively expressing DUO1 (driven by the 35S promoter) show severe seedling patterning defects, twisted and curled leaves and floral defects [21]. These phenotypes demonstrate the importance of restricting high level expression of DUO1 to male germ cells.

Such restriction may partially rely upon degradation of DUO1 mRNA by microRNA159 [21] in certain cell types but promoter elements are also likely to be important. As such, restriction of DUO1 expression to the male germline has been proposed to rely on the repressor protein GRSF due to a putative GRSF binding site in the DUO1 promoter [14]. Mutagenesis of similar sequences in the LGC1 promoter led to ectopic activation of the LGC1 promoter in non-germ line cells in tobacco and Arabidopsis [14]. However, when we specifically mutated the putative GRSF binding site in the DUO1 promoter this did not affect the germline-specific expression of DUO1 (Figure 4A–D). Moreover, sequences in the 150 bp proximal DUO1 promoter, excluding putative GRSF binding sites, were sufficient for germline-specific expression (Figure 4E). Although factors that bind to the lily LGC1 silencer appear to be present in non-germline cells in Arabidopsis [14], the germline-restricted activation of DUO1 does not appear to involve GRSF mediated repression. Since the DUO1 promoter appears to be active only after asymmetric division in the newly formed germ cell and that activation does not depend upon DUO1 itself (see Figure 1), activation of the DUO1 promoter may depend on proximal region-binding transcription factors that are inherited and/or segregated during asymmetric division of the microspore.

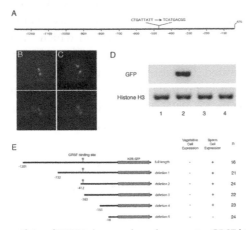

Figure 4. Male germline specificity of DUO1 does not depend on putative GRSF binding sites.
(A) Schematic of the DUO1 promoter region illustrating the mutagenized putative GRSF binding site. (B,C) Expression of H2B::GFP in pollen driven by the native (B) or mutagenized DUO1 (C) promoters. Top panels show GFP signal, lower panels show DAPI staining. (D) RT-PCR analysis of native and mutagenized DUO1 promoter activity in seedlings. PCR was conducted on cDNA from wild type plants (1), control plants transformed with a constituitive HistoneH3 promoter-H2B::GFP fusion (2), and plants transformed with the native (3), or mutagenized (4), DUO1 promoters driving H2B::GFP expression. The primers used were specific for GFP (upper panel) or native Histone H3 transcripts (lower panel). The native or mutagenized DUO1 promoters showed no sporophytic expression of GFP transcripts. (E) Schematic representation of the of the DUO1 promoter 5′ deletion series used to drive expression of H2B::GFP. The first four deletions, including deletion 3 in which the putative GRSF binding site is removed, showed a similar expression pattern to that of the full-length DUO1 promoter, with GFP signal only observed in sperm cell nuclei. The same expression pattern was observed in all independent lines examined (n). GFP expression was not observed in any transformants harbouring the shortest promoter fragment (deletion 5).

Conclusions

We have shown that DUO1 is both necessary and sufficient for the expression of several male germline genes including AtGCS1 that is required for gamete fusion [13], thus DUO1 has a major role in the specification of functional male gametes. DUO1 is not involved in regulating microspore division and is first expressed in germ cells after asymmetric division. DUO1 is also required for the entry of male germ cells into mitosis and for the germline expression of the G2/M regulator AtCycB1;1. Thus, the germ cell programme under DUO1 control has an important role in regulating core cell cycle machinery specifically in the male germline. The discovery of the dual role of DUO1 as a positive regulator in male germline specification and cell cycle progression is a major advance in uncovering the molecular mechanisms involved in plant sexual reproduction. DUO1 is currently the only regulatory factor that has been shown to be required for gamete specification in plants. Recently we described an independent mechanism for male germ cell cycle regulation where the F-BOX protein FBL17 controls germ cell entry into S-phase via the degradation of the CDKA inhibitors KRP6 and 7 [4]. Taken together these data establish a molecular framework for twin sperm cell production in flowering plants (Figure 5).

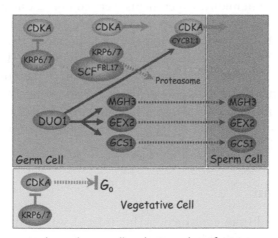

Figure 5. Regulatory events in plant male germ cell production and specification.
Model integrating the role of DUO1 and SCFFBL17 [18] in plant germ cell production and specification. The germline-specific DUO1 protein (blue) activates the expression of several germline specific proteins (red). In parallel, the CDKA inhibitors KRP6 and KRP7 (green) are expressed in the vegetative cell and germ cell after asymmetric division, where they inhibit CDKA activity and S phase progression. The F-box protein FBL17 is then transiently expressed in the germline and forms an SCFFBL17 complex (blue) that targets KRP6/7 for proteasome dependent proteolysis, licensing S-phase progression (green arrow). Further germ cell cycle progression is controlled by the DUO1-dependent G2/M phase expression of the CDKA regulatory subunit AtCYCB1;1 (red). Thus, while SCFFBL17 and DUO1 promote male germ cell proliferation at successive stages of the cell cycle, DUO1 integrates germ cell specification and division to ensure the production of functional twin sperm cells that are essential for double fertilization. Arrows indicate a requirement for the protein rather than direct binding.

Further analysis will shed light on how DUO1 activates its targets, and how DUO1 itself is activated specifically in the male germline. The identification of the role of DUO1 in germ cell specification also provides an exciting platform to develop a detailed regulatory network for male gametogenesis and for comparative studies of the control of sperm cell production. DUO1 homologs are found throughout the land plants from the non-flowering plants Selaginella moellendorffii and Physcomitrella patens (moss) through to the monocots and dicots. Exploring the functional conservation of DUO1 in different species will reveal if DUO1 has a conserved role in male gamete production, in terms of both of germline mitosis and specification, where DUO1 may regulate the expression of a similar suite of genes such as the conserved GCS1 protein. Such studies may shed light on the evolution of regulatory mechanisms in plant germline development and their significance in double fertilization in flowering plants.

Materials and Methods

Plant Material and Transformation

Arabidopsis plants were grown at 21°C with a 16 h-light and 8 h-dark cycle or with 24 h light, with variable humidity. Experiments were conducted in the duo1-1 (in No-0) or the No-0 backgrounds, except for those involving the inducible ectopic expression of mDUO1 and analysis of the DUO1 promoter that were conducted in Col-0. The AtGCS1-AtGCS1::GFP, AtGEX2-GFP and CDG marker lines are also in Col-0. Plants were transformed with Agrobacterium tumefaciens (GV3101) using a standard floral dipping method. Transformants were selected either on Murashige and Skoog (MS) agar containing 50 µg/ml kanamycin or 20 µg/ml hygromycin or on soil with 30 µg/ml BASTA (glufosinate ammonium, DHAI PROCIDA) fed by sub-irrigation.

Vector Construction

Gateway single and multi-site construction (Invitrogen) was used to generate most vectors. DNA was amplified from genomic DNA, cDNA or plasmid DNA by PCR with high fidelity Phusion DNA polymerase (Finnzymes) and primers with suitable attachment site (attB) adapters. Full-length attB sites were added to each fragment in a second high fidelity PCR. For site-directed mutagenesis of the putative GRSF binding site in the DUO1 promoter a two-step recombinant PCR approach was taken. Two overlapping PCR fragments were generated containing the mutated sequence and the two fragments joined in a stitching PCR. PCR fragments were cloned into pDONR vectors (Invitrogen; pDONR207 for

AtCycB1;1 cDNA or pDONR221 for H2B and DUO1 and mDUO1 cDNA, pDONRP4P1R for promoter regions and pDONRP2RP3 for GFP and RFP) via a BP reaction using BP Clonase II (Invitrogen). The product of BP reactions was transformed into alpha-select chemically competent cells (Bioline) and all clones were verified by sequencing.

A multipart LR reaction using LR Clonase plus (Invitrogen) and the destination vector pK7m34GW [32] was used to generate the AtMGH3 marker, AtMGH3-H2B::GFP. This contains the region upstream of the AtMGH3 coding region [7] driving expression of a H2B::GFP fusion protein, with the H2B used to give a nuclear GFP signal. The GCS1-GCS1::GFP marker was constructed by inserting a PCR fragment of GFP into an AflII site in the 16th exon of a Arabidopsis GCS1 genomic DNA fragment in the previously described binary vector [10]. The AtGEX2-GFP marker, with the GEX2 promoter region driving GFP expression was kindly provided by Shelia McCormick [9].

The vectors DUO1-DUO1::mRFP, DUO1-H2B::mRFP and LAT52-DUO1::mRFP were also generated using gateway multisite cloning and the vectors pK7m34GW or pB7m34GW [32]. DUO1-DUO1::mRFP uses the DUO1 promoter region to drive expression of a DUO1::mRFP fusion (used to follow the DUO1 protein during pollen development) while DUO1-H2B::GFP uses the DUO1 promoter to produce a H2B::mRFP fusion protein (used to follow the activity of the DUO1 promoter in duo1 pollen). The LAT52 promoter is active in the vegetative cell [23] so was used to ectopically express DUO1::mRFP in the vegetative cell. Vectors to analyse the DUO1 promoter region were also constructed using gateway multisite cloning.

The DUO1 mRNA contains a functional recognition site for the microRNA miR159 [21], so for inducible expression of DUO1 a miR159 resistant version of the DUO1 cDNA was used containing silent mutations in the miR159 binding site. This was cloned in the vector pMDC7 [33] that contains the XVE estradiol inducible promoter system [22], using a single part LR reaction and LR Clonase II (Invitrogen).

For experiments examining the ability of AtCycB1;1 to complement the duo1 division phenotype the vectors pB2GW7 and pH2GW7 [34] were modified to contain the DUO1 and LAT52 promoters respectively. The 1.2 kb DUO1 and 609 bp LAT52 promoter fragments were amplified from cloned sequences using restriction tagged oligonucleotide primer pairs. A single part gateway reaction was then used to clone AtCycB1;1 into the vectors creating DUO1-CycB1;1 and LAT52-CycB1;1. MGH3-CycB1;1::GFP was generated using a multipart gateway reaction.

Inducible Expression of DUO1

T2 seed from transgenic Col-O plants were grown in MSO plates containing 20 μg/ml hygromycin in standard conditions for 12 days. 25 seedlings were transferred to either control plants containing 0.002% v/v DMSO or induction plates contain 2 μM 17β-estradiol dissolved in DMSO. Plants were returned to the growth room for a further 24 h, before being snap frozen in liquid nitrogen.

RT-PCR Analysis

Pollen from ecotype Landsberg erecta at different stages of development was isolated and RNA extracted as described [31]. For RT-PCR on seedling ectopically expressing mDUO1 and for DUO1 promoter analysis, RNA was extracted from frozen samples using the Qiagen RNeasy Kit. Samples of 750 ng or 1 μg of total RNA for pollen stages and seedlings, respectively, were reverse transcribed in a 20 μl reaction using Superscript II RNase H reverse transcriptase (Invitrogen) and an oligodT primer as per the manufactures instructions. For PCR amplification 1 μl of a 10× (pollen stages) or 5× (seedling) diluted cDNA was used in a 25 μl reaction using Biotaq DNA polymerase (Bioline) and 12.5 pmol of each primer. PCR conditions were: 96°C for 1 min, 30 to 40 cycles at 96°C for 30 s, 55°C for 30 s, 72°C for 40 s followed by 5 min at 72°C. Histone H3 (At4g40040) was used as a control.

Analysis and Imaging of Pollen

Mature pollen was stained with DAPI (4'-6-Diamidino-2-phenylindole) as described previously [35]. Staining for GUS activity was performed as described [36] with inflorescences incubated in GUS buffer (100 mM sodium-phosphate, pH 7; 5 mM EDTA, 0.1% Triton X-100) with 1 mM X-gluc (5-bromo-4-chloro-3-indolyl b-D-glucuronide) and 0.5 mM $K_3Fe[CN]_6$, at 37°C for 1–3 days. Stained inflorescences were then cleared with 70% ethanol. Pollen was dissected out and stained with 0.8 μg/ml DAPI in GUS buffer. Phenotypic analysis of pollen was conducted on a Nikon TE2000-E inverted microscope (Nikon, Japan). Bright field and DIC images were captured with a Nikon-D100 camera (Model MH-18, Japan) and fluorescence images were captured with HAMAMATSU – ORCA-ER digital camera (Model C4742-95, Japan) using Openlab software version 5.0.2. (Improvision).

For confocal laser scanning microscopy (CLSM) pollen from buds at different stages of development was teased out of the anther with a needle and mounted in 0.3 M mannitol and mature pollen was released directly into 0.3 M mannitol.

Pollen was viewed with a Nikon TE2000-E inverted microscope and C1 confocal system using Melles Griot Argon Ion (emission 488 nm) and Melles Griot Helium-Neon (emission 543 nm) lasers, detection filters for GFP and RFP, and EZ-C1 control and imaging software.

Acknowledgements

We thank Sheila McCormick and Peter Doerner for GEX2-GFP and pCDG marker lines, Javier Palatnik for mDUO1 cDNA and Nam-Hai Chua for the estradiol inducible promoter. We acknowledge Prof. Tsuneyoshi Kuroiwa for supporting TM in his work with GCS1::GFP.

Authors' Contributions

Conceived and designed the experiments: LB SH MB DT. Performed the experiments: LB SH MB AS. Analyzed the data: LB SH MB DT. Contributed reagents/materials/analysis tools: TM DT. Wrote the paper: LB DT.

References

1. McCormick S (2004) Control of male gametophyte development. Plant Cell 16: SupplS142–53.

2. Honys D, Oh SA, Twell D (2006) Pollen development, a genetic and transcriptomic view. In: Malho R, editor. The Pollen Tube: A Cellular and Molecular Perspective. 3: 15–45.

3. Strome S, Lehmann R (2007) Germ versus soma decisions: lessons from flies and worms. Science 316: 392–393.

4. Borges F, Gomes G, Gardner R, Moreno N, McCormick S, et al. (2008) Comparative Transcriptomics of Arabidopsis Sperm Cells. Plant Physiol 148: 1168–1181.

5. Engel ML, Chaboud A, Dumas C, McCormick S (2003) Sperm cells of Zea mays have a complex complement of mRNAs. Plant J 34: 697–707.

6. Okada T, Bhalla PL, Singh MB (2006) Expressed sequence tag analysis of Lilium longiflorum generative cells. Plant Cell Physiol 47: 698–705.

7. Okada T, Endo M, Singh MB, Bhalla PL (2005) Analysis of the histone H3 gene family in Arabidopsis and identification of the male-gamete-specific variant AtMGH3. Plant J 44: 557–568.

8. Ingouff M, Hamamura Y, Gourgues M, Higashiyama T, Berger F (2007) Distinct dynamics of HISTONE3 variants between the two fertilization products in plants. Curr Biol 17: 1032–1037.

9. Engel ML, Holmes-Davis R, McCormick S (2005) Green Sperm. Identification of male gamete promoters in Arabidopsis. Plant Physiol 138: 2124–2133.

10. Mori T, Kuroiwa H, Higashiyama T, Kuroiwa T (2006) GENERATIVE CELL SPECIFIC 1 is essential for angiosperm fertilization. Nat Cell Biol 8: 64–71.

11. von Besser K, Frank AC, Johnson MA, Preuss D (2006) Arabidopsis HAP2 (GCS1) is a sperm-specific gene required for pollen tube guidance and fertilization. Development 133: 4761–4769.

12. Hirai M, Arai M, Mori T, Miyagishima S, Kawai S, et al. (2008) Male fertility of malaria parasite is determined by GCS1, a plant-type reproduction factor. Curr Biol 18: 607–613.

13. Liu Y, Tewari R, Ning J, Blagborough AM, Garbom S, et al. (2008) The conserved plant sterility gene HAP2 functions after attachment of fusogenic membranes in Chlamydomonas and Plasmodium gametes. Genes Dev 22: 1051–1068.

14. Haerizadeh F, Singh MB, Bhalla PL (2006) Transcriptional repression distinguishes somatic from germ cell lineages in a plant. Science 313: 496–499.

15. Ingouff M, Sakata T, Li J, Sprunck S, Dresselhaus T, et al. (2009) The two male gametes share equal ability to fertilize the egg cell in Arabidopsis thaliana. Curr Biol 19: R19–R20.

16. Iwakawa H, Shinmyo A, Sekine M (2006) Arabidopsis CDKA;1, a cdc2 homologue, controls proliferation of generative cells in male gametogenesis. Plant J 45: 819–831.

17. Nowack MK, Grini PE, Jakoby MJ, Lafos M, Koncz C, et al. (2006) A positive signal from the fertilization of the egg cell sets off endosperm proliferation in angiosperm embryogenesis. Nat Genet 38: 63–67.

18. Kim HJ, Oh SA, Brownfield L, Hong SH, Ryu H, et al. (2008) Control of plant germline proliferation by SCFFBL17 degradation of cell cycle inhibitors. Nature 455: 1134–1137.

19. Chen Z, Tan JLH, Ingouff M, Sundaresan V, Berger F (2008) Chromatin assembly factor 1 regulates the cell cycle but not cell fate during male gametogenesis in Arabidopsis thaliana. Development 135: 65–73.

20. Rotman N, Durbarry A, Wardle A Yang WC, Chaboud A, et al. (2005) A novel class of MYB factors controls sperm-cell formation in plants. Curr Biol 15: 244–248.

21. Palatnik JF, Wollmann H, Schommer C, Schwab R, Boisbouvier J, et al. (2007) Sequence and expression differences underlie functional specialization of Arabidopsis microRNAs miR159 and miR319. Dev Cell 13: 115–125.

22. Zuo J, Niu QW, Chua NH (2000) An estrogen receptor-based transactivator XVE mediates highly inducible gene expression in transgenic plants. Plant J 24: 265–273.

23. Twell D, Wing R, Yamaguchi J, McCormick S (1989) Isolation and expression of an anther-specific gene from tomato. Mol Gen Genet 217: 240–245.

24. Durbarry A, Vizir I, Twell D (2005) Male germ line development in Arabidopsis duo pollen mutants reveal gametophytic regulators of generative cell cycle progression. Plant Physiol 137: 297–307.

25. Colon-Carmona A, You R, Haimovitch-Gal T, Doerner P (1999) Technical advance: spatio-temporal analysis of mitotic activity with a labile cyclin-GUS fusion protein. Plant J 20: 503–508.

26. Menges M, Murray JAH (2002) Synchronous Arabidopsis suspension cultures for analysis of cell-cycle gene activity. Plant J 30: 203–212.

27. Ito M, Araki S, Matsunaga S, Itoh T, Nishihama R, et al. (2001) G2/M-phase-specific transcription during the plant cell cycle is mediated by c-Myb-like transcription factors. Plant Cell 13: 1891–1905.

28. Planchais S, Perennes C, Glab N, Mironov V, Inze D, et al. (2002) Characterization of cis-acting element involved in cell cycle phase-independent activation of Arath;CycB1;1 transcription and identification of putative regulatory proteins. Plant Mol Biol 50: 111–127.

29. Li C, Potuschak T, Colon-Carmona A, Gutierrez RA, Doerner P (2005) Arabidopsis TCP20 links regulation of growth and cell division control pathways. PNAS 102: 12978.

30. Ehsan H, Reichheld JP, Durfee T, Roe JL (2004) TOUSLED kinase activity oscillates during the cell cycle and interacts with chromatin regulators. Plant Physiol 134: 1488–1499.

31. Honys D, Twell D (2004) Transcriptome analysis of haploid male gametophyte development in Arabidopsis. Genome Biol 5: R85.

32. Karimi M, DeMeyer B, Hilson P (2005) Modular cloning in plant cells. Trends Plant Sci 10: 103–105.

33. Curtis MD, Grossniklaus U (2003) A gateway cloning vector set for high-throughput functional analysis of genes in planta. Plant Physiol 133: 462.

34. Karimi M, Inze D, Depicker A (2002) GATEWAY vectors for Agrobacterium-mediated plant transformation. Trends Plant Sci 7: 193–195.

35. Park SK, Howden R, Twell D (1998) The Arabidopsis thaliana gametophytic mutation gemini pollen1 disrupts microspore polarity, division asymmetry and pollen cell fate. Development 125: 3789–3799.

36. Honys D, Oh SA, Renak D, Donders M, Solcova B, et al. (2006) Identification of microspore-active promoters that allow targeted manipulation of gene expression at early stages of microgametogenesis in Arabidopsis. BMC Plant Biol 6: 31.

CITATION

Originally published under the Creative Commons Attribution License. Brownfield L, Hafidh S, Borg M, Sidorova A, Mori T, Twell D. A Plant Germline-Specific Integrator of Sperm Specification and Cell Cycle Progression. PLoS Genet 5(3): e1000430. 2009. doi:10.1371/journal.pgen.1000430.

Effects of Herbivory on the Reproductive Effort of 4 Prairie Perennials

Erica Spotswood, Kate L. Bradley and Johannes M. H. Knops

ABSTRACT

Background

Herbivory can affect every aspect of a plant's life. Damaged individuals may show decreased survivorship and reproductive output. Additionally, specific plant species (legumes) and tissues (flowers) are often selectively targeted by herbivores, like deer. These types of herbivory influence a plant's growth and abundance. The objective of this study was to identify the effects of leaf and meristem removal (simulated herbivory within an exclosure) on fruit and flower production in four species (Rhus glabra, Rosa arkansana, Lathyrus venosus, and Phlox pilosa) which are known targets of deer herbivory.

Results

Lathyrus never flowered or went to seed, so we were unable to detect any treatment effects. Leaf removal did not affect flower number in the other three species. However, Phlox, Rosa, and Rhus all showed significant negative correlations between seed mass and leaf removal. Meristem removal had a more negative effect than leaf removal on flower number in Phlox and on both flower number and seed mass in Rosa.

Conclusions

Meristem removal caused a greater response than defoliation alone in both Phlox and Rosa, which suggests that meristem loss has a greater effect on reproduction. The combination of leaf and meristem removal as well as recruitment limitation by deer, which selectively browse for these species, is likely to be one factor contributing to their low abundance in prairies.

Background

Herbivory has the potential to impact every stage in a plant's life [1], and thus influences where a plant can grow and its abundance [2]. Different kinds of herbivory have differential impacts on plants. Herbivory can reduce resource availability and subsequently have indirect impacts on plant reproduction [3]. Both meristem damage [4] and leaf damage [3] have been shown to negatively impact components of plant fitness such as survival, flower number, and fruit production [1,4-6].

Herbivores may also feed selectively on specific plant species or tissues, which can lead to increased mortality or slower growth rates of damaged individuals [2]. Insect herbivores can directly limit seed production and lifetime fitness by feeding on inflorescences [7]. Mammalian herbivory has been shown to be strong enough to significantly limit the abundance of a plant species [8-10]. Deer in particular have influenced the composition of plant communities in the northeastern and north-central United States [11,12].

Deer have been shown to reduce the proportional rate of increase in the height of some woody species [13]. It also has been suggested that deer browsing can significantly reduce the growth rate of herbaceous plants [11]. Deer herbivory typically involves the removal of entire leaves and terminal meristems, and reduces the proportion of flowering shoots [11], and has the potential to effect reproductive success of browsed plants. For example, deer browsing reduced the number of flowers and proportion of large fruits produced by the forb, Lactuca canadensis[14]. However, there is little known how browsing influences plant fecundity [11].

The objective of this study was to identify the effects of leaf and meristem damage on fruit and flower production in four species of prairie plants that are known targets of deer herbivory. We simulated herbivory with four unrelated species and asked three questions: (1) Does leaf removal influence plant reproduction? (2) If so, is there a threshold level of leaf removal that must be reached before plant reproduction is influenced? (3) Does a combination of leaf removal and meristem removal have a greater impact on a plant than random leaf removal? We report on our findings for each of these questions.

Results

There was a significant influence of the exclosures on the abundance of three out of the four species, Phlox, Rhus, and Lathyrus (Figure 1). Flower number strongly correlated with leaf biomass for Phlox, Rosa, and Rhus (Figure 2). The number of flowers produced by Phlox (P = 0.6), Rosa (P = 0.13), and Rhus (P = 0.3) was not significantly affected by the leaf removal treatment when accounting for plant size. We did not detect an effect in Lathyrus because the few flowers produced were all aborted. Seed mass positively correlated with both flower number (Figure 3a) and leaf biomass. The strong, collinear relationship between flower number and leaf biomass (Figure 2) prevented us from using both variables in the seed mass analyses. We chose to use flower number as a covariate since it has a more direct impact on seed set and the number of seeds produced by an individual plant.

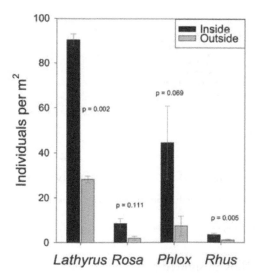

Figure 1. Abundance (mean ± 1 SE) of Lathyrus, Rosa, Phlox, and Rhus inside and outside fenced enclosures. The effect of fencing was compared within each species using oneway ANOVAs (Lathyrus n = 4, Rosa n = 4, Phlox n = 8, Rhus n = 12). Overall, the abundance of all species was significantly higher inside the fence (Type III GLM F = 14.7, df= 7, P < 0.001, R² = 0.837; species F = 21.3, df = 3, P < 0.001; enclosure F = 24.2, df = 1. P < 0.001; interaction F = 6.72, df = 3, p = 0.003).

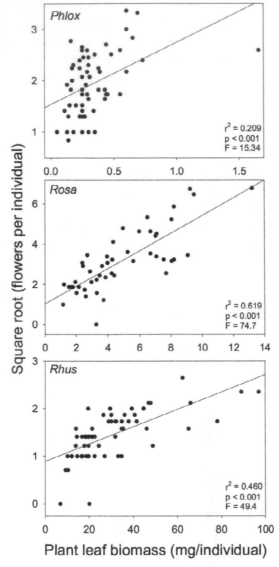

Figure 2. The relationship between flower number and leaf biomass for Phlox (n = 60), Rosa (n = 48) and Rhus (n = 60) respectively.

Flower number per individual was square root transformed.

Phlox, Rosa, and Rhus all showed a significant, negative correlation between leaf removal and seed weight when accounting for flower number (Figure 3b). The more biomass that was removed, the smaller the overall seed mass per individual. There was not a leaf removal frequency threshold that influenced flower or seed production when all the species were examined together (P = 0.1) or when Rhus

(P = 0.3), Phlox (P = 0.1) and Rosa (P = 0.3) were examined individually. We were unable to detect an effect in Lathyrus, which flowered a little, but no single plant went to seed. None of the plants in the study produced fruits and only three fruits were found when the field inside and outside the fence was surveyed.

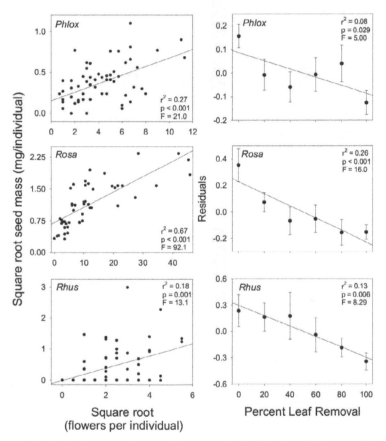

Figure 3. a) The relationship between seed mass and flower number for Phlox (n = 60), Rosa (n = 48) and Rhus (n = 60) respectively.

Both seed mass and flower number were square root transformed, b) The relationship between the residuals (mean ± 1 SE) of seed mass versus flower number and the percent leaf removal for Phlox (n = 60), Rosa (n = 48) and Rhus (n = 60) respectively.

The meristem removal treatment had a significantly negative effect on flower number in both Phlox and Rosa (Figure 4a). The individuals in the meristem removal treatment produced very few flowers when compared to those individuals in the control and leaf removal treatments. Meristem removal also caused the seed mass to be significantly lower in Rosa, though the seed masses of both Phlox and Rhus remained unaffected (Figure 4b).

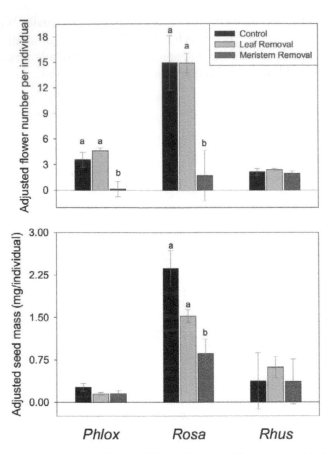

Figure 4. a) Flower number for Phlox, Rosa and Rhus under three different simulated herbivory treatments (control, leaf removal only, or meristem removal).

Values are shown as adjusted means ± 1 SE from Type III GLMs with treatment as the main effect and plant size and leaf removal as covariates. Overall, the treatment effect was significant for Phlox ($F = 14.2$, $P < 0.001$) and Rosa ($F = 6.80$, $P < 0.001$) but not for Rhus ($P = 0.8$). Flower values are based on the average of multiple counts on individual plants (Phlox-June 15, 17 and 21; Rosa-June 15 & 21; Rhus-June 1 & July 1). Flower number was square root transformed for analysis, b) Seed mass for Phlox, Rosa and Rhus under three different simulated herbivory treatments. Values are shown as adjusted means ± 1 SE from Type III GLMs with treatment as the main effect and square root flower number and leaf removal as covariates. Treatment had a significant effect for Rosa ($F = 6.80$, $P < 0.001$) but not for Phlox ($P = 0.2$) or Rhus ($P = 0.6$). Different letters denote significant effects at $P < 0.05$ following Bonferroni correction for multiple comparisons.

Discussion

Three of the four species studied were significantly more abundant within the exclosures than outside of them. This pattern is consistent with other results found for herbaceous species at this [6,15] and other sites [8-10] where mammalian herbivory has been shown to limit overall plant abundance in some species. It is therefore not surprising that deer browsing should effect the overall abundance of species known to be preferred by deer. Less clear is which aspect of herbivory is most important.

Leaf removal did not affect flower production in any of the species, which is consistent with other studies [7,16,17]. Ehrlen demonstrated that flower numbers were predetermined the previous fall by budding in Lathyrus vernus[18]. The same may be true in all of our species because removing leaves did not impact their flower numbers. Nonetheless, high levels of leaf removal did negatively impact the seed weight in all of the study species which produced seeds (Figure 3b). These results suggest that stored resources are available for flowers and seeds before the onset of flowering [19,20] and changes in current year resources have a negligible effect on flower number. However, leaf removal appears to reduce the amount of carbon available for allocation to developing seeds in Phlox, Rosa, and Rhus, which causes a decrease in the overall seed mass produced by an individual plant.

Though we found a negative relationship between seed mass and percent leaf damage in Phlox, Rosa, and Rhus, we did not detect a threshold level of leaf removal that had to be reached before seed mass was impacted. Other studies, which have attempted to quantify the point where defoliation begins to impact reproduction, have yielded widely variable results [1,4-6], though these studies all found significant results at 50% or lower levels of defoliation.

Additionally, defoliation may have differential effects on seeds depending on when it occurs. In this study, all treatments were administered within a few weeks of flowering. One study [5,20] found that when leaves were removed several months before the time of flowering, the plant suffered a large loss in reproductive output. When the same treatment was administered just before flowering, there was no response [5,20]. Timing, then, may be a key in determining how well a plant copes with herbivory [11].

Meristem removal was more harmful to the reproductive output of Phlox and Rosa than leaf removal alone (Figures 4a &4b). With meristem removal, Phlox had fewer flowers than in the control and leaf removal treatments, but its seed mass was not affected. Meristem removal more strongly impacted Rosa, which had fewer flowers and a lower seed mass than either the leaf removal or control groups.

Because Phlox is a small herbaceous plant with terminal flowers, it often suffered complete flower loss and substantial leaf removal under the meristem removal treatment. The individuals in this treatment that did produce seeds sent up a side shoot after the meristem was nipped off. In contrast, Rosa produced many flowers and never suffered a complete flower loss with meristem removal.

The flower loss may have allowed the Phlox to compensate by increasing seed set, which has shown to be resource limited in other species [21], in the remaining flowers. The relationship between seed mass and flower number is much stronger in Rosa than in Phlox (Figure 3a), and Rosa, possibly because of its woody nature, was unable to compensate for the flower loss by generating new shoots and flowers or by increasing seed set in the remaining flowers. Therefore, the significance of this treatment is most likely due to a combination of how many buds remained after meristem removal as well as the allocation of remaining resources for reproduction.

Conclusions

High levels of defoliation reduced total seed weight in Phlox, Rosa, and Rhus, all of which are found in Minnesota prairies. The removal of meristems along with defoliation caused a greater response than defoliation alone in both Phlox and Rosa. This suggests that loss of meristems is more important than defoliation alone in its influence on the reproductive success of these species. All three species studied are preferred by large mammal herbivores (primarily white tailed deer). These results suggest that both defoliation, which limits the resources available for reproduction, and meristem removal may be partly responsible for the comparative rarity of the study species outside fenced enclosures.

Materials and Methods

Study Site and Study Species

The study was conducted at Cedar Creek Natural History Area (CCNHA) in central Minnesota. For a detailed description of the study site, see Tilman [22]. The four species studied include smooth sumac (Rhus glabra), wild rose (Rosa arkansana), bushy vetch (Lathyrus venosus), and phlox (Phlox pilosa). Smooth sumac is a perennial shrub (1–4 m tall). Wild rose is a short woody perennial shrub (1 m or shorter). Lathyrus venosus is a perennial legume (1.5 m or shorter). Phlox is anherbaceous perennial (30 cm or shorter). These species was chosen because they were abundant inside the fenced area and absent or rare outside the fence (see methods below). There is also evidence that Rhus[23], Lathyrus[15], Phlox (Haarstad, personal communication), and Rosa[24] are all browsed by deer. The density of deer in this area has been minimally estimated to be 0.16 deer per ha [25]. This density is similar to other protected areas, where deer herbivory has caused changes in plant composition [25]. Target species were located inside exclosures which kept out large herbivores.

Experimental Design

To compare abundance of the study species inside and outside the fenced enclosures, temporary transects (0.5 × 8 m) were established within and outside of each fenced area. For each species, the total number of individuals along the transects were counted. For Rosa and Lathyrus, two transects on either side of the fence were counted. Phlox was counted in four transects inside the fence and four outside. Rhus transects were established at fenced areas in 2 different fields. Two transects on either side of the fences were counted in each field.

To measure the effects of different levels of defoliation, individuals of each species within the exclosures were randomly selected and tagged. Initial height and number of leaves were recorded. Ten individuals of each species (except Rosa, which only had enough for 8 individuals for each treatment level) were randomly assigned to one of the following treatments: 1) control, no simulated herbivory, 2) 20 % of all leaves removed, 3) 40 % of all leaves removed, 4) 60 % of all leaves removed, 5) 80 % of all leaves removed, 6) 100 % of all leaves removed, or 7) meristem + natural leaf removal (called the meristem removal hereafter). This treatment was designed to simulate deer and rabbit browsing in which the entire top of a plant is often removed. Meristems, leaves and flower buds were all removed from the top of the plant and left at the bottom of the plant. The mass of the leaves removed by the meristem removal was determined and converted to the percent of the plant's total leaf biomass.

Removed leaves were dried at 55 degrees C for one week and then weighed. Following the initial damage treatment, the sites were visited twice a week. Flowers were counted on multiple visits. Seeds were collected and dried, and then weighed to give the total mass of all the seeds collected per individual plant. Mesh bags were placed over Phlox flowers because seeds are small and fall off when they ripen. No such bags were needed for Rosa or Rhus, both of which have large seeds, which are retained on the parent plant.

Statistics

All statistical analysis was performed on SPSS 10.0 for Windows. One-way ANOVAs were used to determine the effect of the enclosures on the abundance of the individual species. Type III GLM analysis was used to test for differences between areas within and outside the enclosure, with abundance as the dependent variable and species, enclosure, and their interaction as the independent variables.

Total leaf biomass was calculated for each plant since larger plants generally produce more biomass and larger and/or more seeds than smaller plants. Using the weight of the leaves collected, the following formula was used to calculate the total leaf biomass per individual:

(dried leaf weight/number of leaves collected) × (total number of leaves on the plant)

This leaf biomass was used to account for plant size in statistical analysis.

Multiple regression was used to examine the relationship between percent leaf removal and flower number with plant size as the covariate. Multiple regression was also used to examine the relationship between percent leaf removal and seed mass with flower number as the covariate. Type III GLMs were run to examine the effect of the different treaments (leaf removal, meristem removal, and controls) on both flower number and seed mass. Plant size was run as a covariate for flower number, and flower number was used as a covariate for seed mass. We also corrected for the actual biomass of the leaves removed since the meristem removal often removed leaves. The level of Type III GLM analysis was also used to test for effects of different levels of leaf removal on flower number and seed mass. Bonferroni tests were performed for multiple comparisons. For all these analyses, seed mass and flower number were square root transformed.

Acknowledgements

Thanks to Bryan Foster, John Haarstad, Troy Mielke, Cini Brown, Lenny Sheps and Joe Craine for their help and support in the field. We extend our thanks to Ian Dickie and Janneke HilleRisLambers for advice and comments on earlier versions of this manuscript. This work was supported by the NSF as part of the Research Experience for Undergraduates supplement grant.

References

1. Crawley MJ: Reduction of oak fecundity by low-density herbivore populations. Nature 1985, 314:163–164.

2. Huntly N: Herbivores and the dynamics of communities and ecosystems. Annual Review of Ecology and Systematics 1991, 22:477–503.

3. Mauricio R, Bowers MD, Bazzaz FA: Pattern of leaf damage affect fitness of the annual plant Raphanus sativus (Brassicaceae). Ecology 1993, 74:2066–2071.

4. Ehrlen J: Demography of the perennial herb Lathyrus vernus. I. Herbivory and individual performance. Journal of Ecology 1995, 83:287–295.

5. Marquis RJ: Leaf herbivores decrease fitness of a tropical plant. Science 1984, 226:537–539.

6. Allison TD: The influence of deer browsing on the reporductive biology of Canada yew (Taxus canadensis Marsh.) I. Directs effects on pollen, ovule and seed production. Oecologia 1990, 83:523–529.

7. Louda SM, Potvin MA: Effect of inflorescence-feeding insects on the demography and lifetime fitness of a native plant. Ecology 1995, 76:229–245.

8. Ross BA, Bray JR, Marshall WH: Effects of long-term deer exclosure on a Pinus resinosa forest in north-central Minnesota. Ecology 1970, 51:1088–1093.

9. Davidson DA: The effects of herbivory and granivory on terrestrial plant succession. Oikos 1993, 68:23–35.

10. Erneberg M: Effects of herbivory and competition on an introduced plant in decline. Oecologia 1999, 118:203–209.

11. Russell FL, Zippin DB, Fowler NL: Effects of white-tailed deer (Odocoileus virginianus) on plants, plant populations and communities: A review. American Midland Naturalist 2001, 146:1–26.

12. McShea WJ, Underwood HB, Rappole JH, eds: The science of overabundance: deer ecology and population management. Washington D.C.: Smithsonian Institution Press; 1997.

13. Inouye RS, Allison TD, Johnson NC: Old field succession on a Minnesota sand plain: Effects of deer and other factors on invasion by trees. Bulletin of the Torrey Botanical Club 1994, 121:266–276.

14. Shelton AL, Inouye RS: Effect of browsing by deer on the growth and reproductive success of Lactuca canadensis (Asteraceae). American Midland Naturalist 1995, 134:332–339.

15. Ritchie ME, Tilman D: Responses of legumes to herbivores and nutrients during succession on a nitrogen-poor soil. Ecology 1995, 76:2648–2655.

16. Paige KN: Overcompensation in response to mammalian herbivory: From mutalistic to antagonistic interactions. Ecology 1992, 73:2076–2085.

17. Maschinski J, Whitham TB: The continuum of plant response to herbivory: the influence of plant association, nutrient availability and timing. American Naturalist 1989, 40:329–336.

18. Ehrlen J: Ultimate fucntions of non-fruiting flowers in Lathyrus vernus. Oikos 1993, 68:45–52.

19. Ehrlen J: Proximate limits to seed production in a herbaceous perennial legume, Lathyrus vernus. Ecology 1992, 73:1820–1831.

20. Marquis RJ: A bite is a bite? Constraints on response to folivory in Piper ariea-num (Piperaceae). Ecology 1992, 73.

21. Ehrlen J, Groendael JV: Storage and the delayed costs of reproduction in the understorey perennial Lathyrus vernus. Journal of Ecology 2001, 89:237–246.

22. Tilman D: Secondary successional and the pattern of plant dominance along experimental nitrogen gradients. Ecological Monographs 1987, 57:189–214.

23. Strauss SY: Direct, indirect, and cumulative effects of three native herbivores on a shared host plant. Ecology 1991, 72:543–558.

24. Ritchie ME, Tilman D, Knops J: Herbivore effects on plant and nitrogen dy-namics in oak savanna. Ecology 1998, 79:165–177.

25. Knops JMH, Ritchie M, Tilman GD: Selective herbivory on a nitrogen fix-ing legume (Lathyrus venosus) influences productivity and ecosystem nitrogen pools of an oak savanna. Ecoscience 2000, 7:166–174.

CITATION

Originally published under the Creative Commons Attribution License. Spotswood E, Bradley KL, and Knops JMH. Effects of Herbivory on the Reproductive Effort of 4 Prairie Perennials. BMC Ecology 2002, 2:2. doi:10.1186/1472-6785-2-2.

Identification of Flowering Genes in Strawberry, a Perennial SD Plant

Katriina Mouhu, Timo Hytönen, Kevin Folta, Marja Rantanen, Lars Paulin, Petri Auvinen and Paula Elomaa

ABSTRACT

Background

We are studying the regulation of flowering in perennial plants by using diploid wild strawberry (Fragaria vesca L.) as a model. Wild strawberry is a facultative short-day plant with an obligatory short-day requirement at temperatures above 15°C. At lower temperatures, however, flowering induction occurs irrespective of photoperiod. In addition to short-day genotypes, everbearing forms of wild strawberry are known. In 'Baron Solemacher' recessive alleles of an unknown repressor, SEASONAL FLOWERING LOCUS (SFL), are responsible for continuous flowering habit. Although flower induction has a central effect on the cropping potential, the molecular control of flowering in strawberries has not been studied and the genetic flowering pathways are still poorly understood. The comparison of everbearing and short-day genotypes of wild strawberry could facilitate our understanding of fundamental molecular mechanisms regulating perennial growth cycle in plants.

Results

We have searched homologs for 118 Arabidopsis flowering time genes from Fragaria by EST sequencing and bioinformatics analysis and identified 66 gene homologs that by sequence similarity, putatively correspond to genes of all known genetic flowering pathways. The expression analysis of 25 selected genes representing various flowering pathways did not reveal large differences between the everbearing and the short-day genotypes. However, putative floral identity and floral integrator genes AP1 and LFY were co-regulated during early floral development. AP1 mRNA was specifically accumulating in the shoot apices of the everbearing genotype, indicating its usability as a marker for floral initiation. Moreover, we showed that flowering induction in everbearing 'Baron Solemacher' and 'Hawaii-4' was inhibited by short-day and low temperature, in contrast to short-day genotypes.

Conclusion

We have shown that many central genetic components of the flowering pathways in Arabidopsis can be identified from strawberry. However, novel regulatory mechanisms exist, like SFL that functions as a switch between short-day/low temperature and long-day/high temperature flowering responses between the short-day genotype and the everbearing 'Baron Solemacher'. The identification of putative flowering gene homologs and AP1 as potential marker gene for floral initiation will strongly facilitate the exploration of strawberry flowering pathways.

Background

Transition from vegetative to reproductive growth is one of the most important developmental switches in plant's life cycle. In annual plants, like Arabidopsis, flowering and consequent seed production is essential for the survival of the population until the following season. To assure timely flowering in various environments, Arabidopsis utilizes several genetic pathways that are activated by various external or internal cues. Light and temperature, acting through photoperiod, light quality, vernalization and ambient temperature pathways, are the most important environmental factors regulating flowering time [1]. Moreover, gibberellin (GA) and autonomous pathways promote flowering by responding to internal cues [2,3]. In contrast to annual plants, the growth of perennials continues after generative reproduction, and the same developmental program is repeated from year to year. Regulation of generative development in these species is even more complex, because other processes like juvenility, winter dormancy and chilling are tightly linked to the control of flowering time.

In Arabidopsis photoperiodic flowering pathway, phytochrome (phy) and cryptochrome (cry) photoreceptors perceive surrounding light signals and reset the circadian clock feedback loop, including TOC1 (TIMING OF CAB EXPRESSION), CCA1 (CIRCADIAN CLOCK ASSOCIATED 1) and LHY (LATE ELONGATED HYPOCOTYL) [4-7]. The central feature in the photoperiodic flowering is the clock generated evening peak of CO (CONSTANS) gene expression [8]. In long-day (LD) conditions, CO peak coincidences with light resulting in accumulation of CO protein in the leaf phloem and consequent activation of the expression of FT (FLOWERING LOCUS T) [9]. FT protein, in turn, moves to the shoot apex, and together with FD triggers floral initiation by activating floral identity gene AP1 (APETALA 1) [10,11]. FT, together with SOC1 (SUPPRESSOR OF OVEREXPRESSION OF CONSTANS 1) and LFY (LEAFY) form also convergence points for different flowering pathways, and therefore are called flowering integrator genes [12].

In winter-annual ecotypes of Arabidopsis, MADS-box gene FLC (Flowering Locus C) prevents flowering by repressing FT and SOC1, and vernalization is needed to nullify its function [13]. The major activator of FLC is FRI (FRIGIDA) [14], but several other proteins, including for example FRL1 (FRIGIDA-LIKE 1) [15], PIE (PHOTOPERIOD INDEPENDENT EARLY FLOWERING 1) [16], ELF7 and ELF8 (EARLY FLOWERING 7 and 8) [17], and VIP3 (VERNALIZATION INDEPENDENCE 3) [18] are also needed to maintain high FLC expression. During vernalization, FLC is down-regulated by VRN2-PRC2 (Vernalization 2 - Polycomb Repressive Complex 2) protein complex containing low temperature activated VIN3 (VERNALIZATION INSENSITIVE3), allowing plants to flower [19,20].

Autonomous and GA pathways respond to endogenous cues to regulate flowering time. The role of the autonomous pathway is to promote flowering by lowering the basal level of FLC transcription [3]. Autonomous pathway consists of few sub-pathways, which include for example RNA processing factors encoded by FCA, FPA, FLK (FLOWERING LOCUS K), FY and LD (LUMINIDEPENDENS) [21], putative histone demethylases LDL1 and LDL2 (LSD1-LIKE 1 and 2) [22], and deacetylases FLD (Flowering locus D) and FVE [23,24]. GA pathway is needed to induce LFY transcription and flowering in short-day (SD) conditions [25].

Strawberries (Fragaria sp.) are perennial rosette plants, belonging to the economically important Rosaceae family. Most genotypes of garden strawberry

(Fragaria × ananassa Duch.) and wild strawberry (F. vesca L.) are Junebearing SD plants, which are induced to flowering in decreasing photoperiod in autumn [26,27]. In some genotypes, flowering induction is also promoted by decreasing temperatures that may override the effect of the photoperiod [27,28]. In contrast to promotion of flowering by decreasing photoperiod and temperature, these "autumn signals" have opposite effect on vegetative growth. Petiole elongation decreases after a few days, and later, around the floral transition, runner initiation ceases and branch crowns are formed from the axillary buds of the crown [29,30]. Crown branching has a strong effect on cropping potential as it provides meristems that are able to initiate inflorescences [31].

In addition to SD plants, everbearing (EB) genotypes are found in garden strawberry and in wild strawberry [29,32]. Environmental regulation of induction of flowering in EB genotypes has been a topic of debate for a long time. Several authors have reported that these genotypes are day-neutral [29,33]. Recent findings, however, show that long-day (LD) accelerates flowering in several EB Fragaria genotypes [34,35]. Interestingly, in wild strawberry genotype 'Baron Solemacher' recessive alleles of SFL gene locus (SEASONAL FLOWERING LOCUS) have been shown to cause EB flowering habit [36]. SFL has not been cloned, but it seems to encode a central repressor of flowering in wild strawberry. Consistent with the repressor theory, LD grown strawberries have been shown to produce a mobile floral inhibitor that is able to move from mother plant to the attached runner plant [37]. GA is one candidate corresponding to this inhibitor, since exogenously applied GA has been shown to repress flowering in strawberries [38,39].

Identification of central genes regulating flowering time and EB flowering habit, as well as those controlling other processes affecting flowering, is an important goal that would greatly accelerate breeding of strawberry and other soft fruit and fruit species of Rosaceae family. In this paper, we have searched Fragaria homologs with the known Arabidopsis flowering time genes by EST sequencing and bioinformatics analysis. Dozens of putative flowering genes corresponding to all known genetic pathways regulating flowering time were identified. The expression analysis of several candidate flowering time genes revealed only few differences between the SD and EB wild strawberries, including the presence or absence of AP1 mRNA in the apices of EB and SD genotypes, respectively. Our data provides groundwork for detailed studies of flowering time control in Fragaria using transcriptomics, functional genomics and QTL mapping.

Results

Environmental Regulation of Flowering in Two EB Genotypes of Wild Strawberry

We studied the effect of photoperiod and temperature on flowering time in two EB genotypes, 'Baron Solemacher', which contains recessive alleles in SFL locus [40,41], and 'Hawaii-4'. Flowering time was determined by counting the number of leaves in the main crown before formation of the terminal inflorescence. In SD genotypes of the wild strawberry, SD (<15 h) or, alternatively, low temperature (~10°C) is needed to induce flowering [27]. In EB genotypes 'Baron Solemacher' and 'Rugen', instead, LD and high temperature has been shown to accelerate generative development [35], but careful analysis of the environmental regulation of flowering induction has so far been lacking.

Both 'Baron Solemacher' and 'Hawaii-4' produced five to six leaves in LD at 18°C before the emergence of the terminal inflorescence showing that they are very early-flowering in favorable conditions (Figure 1A and 1B). In 'Baron Solemacher', low temperature (11°C) or SD treatment for five weeks at 18°C clearly delayed flowering, but low temperature did not have an additional effect on flowering time in SD. Also in 'Hawaii-4', SD and low temperature delayed flowering, but all treatments differed from each other. Compared to the corresponding LD treatment, SD at 18°C doubled the number of leaves, and low temperature (11°C) delayed flowering time by about three leaves in both photoperiods. Thus, flowering induction in these EB genotypes is oppositely regulated by photoperiod and temperature than previously shown for the SD genotypes [27].

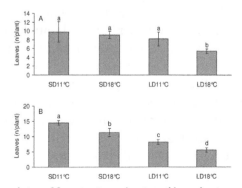

Figure 1. Environmental regulation of flowering in everbearing wild strawberries.

The effect of photoperiod (SD 12 h, LD 18 h) and temperature (11/18°C) on the flowering time of 'Baron Solemacher' (A) and 'Hawaii-4' (B). Seeds were germinated in LD at 18°C, and seedlings were exposed to the treatments for five weeks, when the cotyledons were opened. After treatments, plants were moved to LD at 18°C and flowering time was recorded as number of leaves in the main crown before the terminal inflorescence. Values are mean ± SD. Pairwise comparisons between the treatments were done by Tukey's test, and statistically significant differences (p ≤ 0.05) are denoted by different letters above the error bars.

Construction and Sequencing of Subtracted cDNA Libraries

We constructed two subtracted cDNA libraries from LD grown EB genotype 'Baron Solemacher' and SD genotype, in order to identify differentially expressed flowering time genes in these genotypes. Plants were grown in LD conditions, where the SD genotype stays vegetative and the EB plants show early flowering. Pooled shoot apex sample covering the floral initiation period was collected from the EB genotype, and vegetative apices of the same age were sampled from the SD genotype. Suppression subtractive hybridization (SSH), the method developed for extraction of differentially expressed genes between two samples [42], was used to enrich either flowering promoting or flowering inhibiting transcripts from EB and SD genotypes, respectively.

A total of 1172 ESTs was sequenced from the library enriched with the genes of the SD genotype (SD library subtracted with EB cDNA) and 1344 ESTs from the library enriched with the EB genes (EB library subtracted with cDNA of the SD genotype). 970 SD ESTs [Genbank:GH202443-GH203412] and 1184 EB ESTs [GenBank:GH201259-GH202442] passed quality checking. Pairwise comparison of these EST datasets revealed that there was very little overlap between the libraries. However, general distribution of the sequences to functional categories (FunCat classification) did not reveal any major differences between the two libraries.

BLASTx searches against Arabidopsis, Swissprot and non-redundant databases showed that over 70% of the ESTs gave a match in one or all of the three databases (Table 1). Moreover, tBLASTx comparison with different genomes revealed highest number of hits with Populus trichocarpa (Table 1). We also performed tBLASTx searches against TIGR plant transcript assemblies of Malus × domestica, Oryza sativa and Vitis vinifera and found hits for 64-76% of ESTs in these assemblies. Finally, the comparison of our sequences with a current Fragaria unigene list at the Genome Database for Rosaceae (GDR) showed that 38.2% of our ESTs are novel Fragaria transcripts. Taken together, depending on the analysis, 15-22% of sequences from SD genotype and 22-27% of EB sequences encode novel proteins, or originate from untranslated regions of mRNA. Moreover, the high number of novel Fragaria sequences in our libraries indicates that SSH method efficiently enriched rare transcripts in the libraries.

Table 1. The comparison of F. vesca ESTs with different databases.

		WT		EB	
		number	average length	number	average length
A)	Raw	1172	946	1344	965
	Poor Quality	202	1037	160	1066
	Singletons/ESTs	970	452	1184	451
		number	%	number	%
B)	Arabidopsis	695	72	781	66
	Swissprot	483	50	570	48
	Non-redundant	749	77	852	72
	In all 3 datab.	752	78	862	73
C)	Malus	741	76	874	74
	Oryza	689	71	807	68
	Vitis	666	69	761	64
	Populus	829	85	928	78
D)	No protein hits	218	22	322	27
	No Fragaria hits	370	38	454	38

Average numbers, lengths and percentages of ESTs from EB and SD genotypes. A) numbers and average lengths of raw and poor quality ESTs, and singletons, B) numbers and percentages of BLASTx hits against protein databases, C) numbers and percentages of tBLASTx hits against TIGR plant transcript assemblies of Malus x domestica, Oryza sativa and Vitis vinifera and against Populus genome database, D) numbers and percentages of novel ESTs.

Identification of Flowering Time Genes

Flowering related genes were identified from our libraries by BLASTx searches as described above and fourteen putative flowering time regulators were identified; four gene homologs were present only in EB library, eight in SD library, and two genes in both libraries. In figure 2, we have summarized the Arabidopsis flowering pathways and highlighted the putative homologous genes identified from our EST collection. In general, candidate genes for all major pathways were identified. In addition, 118 Arabidopsis flowering time genes were used as a query to search publicly available GDR Fragaria EST and EST contig databases using tBLASTn. Sequences passing cut-off value of 1e-10 were further analysed by BLASTx algorithm against Arabidopsis protein database, and those returning original Arabidopsis protein were listed. Moreover, sequences that were absent from Fragaria databases were similarly searched from GDR Rosaceae EST database. In these searches, 52 additional Fragaria sequences were identified. Moreover, the total number of 88 homologs of Arabidopsis flowering time genes were found among all available Rosaceae sequences.

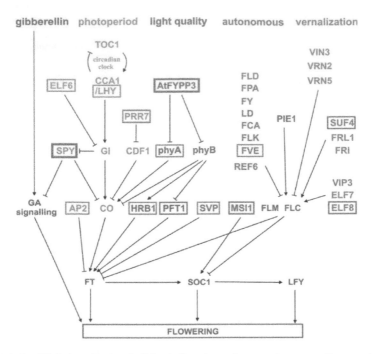

Figure 2. A simplified chart showing Arabidopsis flowering pathways and corresponding gene homologs in Fragaria.

Gene homologs found in cDNA libraries produced from SD and EB genotypes are surrounded by blue and red boxes, respectively. Arrows indicate positive regulation and bars negative regulation.

Most genes of the Arabidopsis photoperiodic pathway were found also in Fragaria, and some of the lacking genes were present among Rosaceae ESTs (Table 2). We found several genes encoding putative Fragaria photoreceptor apoproteins including phyA, phyC, cry2, ZTL (ZEITLUPE) and FKF1 (FLAVIN BINDING KELCH REPEAT F-BOX 1) [43]. Of the central circadian clock genes, homologs of LHY and TOC1 [5,7] were present in our EST libraries and GDR, respectively, but CCA1 [6] was lacking from both Fragaria and Rosaceae databases. Furthermore, a putative Fragaria CO from the flowering regulating output pathway has been cloned earlier [44]. Among the regulators of CO transcription and protein stability, GI (GIGANTEA) [45] was identified from Rosaceae and putative COP1, SPA3 and SPA4 [46,47] from Fragaria. In addition to genes of the photoperiodic pathway, homologs for both known sequences belonging to light quality pathways, PFT1 (PHYTOCHROME AND FLOWERING TIME 1) and HRB1 (HYPERSENSITIVE TO RED AND BLUE 1) [48,49], were found from our EST libraries.

Table 2. The list of genes belonging to the photoperiodic flowering pathway.

Gene	AT gene locus	Biological function	Act./Repr. +/-	Reference	Fragaria	E-value
Photoreceptors and clock input						
PhyA	AT1G09570	Red light photoreceptor	+	[78]	VES-002-C06	5E-33
PhyB	AT2G18790	Red light photoreceptor	-	[79]	nf	
CRY1	AT4G08920	Blue light photoreceptor	+	[79]	nf	
CRY2	AT1G04400	Blue light photoreceptor	+	[79]	DY669844	2E-110
ZTL	AT5G57360	F-box protein/blue light photoreceptor	+	[80]	EX668764	2E-97
FKF1	AT1G68050	F-box protein/blue light photoreceptor	+	[65]	DY671170	2E-54
ELF3	AT2G25920	Unknown	-	[60]	DY675323	3E-33
FYPP3	AT1G50370	Ser/Thr-specific protein phosphatase 2A	-	[81]	BAR-009-A02	1E-56
SRR1	AT5G59560	Unknown	-	[82]	CO817759	1E-10
Circadian clock						
LHY	AT1G01060	Myb domain TF	-	[7]	VES-005-E09	9E-19
CCA1	AT2G46830	Myb domain TF	-	[6]	nf	
TOC1	AT5G61380	Pseudo-response regulator	-	[5]	DY673134	1E-75
LUX	AT3G46640	Myb TF	-	[83]	DY668516	3E-43
ELF4	AT2G40080	Unknown	-	[84]	EX674323	2E-25
GI	AT1G22770	Unknown	+	[45]	nf	
PRR5	AT5G24470	Pseudo-response regulator	+	[85]	DY676242	3E-56
PRR7	AT5G02810	Pseudo-response regulator	+	[85]	VES-013-D12	5E-52
ELF6	AT5G04240	Jumonji/zinc finger-class TF	-	[86]	VES-002-F05	1E-45
Output pathway						
CO	AT5G15840	putative zinc finger TF	+	[8]	DY672035	2E-45
CDF1	AT5G62430		-	[65]	nf	
FT	AT1G65480	Phosphatidylethanolamine binding	+	[11]	nf	
TFL1	AT5G03840	Phosphatidylethanolamine binding	-	[87]	nf	
FD	AT4G35900	bZIP TF	+	[10]	EX675574	2E-14
COP1	AT2G32950	E3 ubiquitin ligase	-	[46]	DY667888	1E-94
SPA1	AT2G46340	WD domain protein	-	[47]	nf	
SPA3	AT3G15354	WD domain protein	-	[47]	DY671873	3E-24
SPA4	AT1G53090	WD domain protein	-	[47]	DY671245	2E-83
RFI2	AT2G47700	Ring domain zinc finger	-	[88]	nf	
HAP3b	AT5G47640	CCAAT-binding TF	+	[89]	EX658204	2E-60

The most important genes belonging to the photoperiodic pathway in *Arabidopsis* and their biological function are presented. Floral activators and repressors are indicated by + and - marks, respectively. Moreover, the presence or absence of homologous sequence in *Fragaria* sequence databases and E-value of BLASTx comparison against *Arabidopsis* are indicated. Sequences found in our libraries are named BAR and VES for everbearing genotype 'Baron Solemacher' and short-day genotype, respectively. Other ESTs and EST contigs are found from Genome Database for Rosaceae http://www.bioinfo.wsu.edu/gdr/. More complete list is available in Additional file 2.

For the vernalization pathway, we were not able to find FLC-like sequences from our EST libraries or public Fragaria or Rosaceae EST databases by tBLASTn searches although we used the FLC and FLC-like sequences from Arabidopsis (MAF1-MAF5, MADS AFFECTING FLOWERING 1-5) and several other plant species as query sequences [13,50,51]. Similarly, also FRI [14] was lacking from Rosaceae ESTs but putative FRL (FRIGIDA-LIKE) [15] sequences were identified in Fragaria. In addition, we identified several gene homologs belonging to the FRI complex as well as other regulatory complexes (SWR1, PAF) involved in promoting the expression of FLC (Table 3) [17,52,53]. Also putative members of FLC repressing PRC2 complex, were present in strawberry ESTs. These include putative VIN3 (VERNALIZATION INSENSITIVE 3) [19,20] that has been identified earlier [54], and putative SWN1 (SWINGER 1), FIE (FERTILIZA-TION INDEPENDENT ENDOSPERM), VRN1 (VERNALIZATION 1) and LHP1 (LIKE HETEROCHROMATIN PROTEIN 1) [19,55,56], which were found in this investigation (Table 3). However, putative VRN2 that is needed for the repression of FLC by PRC2 was not found [19].

Table 3. The list of genes belonging to the vernalization pathway.

Gene	AT gene locus	Biological function	Act./Repr. +/-	Reference	Fragaria	E-value
FLC	AT5G10140	MADS-box TF	-	[13]	nf	
MAFI/FLM	AT1G77080	MADS-box TF	-	[50]	nf	
Fri complex						
FRI	AT4G00650	Unknown, enhancer of FLC	-	[14]	nf	
FRL1	AT5G16320	Unknown, enhancer of FLC	-	[15]	EX686406	4E-45
FRL2	AT1G31814	Unknown, enhancer of FLC	-	[15]	Contig 4768	6E-49
FES1	AT2G33835	CCCH zinc finger protein	-	[53]	nf	
SUF4	AT1G30970	putative zinc finger containing TF	-	[53]	BAR-003-F06	5E-46
Swr complex						
PIE	AT3G12810	ATP-dependent chromatin-remodelling factor	-	[16]	nf	
SEF1/SWC6	AT5G37055	Component of chromatin remodelling complex	-	[52]	DY670674	4E-70
ARP6/ESD1	AT3G33520	Component of chromatin remodelling complex	-	[52]	nf	
ATX1	AT2G31650	Putative SET domain protein	-	[90]	EX687477	4E-71
Paf1 complex						
ELF7	AT1G79730	RNA polymerase 2 associated factor 1 -like	-	[17]	nf	
ELF8	AT2G06210	RNA polymerase 2 associated factor -like	-	[17]	BAR-008-H08	3E-42
VIP4	AT5G61150	RNA polymerase 2 associated factor -like	-	[91]	EX660943	2E-50
VIP3	AT4G29830	RNA polymerase 2 associated factor -like	-	[18]	EX675781	7E-98
EFS/SDG8	AT1G77300	putative histone H3 methyltransferase	-	[53]	nf	
VRN2-PRC2 complex						
VRN2	AT4G16845	Polycomb group zinc finger	+	[92]	nf	
CLF	AT2G23380	Polycomb group protein	+	[93]	nf	
SWN1/EZA	AT4G02020	Polycomb group protein	+	[93]	EX687655	3E-114
FIE	AT3G20740	Polycomb group protein	+	[93]	DY671601	1E-112
VIN3	AT5G57380	PHD domain protein	+	[20]	CO816801	2E-58
LHP1	AT5G17690	epigenetic silencing	+	[56]	DY669633	2E-40
VRN1	AT3G18990	DNA binding protein	+	[55]	DY670727	8E-43

The most important genes belonging to the vernalization pathway in *Arabidopsis* and their biological function are presented. Floral activators and repressors are indicated by + and - marks, respectively. Moreover, the presence or absence of homologous sequence in *Fragaria* sequence databases and E-value of BLASTx comparison against *Arabidopsis* are indicated. Sequences found in our libraries are named BAR and VES for everbearing genotype 'Baron Solemacher' and short-day genotype, respectively. Other ESTs and EST contigs are found from Genome Database for Rosaceae http://www.bioinfo.wsu.edu/gdr/. More complete list is available in Additional file 2.

In addition to the photoperiod and the vernalization pathways, we searched candidate genes for the autonomous and GA pathways. Several sequences corresponding to Arabidopsis genes from both pathways were identified suggesting the presence of these pathways also in Fragaria (Table 4). Among these genes we found homologs for Arabidopsis FVE and SVP which have been shown to control flowering in a specific thermosensory pathway [24,57]. Moreover, some additional flowering time regulators that are not placed to any specific pathway were identified (Table 4).

Table 4. The list of genes belonging to autonomous and gibberellin flowering pathways.

Gene	AT gene locus	Biological function	Act./Repr. +/-	Reference	Fragaria	E-value
Autonomous pathway						
FCA	AT4G16280	RRM-type RNA binding domain containing	+	[94]	nf	
FPA	AT2G43410	RRM-type RNA binding domain containing	+	[95]	nf	
FLK	AT3G04610	KH-type RNA binding domain containing	+	[96]	EX668302	5E-52
FY	AT5G13480	mRNA 3' end processing factor	+	[97]	EX659635	5E-75
SKB1	AT4G31120	Arginine methyltransferase	+	[98]	nf	
FVE	AT2G19520	retinoblastoma associated	+	[24]	VES-001-B03	3E-76
LD	AT4G02560	DNA/RNA binding homeodomain protein	+	[99]	DY670534	3E-49
FLD	AT3G10390	component of histone deacetylase complex	+	[23]	nf	
LDL1/SWP1	AT1G62830	Histone H3-Lys 4 demetylase-like	+	[22]	Contig 2573	2E-27
LDL2	AT3G13682	Histone H3-Lys 4 demetylase-like	+	[22]	DY669828	1E-42
Gibberellin pathway						
GAI	AT1G14920	putative transcriptional repressor	-	[100]	Contig 3276	3E-147
RGA	AT2G01570	putative transcriptional repressor	-	[100]	DQ195503	8E-60
SPY	AT3G11540	O-linked N-acetylglucosamine transferase	-	[101]	BAR-002-C02	2E-93
DDF1	AT1G12610	AP2-like TF	+	[102]	Contig 3158	5E-49
DDF2	AT1G63030	AP2-like TF	+	[102]	nf	
AtMYB33	AT5G06100	MYB TF	+	[25]	DY669997	5E-29
FPF1	AT5G24860	Unknown	+	[103]	Contig 4074	7E-38
Other						
SVP	AT2G22540	MADS-box TF	-	[57]	VES-013-D05	5E-22
AP2	AT4G36920	AP2 TF	-	[104]	VES-008-A07	9E-16
PFT1	AT1G25540	vWF-A domain protein	+	[48]	BAR-002-D08	1E-17
HRB1	AT5G49230	ZZ type zinc finger protein	+	[49]	VES-012-B01	7E-22

The most important genes of *Arabidopsis* autonomous and gibberellin pathways as well as some other floral regulators are presented. The biological function of the genes is indicated, and floral activators and repressors are marked by + and - marks, respectively. Moreover, the presence or absence of homologous sequence in *Fragaria* sequence databases and E-value of BLASTx comparison against *Arabidopsis* are indicated. Sequences found in our libraries are named BAR and VES for everbearing genotype 'Baron Solemacher' and short-day genotype, respectively. Other ESTs and EST contigs are found from Genome Database for Rosaceae http://www.bioinfo.wsu.edu/gdr/. More complete list is available in Additional file 2.

Identification of Floral Integrator Genes in Fragaria

Sequencing of our EST collections did not reveal any homologs for the floral integrator or identity genes such as FT, SOC1, LFY or AP1 [12,58]. A full-length cDNA sequence of SOC1 homolog [GenBank:FJ531999] and a 713 bp 3'-end fragment of putative LFY [GenBank:FJ532000] were isolated using PCR. Closest protein homolog of the putative FvSOC1 was 72% identical Populus trichocarpa MADS5, and the putative FvLFY showed highest amino acid identity (79%) to Malus domestica FL2. Comparison to Arabidopsis showed that AtSOC1 and AtLFY, respectively, were 66% and 75% identical with the corresponding wild strawberry protein sequences (Figure 3A and 3B). FT homolog, instead, was not identified in Fragaria despite of many attempts using degenerate PCR and screening of cDNA library plaques and E.coli clones from a variety of tissues and developmental conditions with the Arabidopsis coding sequence (K. Folta, unpublished). However, a putative FT was found in Prunus and Malus protein databases at NCBI. Among the other genes belonging to the same gene family, homologs of MFT (MOTHER OF FT AND TFL1) and ATC (ARABIDOPSIS

CENTRORADIALIS) [59] were present in GDR Fragaria EST. Moreover, an EST contig corresponding to the floral identity gene AP1 was found. The length of the translated protein sequence of FvAP1 was 284 amino acids, being 30 amino acids longer than the corresponding Arabidopsis sequence. However, FvAP1 EST contig contained an unknown sequence stretch of 81 bp at nucleotide position 596-677. Putative FvAP1 showed highest overall identity (68%) with putative AP1 from Prunus persica (Figure 3C). Moreover, the 5' sequence containing 187 amino acids (the sequence before the unknown part) was 73% identical with the Arabidopsis AP1.

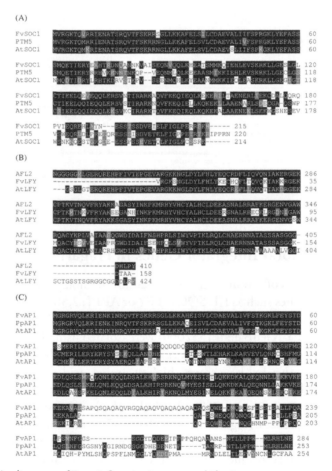

Figure 3. Protein alignments of Fragaria flowering integrator and identity genes. Multiple alignments of Fragaria protein sequences of full length SOC1 (A), partial LFY (B) and full-length AP1 (C) with closest protein homologs and corresponding protein sequence of Arabidopsis thaliana. Alignments were done by ClustalW (A, B) or T-Coffee (C) and modified by Boxshade program. F. vesca AP1 protein sequence was translated from GDR Fragaria EST contig 4941. PTM5 = Populus tremuloides MADS5, AFL2 = Apple FLORICAULA 2, PpAP1 = putative Prunus persica AP1.

Gene Expression Analysis Revealed Few Differences Between EB and SD Genotypes

We compared the expression of selected flowering time genes (Table 5) corresponding to each flowering pathway in the leaf and shoot apex samples of EB and SD genotypes in order to explore the role of different pathways. Only few of the analysed genes were differentially expressed between the genotypes. Floral integrator gene LFY was slightly up-regulated in the shoot apex samples of EB (Table 6). Moreover, PCR expression analysis with two different primer pairs showed that AP1 was specifically expressed in EB apices correlating with the identity of the meristems. Among the genes from different flowering pathways, only two genes, vernalization pathway gene ELF8 [17] and photoperiod pathway gene ELF3 [60], were slightly differentially expressed between the genotypes (Table 6).

Table 5. The list of PCR primers used in real-time RT-PCR.

Gene	Forward primer	Reverse primer
UBI	CAGACCAGCAGAGGCTTATCTT	TTCTGGATATTGTAGTCTGCTAGGG
LFY	CGGCATTACGTTCACTGCTA	CCTGTAACACGCCTGCATC
SOC1	CAGGTGAGGCGGATAGAGAA	AGAGCTTTCCTCTGGGAGAGA
AP1	CGCTCCAGAAGAAGGATAAGG	CATGTGACTGAGCCTGTGCT
AP1	TCTGAAGCACGTAAGGTCTA	ATCCTGATCATAACCTCCAG
LHY	AAAGCTGGAGAAGGAGGCAGTC	CCGAGGATAAGGATTGCTTGGT
ZTL	TGCATGGGGTAGTGAAACAA	CACCTCCGACAGTGACCTTT
FKF1	ACCCACATCGTTTGTGGTCT	ACATCAGGATCCACCAGAGG
ELF3	TCCTCCAAGGAACAAGATGG	CCATTCCCCTGATTTGAGAG
ELF6	TTCGAAGGTCTTGGCAATGG	GCGCCTGAGTTTTATCCAACAC
COL4	GACCGAGAAATCCACTCTGC	CTCTCCGTCCGACAAGTAGC
CO	GACATCCACTCCGCCAAC	GTGGACCCCACCACTATCTG
PFT1	GCGACATGCCAAGGTTAGAATT	TCAGCGCCTCACACTCTTACAC
HRB1	GAATGGTGGACATCAGCAATCC	CCTCCGAAAGAATTGCTCAACA
FYPP3	ACAAAATGGCCCCTCATGTG	TGTGCTATGTGTCCATGGTGGT
FRL	CGCTAGTCAAGGTCGAGGAG	CGACTTCATCTCCATCAGCA
ELF8	GCTCAGAATGCTCCTCCTGT	TGAGTATTGCAGCCACTTGC
VRN5	AGCCCTTGATGTCATCAGCTG	CCGATGAATGGTTGGCTAATG
MSI1	TCTCCACACCTTTGATTGCCA	ACACCATCAGTCTCCTGCCAAG
LHP1	GGAGAGCCAGAACCAGGAG	CTCACCTTCTTCCCCTTCCT
FVE	GATCCAGCAGCAACCAAGTCTC	CCTCTTGGTGCAACAGAAGGAC
SVP	CGTGCTAAGGCAGATGAATGG	TGAAGCACACGGTCAAGACTTC
SPY	TGCGGTGTCAAATTGCATCA	GGCAACACTCAAGATGGATTGC
GA3ox	CCTCACAATCATCCACCAATCC	CGCCGATGTTGATCACCAA
GA2ox	CACCATGCCCAGAGCTTCA	AGGCCAGAGGTGTTGTTGGAT
TFL1	TGCAGAAACAAACGAGTTCGG	CCAAGAGCATCGATCATTTGGT
AP2	CCCGAAATCCTTGATTGTTCC	AACACTGCAATCGAACAACAGC

T$_m$value of the primers is 60 ± 1°C.

Table 6. The expression of selected genes in the wild strawberry.

Gene	MSI1 as a control	FVE as a control
Shoot apex samples		
AP1	Expressed only in EB	Expressed only in EB
LFY	1.8 ± 0.4	1.9 ± 0.3
ELF8	1.5 ± 0.1	1.6 ± 0.1
Leaf samples		
ELF3	1.5 ± 0.1	1.8 ± 0.0

Developmental Regulation of Floral Integrator, Floral Identity, and GA Pathway Genes

We analysed the developmental regulation of AP1, LFY, SOC1, GA3ox and GA2ox transcription in the shoot apices of LD grown plants of EB and SD genotype containing one to four leaves. Ubiquitin, used as a control gene, was stable between different developmental stages, but was amplified ~1 PCR cycle earlier in SD genotype. Thus direct comparison between the genotypes is not possible, but the trends during development are comparable. Three genes, AP1, LFY, and GA3ox, had clear developmental stage dependent expression pattern in EB apices, showing biggest changes after one or two leaf stage (Figure 4). The expression of AP1 was detected in EB apices already at one leaf stage, and its mRNA accumulated gradually reaching 6-fold increase at two leaf stage and 50-fold increase at four leaf stage (Figure 4A). In parallel, transcription of LFY started to increase at 2-leaf stage, but the change in its expression was much smaller (Figure 4B). A floral integrator gene, SOC1, in contrast, did not show clear developmental regulation (Figure 4C). Also GA pathway was co-regulated with AP1 and LFY, since GA biosynthetic gene GA3ox was strongly down-regulated after two leaf stage (Figure 4D). In addition, GA catabolism gene, GA2ox, tended to follow changes in the expression of GA3ox, although the results were not so clear (data not shown). In SD genotype, in contrast, AP1 was absent and other genes did not show clear developmental regulation (Figure 4). In this experiment, control plants of EB genotype flowered very early, after producing 4.7 ± 0.3 leaves to the main crown, whereas plants of SD genotype remained vegetative.

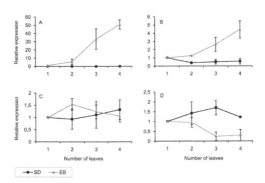

Figure 4. Developmental regulation of gene expression in wild strawberry shoot apices.

The expression of AP1 (A), LFY (B), SOC1 (C) and GA3ox (D) in the SD and EB ('Baron Solemacher') genotype of the wild strawberry. Triplicate shoot apex samples were collected from LD grown plants at one to four leaf stage. Ct values were normalized against a Ubiquitin [GenBank:DY672326] gene to get normalized ΔCt values. The expression differences between one leaf stage and later developmental stages were calculated from the formula $2^{\Delta Ct \text{ later developmental stage}}/2^{\Delta Ct \text{ one leaf stage}}$. The expression values at one leaf stage were artificially set to 1 separately for both genotypes. Values are mean ± SD. Note that Ubiquitin was amplified ~1 cycle earlier in SD genotype, but was stable between different developmental stages. Therefore, expression values between genotypes cannot be directly compared, while the expression levels between the various developmental stages are comparable.

Discussion

Identification of Flowering Genes in Strawberry

Genetic regulation of flowering in strawberry has earlier been studied only by crossing experiments. According to Weebadde et al. [61], everbearing character is a polygenic trait in garden strawberry whereas other studies indicate the presence of a single dominant gene [62]. Different results may arise from different origin of everbearing habit, since at least three different sources have been used in strawberry breeding [32,61,62]. Studies in F. vesca 'Baron Solemacher'have shown that EB flowering habit in this genotype is controlled by recessive alleles of a single locus, called seasonal flowering locus (sfl) [40,41]. Identification of central genes regulating flowering, as well as those controlling other processes that affect flowering (runnering, chilling), is an important goal that would greatly accelerate breeding of strawberry and other soft fruit and fruit species of Rosaceae family.

For comprehensive identification of candidate genes of the strawberry flowering pathways, we searched homologs for 118 Arabidopsis flowering time genes from our own cDNA libraries and from GDR. In total, we were able to identify 66 gene homologs among about 53000 EST sequences. Moreover, gene homologs lacking from Fragaria were further mined from Rosaceae EST collections containing about 410 000 EST sequences. These searches revealed 22 additional putative flowering time genes in Rosaceae. Ongoing genome sequencing projects in apple, peach and wild strawberry will ultimately reveal the currently lacking flowering regulators in these species [63].

Sequences found in Fragaria corresponded to all known Arabidopsis flowering time pathways [2] suggesting that all of these genetic pathways may be present in Fragaria. However, the sequence conservation does not necessarily mean functional conservation, so major candidate genes from different pathways have to be functionally characterized in order to prove the presence of these pathways in strawberry. Few central regulators of flowering time are lacking from Fragaria sequence collections and some of them also from Rosaceae databases. For example, we were not able to identify a homolog for the florigen gene FT [11] in Fragaria regardless of several different attempts. This is probably due to its low expression level and tissue specific expression pattern [64]. Similarly, GI, which links circadian clock and CO [8,65], was absent from the Fragaria sequences. FT and GI homologs were, however, found in apple and Prunus, showing that they are present in Rosaceae. Moreover, consistent with studies in model legumes [66], CCA1 was lacking in Rosaceae, but its redundant paralog, LHY, is represented by few ESTs in Fragaria. CCA1 and LHY are MYB-type transcription factors which repress the expression of TOC1 in the central loop of Arabidopsis circadian clock [67]. Thus, in Fragaria and other species of Rosaceae family, LHY alone may control the expression of TOC1 in the clock core. This contrasts with other species, like Populus, where duplications of the LHY/CCA1 genes contribute to an apparently more complex mode of clock control [68].

Vernalization pathway in Arabidopsis culminates in FLC and FLC-like floral repressors [13,50]. They have been functionally characterized only in Brassicaceae [13,69], although homologous MADS box genes have been recently found from several eudicot lineages by phylogenetic analysis [51]. However, we were not able to identify FLC-like sequences in Rosaceae by using several FLC-like sequences as a query. Similarly, also FRI homologs were lacking from the Rosaceae sequence collections. However, putative homologs of FRI-like genes, FRL1 and FRL2, which are involved in FLC activation in Arabidopsis [15] were found, as well as several other homologs of genes belonging to FLC regulating protein complexes. Despite the presence of these transcripts, the presence of FLC is unclear, since at least PRC2 complex has several target genes [70]. Cloning and characterization of putative FLC-like and FRI genes as well as FT in strawberry would greatly expand our understanding of strawberry flowering pathways, and therefore, it is one of the most important targets of further studies. If these transcripts are present in strawberry, it is likely that the precise control of flowering has placed their expression in specific tissues or contexts where they are not easily detected. However, their presence should be substantiated in analysis of the impending genome sequence. Another important goal is the identification of putative Rosaceae or Fragaria specific flowering time genes. Ultimately, transcriptomics studies and functional analysis of central genes may reveal how different flowering pathways, which may be closely related to Arabidopsis pathways, make seasonal flowering in strawberry.

What is the SFL Gene?

SFL is a single dominant locus that enforces seasonal flowering habit in wild strawberry, and homozygous mutation in this locus leads to continuous flowering habit in at least one genotype, 'Baron Solemacher' [36]. In SD genotypes of wild strawberry, SD or low temperature induce flowering [27] probably by overcoming the function of SFL repressor gene. We showed here that EB genotypes 'Baron Solemacher' and 'Hawaii-4' produce only 5 - 6 leaves to the main crown before the formation of the terminal inflorescence in LD at 18°C. Hence, flowering induction in these conditions occurs soon after germination. In SD (12 h) or at low temperature (11°C) instead, plants formed several leaves more before the inflorescence. This finding shows that, in contrast to SD genotypes, both SD and low temperature restrain flowering induction in these genotypes, confirming earlier suggestions that EB genotypes of wild strawberry are in fact LD plants [35]. Most simple explanation for these opposite environmental responses is that the lack of flowering inhibitor, produced by active SFL gene, unmasks LD induced flowering promotion pathway in 'Baron Solemacher' and possibly in other EB genotypes. Given that both SD and low temperature repress SFL, analogous flowering regulating pathway has not yet been characterized at molecular level.

Our gene expression analysis did not give any hints of the putative location of SFL in wild strawberry flowering pathways. However, homologs of certain flowering repressors can be consireded as candidates for SFL, including the rice CO homolog HD1 (HEADING DATE 1), or Arabidopsis vernalization pathway genes FLC and FRI [13,14,71]. In strawberry, the role of vernalization pathway remains unclear until the presence or absence of FRI or FLC function is confirmed or other targets for this pathway are found. Strawberry CO, instead, has been cloned and mapped in Fragaria reference map, but its position does not match with the genomic location of SFL showing that CO itself is not SFL [44,72]. However, the possibility that some regulator of CO transcription or protein stability could be SFL cannot be ruled out and should be studied further.

Exogenously applied GA inhibits flowering in wild strawberry, and therefore, GA has been suggested to be a floral repressor [38,39]. Similar patterns have been observed and delineate differences between recurrent and non-recurrent roses [73]. However, we did not find clear differences in the expression of GA biosynthetic and catabolism genes, GA3ox and GA2ox, in the shoot apex samples of EB and SD genotypes before the floral initiation had occurred. In contrast, GA3ox was strongly repressed in EB apices after floral initiation and GA2ox showed similar trend. The fact that these changes in GA pathway occurred after two leaf stage suggests that GA signal was regulated during early flower development rather than during floral transition. These data does not support the role of endogenous GA as the regulator of flowering induction, indicating that SFL is not situated in the GA pathway. However, quantitative analysis of GA levels is needed to show whether the observed changes in the expression of GA pathway genes are reflected at the metabolic level.

AP1 is a Potential Marker of Floral Initiation in Strawberry

Gene expression analysis revealed that two putative flowering genes, AP1 and LFY, were co-regulated during floral development in EB wild strawberry. The homolog of floral identity gene AP1 was expressed in the EB apex already at one leaf stage, and its expression was strongly enhanced during later developmental stages. Also LFY mRNA accumulated along with AP1 during floral development in EB genotype, whereas SOC1 did not show a clear trend. The mRNA of SOC1 and LFY were present also in SD genotype, but AP1 transcription was not detected. In

Arabidopsis, LFY and AP1 activate each other's expression constituting a feedback loop [12,58]. Moreover, AP1 is activated by FT-FD heterodimer shortly after flowering induction [10]. Thus, the expression patterns of AP1and LFY in the meristems of EB genotype suggest that flowering induction in these plants occurs before two leaf stage in LD conditions. Consistent with this conclusion, flower initials were clearly visible by stereomicroscope in the meristems at three or four leaf stage, and plants flowered after producing on average 4.7 leaves in the main crown. Based on our results, AP1 can be used as a marker for floral initiation in wild strawberry. However, functional studies are needed to confirm the role of AP1, LFY and SOC1 as floral integrator and identity genes, and this approach is currently going on.

Conclusion

We have explored putative components for the genetic flowering pathways in perennial SD plant wild strawberry by identifying 66 homologs of Arabidopsis flowering time genes. Although few central genes are lacking, these data indicate that all known genetic flowering pathways may be present in Fragaria. This is consistent with the finding that EB genotypes, 'Hawaii-4' and 'Baron Solemacher', show similar environmental regulation of flowering than Arabidopsis summer-annuals. We also studied the expression of selected candidate genes and found that few genes were co-regulated in the shoot apex of the EB genotype during early floral development. Most strikingly, the mRNA of AP1 specifically accumulated in EB genotype, but was absent in SD genotype, showing its usefulness as a marker of floral initiation. Finally, identification of putative flowering time genes reported here enables their transcriptional and functional characterization, as well as genetic mapping, which may give answers for the relative importance of each genetic flowering pathway and lead to cloning of the central repressor gene, SFL. Ultimately, these genetic resources could be utilized in cultivar breeding of various species of Rosaceae family through genetic transformation and marker assisted selection breeding.

Methods

Plant Materials, Growing Conditions and Sampling

Seeds of SD and EB ('Baron Solemacher') genotypes of the wild strawberry (NCGR accession numbers [PI551792] and [PI551507], respectively) were sown on potting soil mixture (Kekkilä, Tuusula, Finland) and grown in a greenhouse under LD conditions (day length min. 18 h), provided by 400 W SON-T lamps

(Airam, Kerava, Finland) and natural sunlight. After two to three leaves had developed per plant, shoot apex samples (tip of the shoot containing the meristem as well as two to three leaf initials) were collected under a stereomicroscope at ten different time points with three days intervals. Samples from each time point were pooled and used for the construction of cDNA libraries and real-time RT-PCR. WT samples contained shoot apices of the main crown, collected from 50 plants per time point. Also in EB genotype, shoot apices of the main crown were collected until the sepal initials became visible in the meristems. After this time point, the apices from one to three side shoots per plant were collected, altogether from 40 plants per sampling. In addition, leaf samples were collected from the same plants at four leaf stage for real-time RT-PCR analysis. Moreover, separate shoot apex samples were collected from WT and EB genotypes at one, two, three and four leaf stages. Control plants were grown in LD and their flowering time was determined by counting the number of leaves in the main crown before the terminal inflorescence. All samples were collected in July - August 2006 - 2008.

Preparation and Sequencing of Subtracted cDNA Libraries

Total RNA from pooled shoot apex samples was extracted with a pine tree method for RNA isolation [74]. The cDNA was synthesized with BD SMART cDNA Synthesis kit (Clontech, Palo Alto, US), amplified with PCR as instructed for subtraction, purified with Chroma Spin-1000 DEPC-H2O Columns (Clontech), extracted with chloroform:isoamylalcohol (24:1) using Phase Loch Gel Heavy 1.5 ml tubes (Eppendorf, Hamburg, Germany), digested with RsaI (Boehringer Mannheim, Mannheim, Germany), and purified with High Pure PCR Product Purification kit (Roche Diagnostics, Indianapolis, US). The cDNAs were subtracted using BD PCR-Select cDNA Subtraction Kit (Clontech) in both forward and reverse directions. The forward and reverse PCR mixtures were digested with RsaI (Boehringer Mannheim) and purified with High Pure PCR Product Purification Kit (Roche). After digestion, A-tailing was done as instructed in the technical manual of pGEM-T and pGEM-T Easy Vector Systems and PCR mixtures were ligated to pGEM-T Easy Vector (Promega, Wisconsin, US), and electroporated to TOP10 cells. The libraries were sequenced at the Institute of Biotechnology, University of Helsinki, as described earlier [75].

Bioinformatics Analysis

Raw EST sequences were quality checked before annotation. Base calling, end clipping and vector removal were performed by CodonCodeAligner-software (CodonCode Corporation, US). After this the ESTs were manually checked and sequences that contained poly-T in the beginning followed by short repetitive sequences were removed. BLASTx was performed against functionally annotated Arabidopsis protein database (v211200, MIPS), Swissprot and non-redundant protein database (NCBI), and Populus trichocarpa genome of DOE Joint Genome Institute [76] using cut-off value 1e-10. tBLASTx was performed against TIGR plant transcript assemblies of Malus x domestica, Oryza sativa and Vitis vinifera [77], and GDR Fragaria and Rosaceae Contigs using cut-off value 1e-10. For MIPS BLAST hits corresponding functional classes and Gene Ontology classes were obtained from Functional Classification Catalogue (Version 2.1) and GO annotation for Arabidopsis thaliana (Version 1.1213).

Homologs of Arabidopsis flowering time genes were searched from GDR Fragaria contig and EST databases using tBLASTx algorithm and Arabidopsis protein sequences as a query. Homologous sequences passing a cut-off value 1e-10 were further analysed by BLASTx algorithm against Arabidopsis protein database, and sequences showing highest sequence homology with the corresponding Arabidopsis genes were selected. The sequences lacking from Fragaria were similarly searched from GDR Rosaceae EST database and from Rosaceae protein database at NCBI.

Photoperiod and Temperature Treatments

For the analysis of environmental regulation of flowering in EB genotypes, seeds of 'Baron Solemacher', and 'Hawaii-4' were germinated in 18 h LD at 18°C. During germination, plants were illuminated using 400 W SON-T lamps (Airam) for 12 h daily (90 ± 10 μmol m^{-2} s^{-1} at plant height plus natural light) and incandescent lamps were used for low-intensity daylength extension (5 ± 1 μmol m^{-2} s^{-1} at plant height). After opening of the cotyledons plants were moved to four treatments, SD and LD (12/18 h) at low or high temperature (11/18°C), for five weeks. In LD, incandescent lamps were used for low-intensity daylength extension (5 ± 1 μmol m^{-2} s^{-1} at plant height) after 12 h main light period. Also photoperiods of 8 and 8 + 8 h (SD/LD) were tested, but because of very slow growth in these light treatments, longer photoperiods were selected (data not shown). SD treatments were carried out at the greenhouse using darkening curtains, while LD treatments (photoperiod 18 h) were conducted in a similar greenhouse compartment without curtains. The experiments were carried out during winter 2007 - 2008, when the natural day length was under 12 h. After treatments, plants were potted to 8 x 8 cm pots, moved to LD (18 h), and flowering time was determined as described above.

Gene Expression Analysis

Total RNA from leaf and shoot apex samples was extracted with a pine tree method [74], and cDNAs were synthesized from total RNA using Superscript III RT kit

(Invitrogen, Carlsbad, US) and $dT_{18}VN$ anchor primers. LightCycler 480 SYBR Green I Master kit (Roche Diagnostics, Indianapolis, US) was used to perform 15 µl real-time RT-PCR reactions in 384-well plates according to manufacturer's instructions by using Light Cycler 480 real-time PCR system (Roche Diagnostics). PCR primers with T_m value of 60°C were used (Table 5). Three biological replicates were analysed for shoot apex samples from different developmental stages (Figure 4), and two biological replicates were used for pooled shoot apex and leaf samples (Table 6).

Authors' Contributions

TH, KM and PE designed all experiments. PE coordinated the study and helped to draft the manuscript. TH run the real-time PCR analysis, performed flowering gene searches from sequence databases, and drafted the manuscript together with KM. KM constructed the subtracted cDNA libraries and performed bioinformatics analysis together with KF. KF also helped to draft the manuscript. MR participated in flowering time analysis and sampling of shoot apices. PA and LP were responsible for the EST sequencing. All authors read and approved the final manuscript.

Acknowledgements

Dr. Michael Brosche is acknowledged for his kind help in the production of subtracted cDNA libraries and M.Sc. Techn. Erkko Airo for his valuable technical help. In addition, Finnish Ministry of Agriculture and Forestry is thanked for financial support.

References

1. Ausín I, Alonso-Blanco C, Martinez-Zapater M: Environmental regulation of flowering. Int J Dev Biol 2005, 49:689–705.

2. Putterill J, Laurie R, Macknight R: It's time to flower: the genetic control of flowering time. Bioessays 2004, 26:363–373.

3. Simpson GG: The autonomous pathway: epigenetic and post-transcriptional gene regulation in the control of Arabidopsis flowering time. Curr Opinion Plant Biol 2004, 7:570–574.

4. Imaizumi T, Kay SA: Photoperiodic control of flowering: not only by coincidence. Trends Plant Sci 2006, 11:550–558.

5. Strayer C, Oyama T, Schultz TF, Raman R, Somers DE, Más P, Panda S, Kreps JA, Kay SA: Cloning of the Arabidopsis clock gene TOC1, an autoregulatory response regulator homolog. Science 2000, 289:768–771.

6. Wang ZY, Tobin EM: Constitutive expression of the CIRCADIAN CLOCK ASSOCIATED 1 (CCA1) gene disrupts circadian rhytms and suppresses its own expression. Cell 1998, 93:1207–1217.

7. Schaffer R, Ramsay N, Samach A, Corden S, Putterill J, Carré IA, Coupland G: The late elongated hypocotyl mutation of Arabidopsis disrupts circadian rhythms and the photoperiodic control of flowering. Cell 1999, 93:1219–1229.

8. Suárez-López P, Wheatley K, Robson F, Onouchi H, Valverde F, Coupland G: CONSTANS mediates between the circadian clock and the control of flowering in Arabidopsis. Nature 2001, 410:1116–1120.

9. Yanovsky MJ, Kay SA: Molecular basis of seasonal time measurement in Arabidopsis. Nature 2002, 419:308–312.

10. Abe M, Kobayashi Y, Yamamoto S, Daimon Y, Yamaguchi A, Ikeda Y, Ichinoki H, Notaguchi M, Goto K, Araki T: FD, a bZIP protein mediating signals from the floral pathway integrator FT at the shoot apex. Science 2005, 309:1052–1056.

11. Corbesier L, Vincent C, Jang S, Fornara F, Fan Q, Searle I, Giakountis A, Farrona S, Gissot L, Turnbull C, Coupland G: FT protein movement contributes to long-distance signalling in floral induction of Arabidopsis. Science 2007, 316:1030–1033.

12. Parcy M: Flowering: a time for integration. Int J Dev Biol 2005, 49:585–593.

13. Searle I, He Y, Turck F, Vincent C, Fornara F, Krober S, Amasino RA, Coupland G: The transcription factor FLC confers a flowering response to vernalization be repressing meristem competence and systemic signalling in Arabidopsis. Genes Dev 2006, 20:898–912.

14. Johanson U, West J, Lister C, Michaels S, Amasino R, Dean C: Molecular analysis of FRIGIDA, a major determinant of natural variation in Arabidopsis flowering time. Science 2000, 290:344–347.

15. Michaels SD, Bezerra IC, Amasino RM: FRIGIDA-related genes are required for the winter-annual habit in Arabidopsis. PNAS 2004, 101:3281–3285.

16. Noh Y, Amasino RS: PIE1, an ISWI family gene, is required for FLC activation and floral repression in Arabidopsis. Plant Cell 2003, 15:1671–1682.

17. He Y, Doyle MR, Amasino RM: PAF1-complex-mediated histone methylation of FLOWERING LOCUS C chromatin is required for the vernalization-responsive, winter-annual habit in Arabidopsis. Genes Dev 2004, 18:2774–2784.

18. Zhang H, Ransom C, Ludwig P, van Nocker S: Genetic analysis of early flowering mutants in Arabidopsis defines a class of pleiotropic developmental regulator required for expression of the flowering-time switch Flowering Locus C. Genetics 2003, 164:347–358.

19. Wood CC, Robertson M, Tanner G, Peacock WJ, Dennis ES, Helliwell CA: The Arabidopsis thaliana vernalization response requires a Polycomb-like protein complex that also includes VERNALIZATION INSENSITIVE 3. PNAS 2006, 103:14631–14636.

20. Sung S, Amasino RM: Vernalization in Arabidopsis thaliana is mediated by the PHD finger protein VIN3. Nature 2004, 427:159–164.

21. Quesada V, Dean C, Simpson GG: Regulated RNA processing in the control of Arabidopsis flowering. Int J Dev Biol 2005, 49:773–780.

22. Jiang D, Yang W, He Y, Amasino RM: Arabidopsis relatives of the human lysine-specific demethylase 1 repress the expression of FWA and FLOWERING LOCUS C and thus promote the floral transition. Plant Cell 2007, 19:2975–2987.

23. He Y, Michaels SD, Amasino RM: Regulation of flowering time by histone acetylation in Arabidopsis. Science 2003, 302:1751–1754.

24. Ausín I, Alonso-Blanco C, Jarillo JA, Ruiz-García L, Martínez-Zapater JM: Regulation of flowering time by FVE, a retinoblastoma-associated protein. Nat Genet 2004, 36:162–166.

25. Gogal GFW, Sheldon CC, Gubler F, Moritz T, Bagnall DJ, MacMillan CP, Li SF, Parish RW, Dennis ES, Weigel D, King RW: GAMYB-like genes, flowering, and gibberellin signaling in Arabidopsis. Plant Physiol 2001, 127:1682–1693.

26. Jonkers H: On the flower formation, the dormancy and the early forcing of strawberries. In Thesis. Mededelingen van de Landbouwhogeschool, Wageningen; 1965.

27. Heide O, Sønsteby A: Interactions of temperature and photoperiod in the control of flowering of latitudinal and altitudinal populations of wild strawberry (Fragaria vesca). Physiol Plant 2007, 130:280–289.

28. Heide O: Photoperiod and temperature interactions in growth and flowering of strawberry. Physiol Plant 1977, 40:21–26.

29. Guttridge CG: Fragaria × ananassa. In CRC Handbook of Flowering. Volume III. Edited by: Halevy A. Boca Raton: CRC Press; 1985:16–33.

30. Konsin M, Voipio I, Palonen P: Influence of photoperiod and duration of short-day treatment on vegetative growth and flowering of strawberry (Fragaria × ananassa Duch.). J Hort Sci Biotech 2001, 76:77–82.

31. Hytönen T, Palonen P, Mouhu K, Junttila O: Crown branching and cropping potential in strawberry (Fragaria × ananassa, Duch.) can be enhanced by day-length treatments. J Hort Sci Biotech 2004, 79:466–471.

32. Darrow GM: The strawberry. History, breeding and physiology. New York: Holt, Rinehart and Winston; 1966.

33. Durner EF, Barden JA, Himelrick DG, Poling EB: Photoperiod and temperature effects on flower and runner development in day-neutral, junebearing and everbearing strawberries. J Amer Soc Hort Sci 1984, 109:396–400.

34. Sønsteby A, Heide OM: Long-day control of flowering in everbearing strawberries. J Hort Sci Biotech 2007, 82:875–884.

35. Sønsteby A, Heide OM: Long-day rather than autonomous control of flowering in the diploid everbearing strawberry Fragaria vesca ssp. semperflorens. J Hort Sci Biotech 2008, 83:360–366.

36. Albani M, Battey NH, Wilkinson MJ: The development of ISSR-derived SCAR markers around the SEASONAL FLOWERING LOCUS (SFL) in Fragaria. Theor Appl Gen 2004, 109:571–579.

37. Guttridge CG: Further evidence for a growth-promoting and flower-inhibiting hormone in strawberry. Annals Bot 1959, 23:612–621.

38. Thompson PA, Guttridge CG: Effect of gibberellic acid on the initiation of flowers and runners in the strawberry. Nature 1959, 184:72–73.

39. Guttridge CG, Thompson PA: The effects of gibberellins on growth and flowering of Fragaria and Duchesnea. J Exp Bot 1963, 15:631–646.

40. Brown T, Wareign PF: The genetic control of the everbearing habit and three other characters in varieties of Fragaria vesca. Euphytica 1965, 14:97–112.

41. Battey N, Miere P, Tehranifar A, Cekic C, Taylor S, Shrives K, Hadley P, Greenland A, Darby J, Wilkinson M: Genetic and environmental control of flowering in strawberry. In Genetic and Environmental Manipulation of Horticultural Crops. Edited by: Cockshull KE, Gray D, Seymour GB, Thomas B. Wallingford, UK, Cab International; 1998:111–131.

42. Diatchenko L, Lau YF, Campbell AP, Chenchik A, Moqadam F, Huang B, Lukyanov S, Lukyanov K, Gurskaya N, Sverdlov ED, Siebert PD: Suppression subtractive hybridization: a method for generating differentially regulated or tissue-specific cDNA probes and libraries. PNAS 1996, 93:6025–6030.

43. Thomas B: Light signals and flowering. J Exp Bot 2006, 57:3387–3393.

44. Stewart P: Molecular characterization of photoperiodic flowering in strawberry (Fragaria sp.). PhD thesis. University of Florida; 2007.

45. Fowler S, Lee K, Onouchi H, Samach A, Richardson K, Morris B, Coupland G, Putterill J: GIGANTEA: a circadian clock-controlled gene that regulates photoperiodic flowering in Arabidopsis and encodes a protein with several possible membrane-spanning domains. EMBO J 1999, 18:4679–4688.

46. Jang S, Marchal V, Panigrahi KCS, Wenkel S, Soppe W, Deng X, Valverde F, Coupland G: Arabidopsis COP1 shapes the temporal pattern of CO accumulation conferring a photoperiodic flowering response. EMBO J 2008, 27:1277–1288.

47. Laubinger S, Marchal V, Le Gourrierec J, Wenkel S, Adrian J, Jang S, Kulajta C, Braun H, Coupland G, Hoecker U: Arabidopsis SPA proteins regulate photoperiodic flowering and interact with floral inducer CONSTANS to regulate its stability. Development 2006, 133:3213–3222.

48. Cerdán PD, Chory J: Regulation of flowering time by light quality. Nature 2003, 423:881–885.

49. Kang X, Zhou Y, Sun X, Ni M: HYPERSENSITIVE TO RED AND BLUE 1 and its C-terminal regulatory function control FLOWERING LOCUS T expression. Plant J 2007, 52:937–948.

50. Scortecci KC, Michaels SD, Amasino RM: Identification of a MADS-box gene, FLOWERING LOCUS M, that repress flowering. Plant J 2001, 26:229–236.

51. Reeves PA, He Y, Schmitz RJ, Amasino RM, Panella LW, Richards CM: Evolutionary conservation of the FLOWERING LOCUS C-mediated vernalization response: evidence from the sugar beet (Beta vulgaris). Genetics 2007, 176:295–307.

52. Choi K, Park C, Lee J, Oh M, Noh B, Lee I: Arabidopsis homologs of components of the SWR1 complex regulate flowering and plant development. Development 2007, 134:1931–1941.

53. Kim KS, Choi K, Park C, Hwanga H, Lee I: SUPPRESSOR OF FRIGIDA4, encoding a C2H2-type zinc finger protein, represses flowering by transcriptional activation of Arabidopsis FLOWERING LOCUS C. Plant Cell 2006, 18:2985–2998.

54. Folta KM, Staton M, Stewart PJ, Jung S, Bies DH, Jesudurai C, Main D: Expressed sequence tags (ESTs) and simple sequence repeat (SSR) markers from octoploid strawberry (Fragaria × ananassa). BMC Plant Biol 2005, 5:12.

55. Levy YY, Mesnage S, Mylne JS, Gendall AR, Dean C: Multiple roles of Arabidopsis VRN1 in vernalization and flowering time control. Science 297:243–246.

56. Mylne JS, Barrett L, Tessadori F, Mesnage S, Johnson L, Bernatavichute VN, Jacobsen SE, Fransz P, Dean C: LHP1, the Arabidopsis homologue of HET-

EROCHROMATIN PROTEIN1, is required for epigenetic silencing of FLC. PNAS 2006, 103:5012–5017.

57. Lee JH, Yoo SJ, Park SH, Hwang I, Lee JS, Ahn JH: Role of SVP in the control of flowering time by ambient temperature in Arabidopsis. Genes Dev 2007, 21:397–402.

58. Wagner D, Sablowski RWM, Meyerowitz EM: Transcriptional activation of APETALA1 by LEAFY. Science 1999, 285:582–584.

59. Turck F, Fornara F, Coupland G: Regulation and identity of florigen: FLOWER-ING LOCUS T moves central stage. Annu Rev Plant Biol 2008, 59:573–594.

60. Zagotta MT, Hicks KA, Jacobs CI, Young JC, Hangarter RP, Meeks-Wagner D: The Arabidopsis ELF3 gene regulates vegetative photomorphogenesis and the photoperiodic induction of flowering. Plant J 1996, 10:691–702.

61. Weebadde CK, Wang D, Finn CE, Lewers KS, Luby JJ, Bushakra J, Sjulin TM, Hancock JF: Using a linkage mapping approach to identify QTL for day-neu-trality in the octoploid strawberry. Plant Breed 2008, 127:94–101.

62. Ahmadi H, Bringhurst RS, Voth V: Modes of inheritance of photoperiodism in Fragaria. J Amer Soc Hort Sci 1990, 115:146–452.

63. Shulaev V, Korban SS, Sosinski B, Abbott AG, Aldwinckle HS, Folta KM, Iez-zoni A, Main D, Arús P, Dandekar AM, Lewers K, Brown SK, Davis TM, Gar-diner SE, Potter D, Veilleux RE: Multiple models for Rosaceae genomics. Plant Physiol 2008, 147:985–1003.

64. Takada S, Goto K: Terminal flower 2, an Arabidopsis homolog of heterochro-matin protein 1, counteracts the activation of Flowering locus T by Constans in the vascular tissues of leaves to regulate flowering time. Plant Cell 2003, 15:2856–2865.

65. Sawa M, Nusinow DA, Kay SA, Imaizumi T: FKF1 and GIGANTEA com-plex formation is required for day-length measurement in Arabidopsis. Science 2007, 318:261–265.

66. Hecht V, Foucher F, Ferrandiz C, Macknight R, Navarro C, Morin J, Vardy ME, Ellis N, Beltran J, Rameau C, Weller JL: Conservation of Arabidopsis flowering genes in model legumes. Plant Physiol 2005, 137:1420–1434.

67. Alabadi D, Oyama T, Yanovsky MJ, Harmon FG, Mas P, Kay SA: Reciprocal regulation between TOC1 and LHY/CCA1 within the Arabidopsis circadian clock. Science 2001, 293:880–883.

68. Böhlenius H: Control of flowering time and growth cessation in Arabidopsis and Populus trees. PhD thesis. Swedish University of Agricultural Sciences, Umeå; 2008.

69. Wang R, Farrona S, Vincent C, Joecker A, Schoof H, Turck F, Alonso-Blanco C, Coupland G, Albani MC: PEP1 regulates perennial flowering in Arabis alpina. Nature 2009, 459:423–427.

70. Zhang X, Clarenz O, Cokus S, Bernatavichute YV, Pellegrini M, Goodrich J, Jacobsen SE: Whole genome analysis of histone H3 lysine 27 trimethylation in Arabidopsis. PloS Biol 2007, 5:e129.

71. Yano M, Katayose Y, Ashikari M, Yamanouchi U, Monna L, Fuse T, Baba T, Yamamoto K, Umehara Y, Nagamura Y, Sasaki T: Hd1, a major photoperiod sensitivity quantitative trait locus in rice, is closely related to the Arabidopsis flowering time gene CONSTANS. Plant Cell 2000, 12:2473–2484.

72. Sargent DJ, Clarke J, Simpson DW, Tobutt KR, Arús P, Monfort A, Vilanova S, Denoyes-Rothan B, Rousseau M, Folta KM, Bassil NV, Battey NH: An enhanced microsatellite map of diploid Fragaria. Theor Appl Genet 2006, 112:1349–1359.

73. Roberts AV, Blake PS, Lewis R, Taylor JM, Dunstan DI: The effect of gibberellins on flowering in roses. J Plant Growth Regul 1999, 18:113–119.

74. Monte D, Somerville S: Pine tree method for isolation of plant RNA. In DNA microarrays: a molecular cloning manual. Edited by: Bowtell D, Sambrook J. New York: Cold Spring Harbour Laboratory Press; 2002:124–126.

75. Laitinen RAE, Immanen J, Auvinen P, Rudd S, Alatalo E, Paulin L, Ainasoja M, Kotilainen M, Koskela S, Teeri TH, Elomaa P: Analysis of the floral transcriptome uncovers new regulators of organ determination and gene families related to flower organ differentiation in Gerbera hybrida (Asteraceae). Genome Res 2005, 15:475-486.

76. Tuskan GA, DiFazio S, Jansson S, Bohlmann J, Grigoriev I, et al.: The genome of black cottonwood, Populus trichocarpa (Torr. & Gray). Science 2006, 313:1596–1604.

77. Childs KL, Hamilton JP, Zhu W, Ly E, Cheung F, Wu H, Rabinowicz PD, Town CD, Buell CR, Chan AP: The TIGR plant transcript assemblies database. Nucleic Acids Res 2007, 35:D846-D851.

78. Mockler T, Yang H, Yu X, Parikh D, Cheng Y, Dolan S, Lin C: Regulation of photoperiodic flowering by Arabidopsis photoreceptors. PNAS 2003, 100:2140–2145.

79. Guo HW, Yang WY, Mockler TC, Lin CT: Regulation of flowering time by Arabidopsis photoreceptors. Science 1998, 279:1360–1363.

80. Kim W, Fujiwara S, Suh S, Kim J, Kim Y, Han L, David K, Putterill J, Nam HG, Somers DE: ZEITLUPE is a circadian photoreceptor stabilized by GIGANTEA in blue light. Nature 2007, 449:356–360.

81. Kim D, Kang J, Yang S, Chung K, Song P, Park C: A phytochrome-associated protein phosphatase 2A modulates light signals in flowering time control in Arabidopsis. Plant Cell 2002, 14:3043–3056.

82. Staiger D, Allenbach L, Salathia N, Fiechter V, Davis SJ, Millar AC, Chory J, Fankhauser C: The Arabidopsis SRR1 gene mediates phyB signaling and is required for normal circadian clock function. Genes Dev 2003, 17:256–268.

83. Hazen SP, Schultz TF, Pruneda-Paz JL, Borevitz JO, Ecker JR, Kay SA: LUX ARRHYTHMO encodes a myb domain protein essential for circadian rhythms. PNAS 2005, 102:10387–10392.

84. Doyle MR, Davis SJ, Bestow RM, McWatters HG, Kozma-Bognar L, Nagy F, Millar AJ, Amasino MR: The ELF4 gene controls circadian rhythms and flowering time in Arabidopsis thaliana. Nature 2002, 419:74–77.

85. Nakamichi N, Kita M, Niinuma K, Ito S, Yamashino T, Mizoguchi T, Mizuno T: Arabidopsis clock-associated pseudo-response regulators PRR9, PRR7 and PRR5 coordinately and positively regulate flowering time through the canonical CONSTANS-dependent photoperiodic pathway. Plant Cell Physiol 2007, 48:822–832.

86. Noh B, Lee S, Kim H, Yi G, Shin E, Lee M, Jung KMR, Doyle KMR, Amasino RM, Noh Y: Divergent roles of a pair of homologous jumonji/zinc-finger-class transcription factor proteins in the regulation of Arabidopsis flowering time. Plant Cell 16:2601–2613.

87. Hanzawa Y, Money T, Bradley D: A single amino acid converts a repressor to an activator of flowering. PNAS 2005, 102:7748–7753.

88. Chen M, Ni M: RFI2, a RING-domain zinc finger protein, negatively regulates CONSTANS expression and photoperiodic flowering. Plant J 2006, 46:823–833.

89. Cai X, Ballif J, Endo S, Davis E, Liang M, Chen D, DeWald D, Kreps J, Zhu T, Wu Y: A putative CCAAT-binding transcription factor is a regulator of flowering timing in Arabidopsis. Plant Physiol 2007, 145:98–105.

90. Pien S, Fleury DF, Mylne JS, Crevillen P, Inzé D, Avramova Z, Dean C, Grossniklaus U: ARABIDOPSIS THITHORAX1 dynamically regulates FLOWERING LOCUS C activation via histone 3 lysine 4 trimethylation. Plant Cell 2008, 20:580–588.

91. Zhang H, van Nocker S: The VERNALIZATION INDEPENDENCE 4 gene encodes a novel regulator of FLOWERING LOCUS C. Plant J 2002, 31:663–673.

92. Gendall AR, Levy YY, Wilson A, Dean C: The VERNALIZATION 2 gene mediates the epigenetic regulation of vernalization in Arabidopsis. Cell 2001, 107:525–535.

93. Chanvivattana Y, Bishopp A, Schubert D, Stock C, Moon Y, Sung ZR, Goodrich J: Interaction of Polycomb-group proteins controlling flowering in Arabidopsis. Development 2004, 131:5263–5276.

94. MacKnight R, Bancroft I, Page T, Lister C, Schmidt R, Love K, Westphal L, Murphy G, Sherson S, Cobbett C, Dean C: FCA, a gene controlling flowering time in Arabidopsis thaliana encodes a protein containing RNA binding domains. Cell 1997, 89:737–745.

95. Schomburg FM, Patton DA, Meinke DW, Amasino RM: FPA, a gene involved in floral induction in Arabidopsis thaliana, encodes a protein containing RNA-recognition motifs. Plant Cell 2001, 13:1427–1436.

96. Lim MH, Kim J, Kim YS, Chung KS, Seo YH, Lee I, Kim J, Hong CB, Kim HJ, Park CM: A new Arabidopsis thaliana gene, FLK, encodes a RNA binding protein with K homology motifs and regulates flowering time via FLOWERING LOCUS C. Plant Cell 2004, 16:731–740.

97. Simpson GG, Dijkwel PP, Quesada V, Henderson I, Dean C: FY is a RNA 3'end-processing factor that interacts with FCA to control the Arabidopsis thaliana floral transition. Cell 2003, 113:777–787.

98. Wang X, Zhang Y, Ma Q, Zhang Z, Xue Y, Bao S, Chong K: SKB1-mediated symmetric dimethylation of histone H4R3 controls flowering time in Arabidopsis. EMBO J 2007, 26:1934–1941.

99. Lee I, Aukerman MJ, Gore SL, Lohman KN, Michaels SD, Weaver LM, John MC, Feldmann KA, Amasino RM: Isolation of LUMINIDEPENDENS: a gene involved in the control of flowering time in Arabidopsis thaliana. Plant Cell 1994, 6:75–83.

100. Cheng H, Qin L, Lee S, Fu X, Richards DE, Cao D, Luo D, Harberd NP, Peng J: Gibberellin regulates Arabidopsis floral development via suppression of DELLA protein function. Development 2004, 131:1055–1064.

101. Tseng TS, Salomé PA, McClung CR, Olszewski NE: SPINDLY and GIGANTEA interact and act in Arabidopsis thaliana pathways involved in light responses, flowering and rhythms in leaf movements. Plant Cell 2004, 16:1550–1563.

102. Magome H, Yamaguchi S, Hanada A, Kamiya Y, Oda K: Dwarf and delayed-flowering 1, a novel Arabidopsis mutant deficient in gibberellin biosynthesis because of overexpression of putative AP2 transcription factor. Plant J 2004, 37:720–729.

103. Kania T, Russenberger D, Peng S, Apel K, Melzer S: FPF1 promotes flowering in Arabidopsis. Plant Cell 1997, 9:1327–1338.

104. Aukerman MJ, Sakai H: Regulation of flowering time and floral organ identity by a microRNA and its APETALA2-like target genes. Plant Cell 2003, 15:2730–2741.

CITATION

Originally published under the Creative Commons Attribution License. Mouhu K, Hytönen T, Folta K, Rantanen M, Paulin L, Auvinen P, Elomaa P. Identification of flowering genes in strawberry, a perennial SD plant. BMC Plant Biology 2009, 9:122 doi:10.1186/1471-2229-9-122.

Changes in Tree Reproductive Traits Reduce Functional Diversity in a Fragmented Atlantic Forest Landscape

Luciana Coe Girão, Ariadna Valentina Lopes,
Marcelo Tabarelli and Emilio M. Bruna

ABSTRACT

Functional diversity has been postulated to be critical for the maintenance of ecosystem functioning, but the way it can be disrupted by human-related disturbances remains poorly investigated. Here we test the hypothesis that habitat fragmentation changes the relative contribution of tree species within categories of reproductive traits (frequency of traits) and reduces the functional diversity of tree assemblages. The study was carried out in an old and severely fragmented landscape of the Brazilian Atlantic forest. We used published information and field observations to obtain the frequency of tree species and individuals within 50 categories of reproductive traits (distributed in four major classes: pollination systems, floral biology, sexual systems, and reproductive systems) in 10 fragments and 10 tracts of forest interior (control plots). As hypothesized, populations in fragments and control plots differed substantially in the representation of the four major classes of reproductive traits (more than 50% of the categories investigated). The most conspicuous differences were the lack of three pollination systems in fragments-pollination by birds, flies and non-flying mammals-and that fragments had a higher frequency of both species and individuals pollinated by generalist vectors. Hermaphroditic species predominate in both habitats, although their relative abundances were higher in fragments. On the contrary, self-incompatible species were underrepresented in fragments. Moreover, fragments showed lower functional diversity (H' scores) for pollination systems (–30.3%), floral types (–23.6%), and floral sizes (–20.8%) in comparison to control plots. In contrast to the overwhelming effect of fragmentation, patch and landscape metrics such as patch size and forest cover played a minor role on the frequency of traits. Our results suggest that habitat fragmentation promotes a marked shift in the relative abundance of tree reproductive traits and greatly reduces the functional diversity of tree assemblages in fragmented landscapes.

Introduction

Functional diversity can be defined as a variety of life-history traits presented by an assemblage of organisms [1], [2] and it has been postulated to be critical for the maintenance of ecosystem processes and properties [3]. For example, previous empirical work has suggested that ecosystems with a high diversity of functional traits have greater efficiency of water, nutrient, and light use, as well as higher productivity [3], [4]. In addition, they may also be more resilient [5] and resistant to biological invasions and to biodiversity loss [6], [7]. Nevertheless, most studies on functional diversity in plant communities have focused on the importance of traits associated with plant physiology. Consequently, we know little regarding the functional diversity of other traits that also affect both community structure and ecosystem functioning, such as those related to plant-animal interactions [1], [2].

Habitat loss and fragmentation (hereafter habitat fragmentation) have been shown to dramatically alter tree communities in tropical forests [8]–[12]. Fragments usually exhibit reduced species richness and diversity, particularly near edges. This reduction in species diversity is due in large part to loss of species that are "shade-tolerant" [8], [12], [13], restricted to the forest understory [10], have large-seeds [14], [15], or are dispersed by vertebrates [12], [16]–[19].

Furthermore, fragments tend to become dominated, both in terms of species richness and individual abundance, by pioneer trees [8], [19]. Because tropical pioneer trees usually share a similar set of life-history traits irrespective of their taxonomic affinities [20]–[22], this biased ratio of pioneers to shade-tolerant plants may reduce the functional diversity of tree assemblages in fragments.

More than 90% of the extant angiosperms are animal-pollinated [23], therefore pollination is considered an essential ecosystem process whose outcome can have major consequences for the maintenance of biodiversity [24], [25]. Indeed, a broad body of empirical evidence has found that the disruption of plant-pollinator interactions by habitat fragmentation can detrimentally affect plant reproductive success [26]–[29]. Potentially, changes in plant-pollinator interactions and pollinator abundance/composition can affect seed dispersal and seedling recruitment and consequently reduce plant population size or even promote local extinction [26], [27], [30]. Nevertheless, patterns and process regarding changes in reproductive functional diversity in fragmented tropical landscapes remain poorly investigated.

Because the long lifespan of tropical trees [31], hypotheses addressing disruptions of functional diversity driven by changes in tree composition can be properly tested in landscapes that were disturbed long enough ago to permit demographic shifts to have occurred, such as fragmented landscapes with longer histories of human occupation. The Atlantic forest of Brazil is a biodiversity hotspot that has been reduced to less than 8% of its original distribution due to forest clearing and fragmentation that dates to the 16th century [32]. In some regions (e.g. Brazil's northeast), over 90% of fragments are smaller than 50 ha and are immersed in a homogeneous and hostile matrix of sugar cane fields [33]. These archipelagos of small fragments and forest edge habitat are currently dominated by a small subset of pioneer trees, retain less than half of the tree species richness of the forest interior [19], and receive an impoverished seed rain biased towards smaller seeds [34]. This scenario offers an excellent opportunity to investigate long-term fragmentation-related changes in tree assemblages and how they influence functional diversity.

Here we test the hypothesis that the habitat fragmentation changes the frequency of tree species and individuals within categories of reproductive traits and consequently reduces the functional diversity of tree assemblages in a fragmented landscape of the Brazilian Atlantic forest. We begin by comparing the pollination systems, floral biology, sexual systems, and reproductive systems of trees in forest fragments and tracts of forest interior (control plots). We then compare the diversity of these traits in these two habitats based on the relative contribution of both species and individuals. Finally, we discuss potential mechanisms driving the patterns we observed. We conclude that habitat fragmentation promotes a marked shift in the relative abundance of tree reproductive traits, including the lack of some specialized pollination systems and a parallel increase in the frequency of generalist ones. Collectively, shifts in reproductive traits promote a conspicuous reduction in the functional diversity of tree assemblages in fragmented landscapes, which may strongly influence forest dynamics and the persistence of biodiversity.

Materials and Methods

Study Site and Landscape Attributes

This study was conducted in the State of Alagoas in northeastern Brazil on the property of Usina Serra Grande (8°58'50"S, 36°04'30"W), a large, privately-owned sugar producer. This landholding has approximately 9,000 ha of forest included in a unique biogeographic region of the Atlantic forest known as the Pernambuco Center of Endemism [sensu 35] or the Atlantic forest of Northeast Brazil, the most threatened sector of the South American Atlantic forest [17]. We selected a large (666.7 km2), severely fragmented landscape within this property containing 109 forest fragments (total forest cover = 9.2%), including the 3,500-ha Coimbra forest–the largest and best preserved remnant in this region [19]. All fragments are entirely surrounded by a uniform matrix of sugar-cane monoculture (Figure 1). In addition to the Coimbra forest, the patches ranged in size from 1.67–295.7 ha.

Figure 1. Study landscape at the Atlantic forest of northeast Brazil.

(A) Northeastern Brazil, where this study was conducted. (B) Distribution of the Atlantic forest of northeast Brazil (= Pernambuco Center of Endemism), note original (grey) and current (black) distribution of this forest in the region; white rectangle represents the study landscape (amplified in C). (C) Study landscape with the fragments used in this study (dark grey polygons), including the 3,500 ha Coimbra forest (lower right). Light grey and white areas represent remaining forest fragments (not sampled) and sugar-cane cultivation, respectively.

Our study landscape consists of a low-altitude plateau (300–400 m above sea level) containing two similar classes of dystrophic soils with high clay fractions: yellow-red latosols and yellow-red podzols (according to the Brazilian soil classification system [36]). Annual rainfall is ~2000 mm, with a 3-month dry season (<60 mm/month) from November to January. Forests in this landscape consist of lowland terra firme forest (<400 m a.s.l.) [37], with the Fabaceae, Lauraceae, Sapotaceae, Chrysobalanaceae and Lecythidaceae accounting for most tree species (≥10 cm DBH) [38], [39]. Sugar cane cultivation in this landscape, which dates to the early 19th century, and possibly as early as the 18th century [see 40], provided the strongest incentive for clearing large tracts of pristine old-growth forests. Remaining forest fragments have been protected against fire and logging to ensure watershed protection and water supply for sugar cane irrigation (C. Bakker, pers. communication). This protection has guaranteed the stability of forest fragment borders and the occurrence of both pioneer and shade-tolerant adult trees along forest edges as evidenced by local patterns of seed rain [15]. The Serra Grande landscape therefore provides a rare and interesting opportunity for Atlantic forest fragmentation studies.

Tree Species Surveys and Habitat Classification

We compared the frequencies and the functional diversity of tree reproductive traits in 10 of the forest fragments (range = 3.4–295.7 ha) and 10 'forest interior plots' [sensu 41] located in the region's largest remnant (Coimbra forest; here adopted as the control site) using floristic data from previously conducted botanical surveys. Although we are aware that the Coimbra forest does not represent a true 'continuous forest,' it is the largest remaining Atlantic forest patch in Northeast Brazil [see 19] and is more than twice as large as the largest fragment analyzed by Ranta et al. [33] in this same center of endemism. In addition, the Coimbra forest still retains the full complement of ecological groups occurring in more continuous tracts of Atlantic forest, such as large-seeded trees and frugivorous vertebrates [16], [17], [42], [43]. It is therefore representative of the largest tracts of forest remaining in the hotspot, making its core area [sensu 41] the best possible control site for assessing persistent and long-term effects of habitat loss and fragmentation.

The tree surveys, upon which we randomly selected our fragments and control plots, were carried out from 2003–2005 by Oliveira et al. [19] and Grillo [39] as part of a regional plant survey. Briefly, all trees ≥10 cm DBH were measured, marked, and identified in one 0.1-ha plot per fragment. Plots were located in the geographic center of fragments to standardize procedures and minimize edge effects [44]. Depending on the size of the fragment, plots were 60.5–502.77 m from nearest edge. The ten control plots, also measuring 0.1-ha, were haphazardly located in the interior of Coimbra forest at distances 200–1012.73 m from nearest edge, in locations consisting of old-growth forest with no detectable edge effects (i.e. forest interior [sensu 41]). Vouchers collected by Oliveira et al. [19] and Grillo [39] are deposited in the Herbarium UFP (No. 34.445 to 36.120), and the checklist of the flora of Usina Serra Grande (ca. 650 plant species) is available at www.cepan.org.br and in Pôrto et al. [45]. Since 2001, the number of botanical investigations carried out in our study landscape has increased [e.g. 15], [19], [34], [38], [39], [45], [46], providing detailed knowledge about the taxonomy and life-history traits of the woody flora.

Reproductive Traits of Tree Species

Floristic surveys revealed a total of 629 individuals from 77 tree species in the forest fragments (32 families, 58 genera), whereas 878 individuals from 119 species (37 families, 87 genera) were recorded in the control plots. Pooling the data from all sites resulted in 1507 individuals from 156 species (41 families, 105 genera). For each species we identified the following "reproductive traits": pollination system, floral biology, sexual system, and reproductive system (Table 1). Classification of species into each category was based on (1) floras and botanical monographs [e.g. 47], [48]–[50], including several issues of Flora Neotropica; (2) web searches including only published and referenced data; (3) field observations and a survey of specimens from the UFP and IPA Herbaria; and (4) personal knowledge and previously published observations [see 51 for a review]. For each fragment and control plot we then calculated the proportion of tree species and individuals within the 50 categories that comprise the four major classes of reproductive traits (Table 1). Although not all categories could be identified for a few of the species (see results), it is unlikely that this biases the qualitative outcome of our analyses because habitats were compared in terms of frequency of species and individuals within categories.

Table 1. Tree reproductive traits with their respective categories adopted in this study.

Reproductive traits	Categories*
1. Pollination system[1]	bats; bees; beetles; birds; butterflies; diverse small insects (DSI); flies; moths (excluding hawkmoths); Sphingids (hawkmoths); non-flying mammals; wasps; wind
2. Floral biology	
Size[2]	inconspicuous (≤4 mm); small (>4≤10 mm); medium (>10≤20 mm); large (>20≤30 mm); very large (>30 mm)
Reward[1]	brood or mating places/floral tissues (BMFT); nectar; oil; pollen; nectar/pollen; without resource (other than deceit flowers)
Type[3]	bell/funnel; brush; camera; flag; gullet; inconspicuous (attributed to very small flowers, ≤4 mm); open/dish; tube
Anthesis period[1]	diurnal; nocturnal
3. Sexual system[4] (morphological expression)	andromonoecious; dioecious; hermaphrodites (distinguishing those heterostylous); heterostylous; monoecious
4. Reproductive system[4,5]	agamospermic; self-compatible; self-incompatible; outcrossing (self-incompatible+dioecious species)

Explanatory Variables

Because a number of patch and landscape-scale environmental variables may affect the structure of tree assemblages in tropical forests [8], [52], we also considered the effects of soil type, distance to the nearest forest edge, forest fragment size, the spatial distribution of plots (i.e. plot location in the landscape), and the amount of forest cover retained in the surrounding landscape (hereafter forest cover) as independent variables for the frequency of reproductive traits in the tree assemblages. Forest cover is positively correlated with overall connectivity between patches [53] and was quantified as the percentage of forest within a 1-km width buffer set from the border of each fragment. Patch and landscape metrics were quantified using a combination of three Landsat and Spot images acquired in 1989, 1998, and 2003, a set of 160 aerial photos (1:8,000) taken from commissioned helicopter overflights on April 2003, a soil map by IBGE [36], and a soil map provided by the Usina Serra Grande Agriculture Office. Analyses were conducted using ArcView 3.2 and Erdas Imagine 8.4.

Functional Diversity of Reproductive Traits

Here we operationally define a functional group as a set of tree species within the same category of reproductive trait, i.e. a set of species sharing a life-history trait as previously adopted elsewhere [1]. To calculate the functional diversity of reproductive traits in forest fragment and control plots, we used Shannon's (log base 2) and Simpson's indices [54]. We used both indices to elucidate the contribution of both the richness of categories and the evenness to diversity scores (note that the use of evenness-based indices for estimating functional diversity has been recommended by some authors [55]–[57]). We calculated these indices twice for each of the 20 plots: first, using categories as the equivalent of species, and the number of tree species within each category as the equivalent of individuals; and second using categories as the equivalent of species and the number of individual within each category.

Statistical Analysis

Differences in (1) the average percentage of species and individuals within each category of reproductive trait, and in (2) the average functional diversity of reproductive traits between the control area and fragments were compared with t or Mann-Whitney tests [58]. General linear models (GLM) were used to detect any effect of explanatory variables on the frequency of traits in tree assemblages by first examining the effects of habitat type (fragments vs. control plots), soil type and distance to the nearest edge considering all 20 plots in the two habitats, and then the effects of forest fragment size and surrounding forest cover considering the 10 fragments (since these patches and landscape metrics had no variance in Coimbra forest). Normality of all response variables were checked using Lilliefors tests; for GLMs the percentage-expressed dependent variables were arcsine transformed as suggested by Sokal & Rohlf [58].

Additionally, to examine the effect of habitat and soil type on species similarity between plots these variables were considered as factors in Analysis of Similarities (ANOSIM) tests [59]. Plots were ordered according to their Bray-Curtis dissimilarities of species composition [54]. Species abundance were square-root transformed and standardized [59] to avoid any bias resulting from very abundant species and differences in the sample size of individuals recorded within each plot. We also performed a Mantel test with Weighted Spearman rank correlations to address the effect of plot geographic location on levels of taxonomic similarity. Straight-line distances between plots were ln-transformed, as suggested by Condit et al. [60] and Jones et al. [61]. The Mantel test was carried out considering a group of 20 fragments and 75 0.1-ha plots from which information on tree species composition is available [19], [39]. Here we assume that the lack of significant relationships between soil type, plot location and plot floristic similarity discard soil and plot location as variables driving the frequency of tree reproductive traits in the landscape. All analyses were carried out using SYSTAT 6.0 [62], PRIMER v. 5 [63], and PC-ORD 4.36 [64].

Results

Reproductive Traits of Tree Species

Fragments and control plots differed significantly in more than 50% of the categories of reproductive traits investigated, but differences were much more notable when evaluating individuals within categories (over 60% of the categories differed) than species (ca. 40%).

For pollination systems, fragments and control plots markedly differed in 50% of all categories (6 out of 12 categories) (Table 2; Figure 2A). The most conspicuous differences concerning species richness within categories of pollination systems can be summarized in four aspects. First, fragments lacked three categories of pollination systems–pollination by birds, flies and non-flying mammals. Second, scores for hawkmoth- and bat-mediated pollination in fragments were about half of the scores recorded in the control plots. Third, when comparing pollination by vertebrates as whole (birds, bats, and non-flying mammals) fragments had a ca. threefold decreased frequency than control plots. Finally, fragments had a 33% increase in the proportion of tree species pollinated by diverse small insects (DSI) in comparison to control plots (Table 2). The proportion of tree individuals within categories of pollination systems showed similar trends (Figure 2A), although for some categories the differences between fragments and control plots were even more dramatic than for species richness (e.g. hawkmoth and vertebrate pollination). Fragments and control plots also differed dramatically when pollination systems were pooled into two categories of pollen vectors-generalists and specialists [sensu 65]. In summary, fragments had proportionately more tree species pollinated by generalist vectors (66.43±14.08%) than control plots (58.18±7.87%; t = 1.616; d.f. = 18; P = 0.06); the relative abundance individuals pollinated by generalists was also higher in fragments than control plots (71.71±16.5% vs. 46.10±15.53, U = 13.0; P = 0.0052).

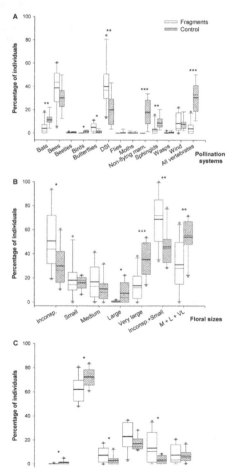

Figure 2. Effect of habitat fragmentation on pollination systems, floral sizes and floral rewards.
Percentage of tree individuals within categories of pollination systems (A; N = 137 spp.), floral sizes (B; N = 105 spp.), and floral rewards (C; N = 124 spp.) at 10 fragments and 10 control plots of an Atlantic forest landscape, northeastern Brazil. Frequencies represented by boxes that are significantly different are indicated with asterisks: *P<0.05; **P<0.01; ***P<0.001.

Table 2. Percentages (mean±SD) of tree species within categories of reproductive traits in forest fragments (N = 10) and control plots (N = 10) in a fragmented landscape of Atlantic forest, northeastern Brazil (data on the reproductive traits for the species are available upon request).

Reproductive traits	Categories	Fragments	Control plots
		%$_{mean}$ species ± SD	
Pollination systems	Bats	5.79±7.85 a	14.04±3.74 b**
N = 137 spp.	Bees	37.77±12.72 a	37.95±7.68 a
	Beetles	0.97±2.15 a	1.20±1.58 a
	Birds	0.00 a	1.75±1.64 b*
	Butterflies	5.59±4.00 a	0.83±1.46 b**
	Diverse small insects	33.62±8.99 a	22.44±8.05 b**
	Flies	0.00 a	0.29±0.90 a
	Moths	1.13±2.39 a	0.42±1.32 a
	Non-flying mammals	0.00 a	3.32±1.00 b***
	Sphingids	4.76±5.65 a	8.94±3.60 b*
	Wasps	1.17±2.49 a	2.22±2.68 a
	Wind	9.22±7.33 a	6.62±3.72 a
	All Vertebrates	5.79±7.85 a	19.10±4.70 b**
Floral sizes (mm)	Inconspicuous (≤4)	42.39±19.01 a	37.31±8.59 a
N = 105 spp.	Small (>4≤10)	21.22±11.23 a	25.31±6.82 a
	Medium (>10≤20)	15.02±6.50 a	10.48±6.18 a
	Large (>20≤0)	1.93±3.26 a	5.77±5.57 a
	Very Large (>30)	19.44±11.51 a	21.12±4.77 a
	Inconspicuous+Small	63.61±15.66 a	62.63±8.61 a
	Medium+Large+Very large	36.39±15.66 a	37.37±8.61 a
Floral rewards	Brood or mating places/floral tissues	1.02±2.28 a	2.88±2.60 a
N = 124 spp.	Nectar	62.51±8.16 a	65.50±6.51 a
	Oil	5.68±2.40 a	2.94±3.04 b*
	Pollen	24.03±8.99 a	24.24±5.16 a
	Nectar/pollen	7.79±4.90 a	3.90±3.08 b*
	Without	6.75±5.52 a	4.44±2.38 a
Floral types	Bell/funnel	3.39±3.74 a	1.72±2.42 a
N = 111 spp.	Brush	8.36±7.51 a	22.34±6.80 b***
	Camera	9.29±7.13 a	10.03±4.40 a
	Flag	3.13±5.17 a	11.75±5.21 b**
	Gullet	9.40±7.40 a	0.32±1.02 b**
	Inconspicuous	36.40±19.70 a	24.61±7.56 b*
	Open/dish	22.97±8.94 a	18.68±5.55 a
	Tube	7.06±5.96 a	10.55±2.80 a
	Inconspicuous+Open	59.37±13.44 a	43.29±6.31 b**
	All non-inconspicuous or open	40.63±13.44 a	56.71±6.31 b**
Anthesis period	Diurnal	91.83±9.21% a	80.42±6.44 b**
N = 116 spp.	Nocturnal	8.17±9.21% a	19.58±6.44% b**
Sexual systems	Andromonoecious	0.91±1.92 a	0.00±0.00 a
N = 129 spp.	Dioecious	27.95±7.94 a	31.80±5.48 a
	Hermaphrodite	65.55±10.80 a	60.28±6.34 a
	Heterostylous	0.45±1.44 a	0.63±1.37 a
	Monoecious	5.14±5.05 a	7.29±4.15 a
	All non-hermaphrodite	34.45±10.80 a	39.72±6.34 a
Reproductive systems	Agamospermic	0.92±2.92 a	2.74±5.79 a
N = 79 spp.	Self-compatible	15.51±7.52 a	5.86±9.82 b*
	Self-incompatible (SI)	51.77±9.27 a	63.44±14.95 b*
	Outcrossing (SI+Dioecious)[1]	83.57±9.50 a	91.39±14.66 b*

The proportion of species within categories of floral size was similar in fragments and control plots (Table 2). However, fragments had twice as many individuals with inconspicuous flowers than control plots (50.75±25.44% vs.

29.99±15.86%; Figure 2B). An opposite trend was observed for large and very large flowers, fragments with more than a 10-fold lower proportion of individuals with large flowers (0.5±0.84%) and almost a three-fold decrease of the very large ones (13.74±11.77%) in comparison to control plots (7.54±8.58% and 35.4±13.54%, respectively). By grouping the five categories of flower size into two [i.e. inconspicuous/small (≤10 mm) and medium/very large (>10 mm)] results were similar. Fragments showing a prevalence of individuals with inconspicuous/small flowers (68.85±21.43%) in contrast with control plots (45.98±16.04%), and a significant lower proportion of individuals with medium to very large flowers ones (31.15±21.43%) than control sites (54.02±16.04%) (Figure 2B).

Nectar was the most frequent floral reward observed in tree species of fragment and control sites, however, these habitats differed in two of the other five categories of floral rewards adopted in this study (Table 2). Nectar/pollen-flower species were twice as higher in fragments than in control plots, and fragments had also higher frequency of species with oil-flowers in comparison with control plots (Table 2). Similar patterns were observed with respect to the proportions of individuals within categories of floral rewards in each habitat, but, additionally, fragments faced a slight and statistically significant reduction on the proportion of individuals with BMFT flowers (0.19±0.41%) in contrast with control plots (1.25±1.56%) (Figure 2C).

As expected, fragments and control plots largely differed in terms of floral types considering the proportion of both species and individuals. Noticeable differences refer to significantly lower scores of species with flag and brush flowers, and higher scores of inconspicuous flowers in fragments in comparison with control plots (Table 2). Similar patterns were detected by analyzing the eight categories of floral types based on reward accessibility: (1) inconspicuous+open/dish flowers (easily accessible resource [sensu 66]), and (2) non-inconspicuous/open (concealed resource, at least some degree of hiddenness [sensu 66]). Under this approach, fragments showed a prevalence of species with inconspicuous/open type, which was significantly higher than in control sites. In terms of relative abundance of tree species within floral types categories, figures described fragments facing the same patterns observed to species regarding flag, inconspicuous (with even stronger differences), and brush flowers. Additionally, fragments showed lower proportions of individuals bearing camera and tube flowers in contrast with control areas (Figure 3A). Similarly, when observing proportions of individuals within categories of floral types according to reward accessibility, fragments had significant higher frequency of individuals with flowers of the inconspicuous/open type than control plots (Figure 3A), differences being yet more expressive than for species richness. Moreover, fragments revealed to be particularly impoverished in terms of tree species with nocturnal anthesis, showing a frequency more than two times lower (8.17±9.21%) than control plots (19.58±6.44%) (t = –3.211; d.f. = 18; P = 0.002). Difference was even more marked when the relative abundance of tree species with nocturnal anthesis is analyzed (4.93±6.67% in fragments vs. 21.18±11.41% in control plots) (t = –3.889; d.f. = 18; P = 0.001).

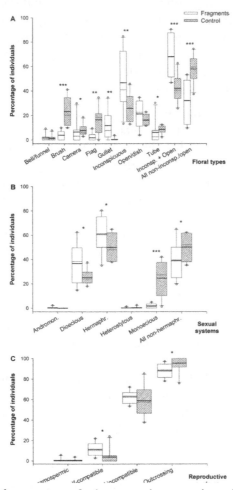

Figure 3. Effect of habitat fragmentation on floral types, sexual systems and reproductive systems.

Percentage of tree individuals within categories of floral types (A; N = 111 spp.), sexual system (B; N = 129 spp.), and reproductive system (C; N = 79 spp.) at 10 fragments and 10 control plots of an Atlantic forest landscape, northeastern Brazil. Frequencies represented by boxes that are significantly different are indicated with asterisks: *P<0.05; **P<0.01; ***P<0.001.

Both habitats, fragments and control, were dominated by hermaphrodite species and showed similar scores for species within the five categories of sexual systems (Table 2). However, habitats were absolutely contrasting with respect to the frequency of individuals, as fragments were dominated by hermaphrodite individuals (61.05±15.33%), whereas non-hermaphrodite systems were prevalent (50.12±10.66%) among individuals of the control plots. Particularly expressive, as well, was the lower representation of monoecious individuals in the fragments— more than 12-times lower (1.72±1.84%) than control plots (24.51±13.92%) (Figure 3B). Fragments also had a slight but statistically significant decrease in the proportion of self-incompatible and overall obligatory outcrossing species (self-incompatible+dioecious; Table 2). In terms of the relative abundance of tree species within categories of reproductive system, fragments had significant lower scores of outcrossing individuals (87.82±6.84% vs. 95.40±8.54 in control plots) and highest frequency of self-compatible ones (11.59±6.56% vs. 3.90±8.03 in control sites) (Figure 3C).

Explanatory Variables

GLMs did not reveal any significant influence of soil type on the proportion of traits in tree assemblages. Habitat was consistently the strongest explanatory variable for the proportion of tree species and individuals within categories of reproductive traits, explaining between 19.4% and 69.4% of their variation, influencing 38 categories (Table 3). GLMs also detected 10 categories of reproductive traits that were influenced by log-distance to edge (considering forest fragments and control plots), two categories influenced by log-fragment area, and eight affected by forest cover (considering forest fragments only) (Table 3). These three fragmentation-related variables explained between 20.7% and 68.6% of the variation on reproductive traits in forest fragments and control plots (Table 3). Additionally, ANOSIM revealed no significant correlation between soil type and level of taxonomic similarity between plots (R = 0.024; P = 0.54), but detected a stronger effect of habitat type (R = 0.95; P = 0.001). A Mantel test failed to uncover any spatial effects on the taxonomic similarity among plots (Rho = 0.155; P = 0.9).

Table 3. Scores from General Linear Models applied to the proportion of tree species and individuals within categories of reproductive traits (48 categories for species, 48 categories for individuals) in forest fragments (N = 10) and control plots (N = 10) in a fragmented landscape of Atlantic forest, northeastern Brazil.

Habitat/explanatory variables	Traits analyzed	Traits affected	P values	R² range
Fragments+control plots				
Habitat	96	38	<0.0001–0.04	19.4–69.4%
Soil	96	0		0
Log-distance to edge	96	10	0.008–0.044	20.7–46.5%
Total		48		
Fragments				
Log-fragment area	96	2	0.014–0.018	52.7–55.4%
Forest cover	96	8	0.003–0.046	39.9–68.6%
Total		10		

Functional Diversity of Reproductive Traits

When using the number of reproductive categories (see Table 1) and the species richness per category, fragments were significantly less diversified (H') with respect to pollination systems (–18.4%) and floral types (–12.65%) in comparison with control plots (Table 4). Simpson's values also evidenced fragments with significant lower functional diversity of pollination systems (Table 4). Differences were much more expressive, both biologically and statistically, when using number of categories (as equivalent of species) and number of individuals within categories for calculating diversity indices. In this case, fragments were significantly less diversified (H' scores) not only in terms of pollination systems (–30.3%) and floral types (–23.6%), but they also presented significant lower functional diversity of floral sizes (–20.8%) in contrast with control plots (Table 4). Simpson's values also evidenced fragments with significant reduced functional diversity of pollination systems (–20.7%) and floral types (–19.62%) (Table 4). Based on Simpson's index, fragments were slightly more diversified than control plots in terms of floral rewards, however, when applying Bonferroni correction, values for floral rewards were not significantly different any more (Table 4).

Table 4. Functional diversity (mean±SD) of pollination systems, floral size, floral type and floral reward categories in tree assemblages of forest fragments (N = 10) and control plots (N = 10) in a fragmented landscape of Atlantic forest, northeastern Brazil.

Functional Diversity	Treatments (N = 10 plots/treatment)	Pollination systems (mean±SD)	Floral sizes (mean±SD)	Floral types (mean±SD)	Floral rewards (mean±SD)
Categories and species					
Shannon's (H')	Fragments	1.965±0.341 a	1.752±0.414 a	2.169±0.429 a	1.386±0.189 a
	Control	2.407±0.213 b***	1.983±0.169 a	2.483±0.168 b**	1.323±0.238 a
Simpson's (1-D)	Fragments	0.732±0.073 a	0.713±0.121 a	0.782±0.126a	0.562±0.069a
	Control	0.781±0.047 b*	0.758±0.053 a	0.843±0.031a	0.521±0.073a
Categories and individuals					
Shannon's H'	Fragments	1.672±0.358 a	1.485±0.567 a	1.810±0.506 a	1.332±0.242 a
	Control	2.398±0.207 b***	1.875±0.161 b*	2.369±0.244 b**	1.167±0.258 a
Simpson's (1-D)	Fragments	0.613±0.130 a	0.566±0.220 a	0.635±0.181 a	0.528±0.100 a
	Control	0.773±0.050 b**	0.695±0.058 a	0.790±0.048 b**	0.437±0.090 b*
Total no. of categories	Fragments	5.4±1.43	4.0±0.94	5.7±1.16	3.9±0.74
	Control	8.0±0.82	4.6±0.52	6.4±0.84	4.1±0.74
Total no. of species	Fragments	18.3±5.81	12.8±4.26	13.4±4.17	17.3±5.48
	Control	32.9±10.54	22.0±7.94	23.3±7.8	28.0±8.10

Discussion

Patterns and Underlying Mechanisms

Our findings suggest that habitat fragmentation promotes marked changes in both the presence and relative abundance of the reproductive traits of tree species, resulting in a reduced functional diversity of tree assemblages in forest fragments. Moreover, small forest patches in severely-fragmented landscapes may be strongly impoverished in terms of the number of species and individuals with particular pollination systems (e.g. pollination by bats, birds, non-flying mammals, Sphingids) and may be dominated by tree species pollinated by generalists. Finally, strategies that are more dependent on long-distance pollen movement and animal-mediated services, such as self-incompatibility, may be negatively affected. These statements are supported by the fact that the differences we found between fragments and control plots could not be explained by soil type or the relative spatial position of the plots in the landscape. Although the distribution of tropical trees has been found to be influenced by variation in soil types [52], [67], there is no evidence that this also influences the spatial distribution of ecological groups (based on reproductive traits, regeneration strategy, etc.) in terra firme forests [8], [68].

An increasing body of evidence has shown that as fragments become older, tree assemblages become drastically altered [12], [69]–[71]. Plant assemblages in small fragments (<10 ha) and forest edges are impoverished (scores of alpha diversity reduced by a half) and biased in taxonomic and ecological terms towards pioneer species. These patch-level findings suggest that fragmented landscapes tend to retain just a small subset of species from the original biota. Despite the recent findings on this topic, our study is one of the first to document a marked shift on the signature of tree assemblages inhabiting a fragmented landscape with respect to the frequency of reproductive-related traits and its functional diversity. Similar results were reported by Chazdon et al. [72] for tree assemblages in second-growth, logged, and old-growth forests in Costa Rica. They found lower relative abundance of mammal-pollinated trees in second-growth forests in comparison to old-growth ones, as well as a higher relative abundance of hermaphroditic trees in second-growth forests. In addition, Murcia [27] suggested fragmented forests tended to have an increased frequency of self-compatible hermaphrodites at the expense of other sexual systems. Our findings are consistent with these results, as well as recent ones indicating self-incompatible systems are more negatively affected than self-compatible ones following habitat loss and fragmentation [12], [25], [29].

Two fragmentation-related processes may be the principal mechanisms driving the changes in reproductive traits and functional diversity we observed: 1) the proliferation of pioneer species with a concomitant decline in the abundance of shade-tolerant trees and 2) depressed population sizes of animal pollinators, which over time led to changes in tree abundance in forest fragments. In tropical forests, myriad processes triggered by the creation of forest edges promote a proliferation of short-lived pioneers [8] and the local extirpation of shade-tolerant trees, including canopy and understory species [10], [19], emergent trees [73] and large-seeded trees [15], [34]. In our study site, pioneer species represent over 80% of all tree species and individuals recorded in the fragments, whereas they represent less 50% in core areas [19], [39]. Furthermore, recent surveys in this site have documented an outstanding predominance of pioneer species in seed rain [15] and seedling assemblages [34], [74] which suggests that pioneer dominance may represent a more pervasive, long-term feature of old and severally fragmented landscapes.

Assuming that pioneer plants are r-strategists and shade-tolerant (climax) ones are K- strategists [21], it is reasonable to expect that these two species groups differ in terms of reproductive traits, sexual systems, and reproductive systems. Some of our findings, such as higher scores of pollination by DSI and flowers with easily accessible resources (inconspicuous+open/dish flowers) in fragments, may simply reflect the dominance of pioneer trees in this habitat as these traits appear to be more frequent among pioneers (65% of the DSI-pollinated species and over 68% of the species with inconspicuous/open/dish flowers are pioneers). On the other hand, a trait such as pollination by bats that was significantly more frequent in control plots (richness and abundance) is also positively associated with a subset of shade-tolerant species–75% of the bat-pollinated species are shade-tolerant (e.g. Bauhinia, Hymenaea-Fabaceae; Manilkara-Sapotaceae; Quararibea-Malvaceae sensu APG II [75]). Because the pioneer species recorded in the fragments - including both short- and long-lived pioneers - belong to 16 orders and eight superordinal clades (sensu APG II [75]), the patterns documented here cannot be explained by phylogenetic clustering among pioneers. Even pioneer species that were recorded exclusively in forest fragments belong to four families in four orders and three superordinal clades. Unfortunately, because of the large number of categories for each reproductive trait and the low number of tree species within each category, it was not possible to properly test trait-associated differences between pioneer and shade-tolerant tree species.

In tropical forests, 98–99% of the flowering plant species (and 97.5% of the trees) rely on biotic vectors such as insects and vertebrates for successful pollination [76], [77], and it has been broadly assumed that plant-pollinator interactions are largely detrimentally affected by habitat loss and fragmentation [26]–[29], [78]–[81]. Some of the changes we documented in our fragments are therefore expected, particularly the lack or reduced occurrence of some pollination systems [27], [28], [82]. For instance, fragmented habitats may support less pollinators than continuous habitats due to limited resource availability for pollinators (area-related effects on animal populations). In turn, plants can have a depressed reproductive output as consequence of changes in pollinator diversity, composition, or behavior [25], [28], i.e., reproductive impairment driven by pollination limitation [sensu 29]. Studies on pollinator diversity carried out in our landscape have documented a decreased diversity of nectarivorous bats [83] and hawkmoths [84] in small fragments. However, empirical evidence to determine which pollination-related traits and plant-pollinator mutualisms are particularly susceptible to habitat disturbance is still scarce [28]. Our results suggest that the reduced number of tree species and individuals pollinated by bats and Sphingids in fragments and the absence of fly-, bird-, and non-flying-mammal-pollinated trees, together with the changes in floral traits and sexual systems, may be a higher order effect promoted by habitat fragmentation.

Implications of Reduced Functional Diversity

The reduced reproductive functional diversity documented in our study landscape's fragments resulted primarily from the lack or skewed representation of some pollination systems, floral types, and floral size categories in terms of both species and individual (see Table 3). In other words, tree assemblages in this habitat appear to carry a narrower range of floral traits and pollination systems in comparison to patches of forest interior, particularly for pollinators such as mammals and hawkmoths (reduced support capacity). Regardless the underlying mechanism, this narrow range may (1) promote the collapse of pollinator populations; (2) restrict the ecological range of plant and animal groups able to colonize remaining patches of forest or even turn fragments into sink habitats for both plants and their pollinators; and (3) alter the course of natural regeneration or the dynamics of forest fragments toward the establishment of impoverished assemblages in terns of species richness, ecological composition and trophic structure.

Unfortunately, few studies have addressed shifts on the diversity of plant reproductive traits in human-disturbed habitats, especially those traits associated with plant-pollinator interactions [1], [72]. Studies linking these shifts to functional diversity are even more scarce [1], [2], despite the fact that pollination processes influence biodiversity maintenance and ecosystem functioning. Fontaine et al. [2], for example, argued that even simple structured plant-pollinator communities may have their persistence threatened due to reduced functional diversity, thereby suggesting that functional diversity of pollination networks is critical to avoid biodiversity loss.

In summary, it is reasonable to propose as a working hypothesis that the persistence of biodiversity and consequently the long-term conservation value of isolated tropical forest fragments may be negatively affected by reduced functional diversity to such an extent yet not anticipated by conservation biologists. Collectively, the proliferation of pioneer species, extirpation of shade-tolerant trees, and reduced functional diversity have the potential to disrupt some trophic interactions [e.g. 85]; even landscapes such as ours that were fragmented long ago and are dominated by pioneers may face future biodiversity loss. We believe it would be beneficial for future research to 1) validate and assess the generality of both the patterns and the underlying mechanisms observed here and 2) address more ecosystem level effects driven by reduced functional diversity in fragmented landscapes, such as changes in biodiversity persistence, primary productivity, nutrient cycling, succession, and ecosystem resilience.

Acknowledgements

We are very grateful to Alexandre Grillo and Marcondes Oliveira (UFPE-Brazil) for their immeasurable cooperation and for permitting us to use data from their botanical surveys; Luís Antônio Bezerra and Clodoaldo Bakker for permitting this research on the private property of the Usina Serra Grande; Carlos Peres (University of East Anglia-England) for discussions and help with analyses; André Santos (UFPE-Brazil) and Antônio Aguiar-Neto (UF-USA) for discussions and help in several phases of this study; Jérôme Chave (Academic Editor) and an anonymous reviewer for valuable suggestions on the manuscript.

Authors' Contributions

Conceived and designed the experiments: AL MT. Performed the experiments: AL LG. Analyzed the data: AL LG MT EB. Wrote the paper: AL LG MT EB.

References

1. Mayfield MM, Boni ME, Daily GC, Ackerly D (2005) Species and functional diversity of native and human-dominated plant communities. Ecology 86: 2365–2372.

2. Fontaine C, Dajoz I, Meriguet J, Loreau M (2006) Functional diversity of plant-pollinator interaction webs enhances the persistence of plant communities. PLoS Biol 4: 129–135.

3. Tilman D, Knops J, Wedin D, Reich P, Ritchie M, et al. (1997) The influence of functional diversity and composition on ecosystem processes. Science 277: 1300–1302.

4. Hooper DU, Vitousek PM (1997) The effects of plant composition and diversity on ecosystem processes. Science 277: 1302–1305.

5. Tilman D, Downing JA (1994) Biodiversity and stability in grasslands. Nature 367: 363–365.

6. Prieur-Richard AH, Lavorel S (2000) Invasions: the perspective of diverse plant communities. Austral Ecol 25: 1–7.

7. Mason NWH, MacGillivray K, Steel JB, Wilson JB (2003) An index of functional diversity. J Veg Sci 14: 571–578.

8. Laurance WF, Nascimento HEM, Laurance SG, Andrade AC, Fearnside PM, et al. (2006) Rain forest fragmentation and the proliferation of successional trees. Ecology 87: 469–482.

9. Tabarelli M, Silva MJC, Gascon C (2004) Forest fragmentation, synergisms and the impoverishment of neotropical forests. Biodivers Conserv 13: 1419–1425.

10. Tabarelli M, Mantovani W, Peres CA (1999) Effects of habitat fragmentation on plant guild structure in the montane Atlantic forest of southeastern Brazil. Biol Conserv 91: 119–127.

11. Metzger JP (2000) Tree functional group richness and landscape structure in a Brazilian tropical fragmented landscape. Ecol Appl 10: 1147–1161.

12. Laurance WF, Nascimento HEM, Laurance SG, Andrade A, Ribeiro J, et al. (2006) Rapid decay of tree-community composition in Amazonian forest fragments. Proc Natl Acad Sci USA 103: 19010–19014.

13. Laurance WF, Ferreira LV, Rankin-De Merona JM, Laurance SG, Hutchings RW, et al. (1998) Effects of forest fragmentation on recruitment patterns in Amazonian tree communities. Conserv Biol 12: 460–464.

14. Terborgh J, Lopez L, Nunez P, Rao M, Shahabuddin G, et al. (2001) Ecological meltdown in predator-free forest fragments. Science 294: 1923–1926.

15. Melo FPL, Dirzo R, Tabarelli M (2006) Biased seed rain in forest edges: Evidence from the Brazilian Atlantic forest. Biol Conserv 132: 50–60.

16. Galetti M, Donatti CI, Pires AS, Guimaraes PR, Jordano P (2006) Seed survival and dispersal of an endemic Atlantic forest palm: the combined effects of defaunation and forest fragmentation. Bot J Linnean Soc 151: 141–149.

17. Silva JMC, Tabarelli M (2000) Tree species impoverishment and the future flora of the Atlantic forest of northeast Brazil. Nature 404: 72–74.

18. Cordeiro NJ, Howe HF (2001) Low recruitment of trees dispersed by animals in African forest fragments. Conserv Biol 15: 1733–1741.

19. Oliveira MA, Grillo AS, Tabarelli M (2004) Forest edge in the Brazilian Atlantic forest: drastic changes in tree species assemblages. Oryx 38: 389–394.

20. Richards PW (1996) The tropical rain forest - an ecological study. Cambridge: Cambridge University Press.

21. Turner IM (2001) The ecology of trees in the tropical rain forest. Cambridge: Cambridge University Press.

22. Whitmore TC (1990) An introduction to tropical rain forests. Oxford: Clarendon Press.

23. Buchmann SL, Nabhan GP (1996) The forgotten pollinators. Whashington D.C.: Island Press.

24. Kevan PG (1999) Pollinators as bioindicators of the state of the environment: species, activity and diversity. Agric Ecosyst Environ 74: 373–393.

25. Vamosi JC, Knight TM, Steets JA, Mazer SJ, Burd M, et al. (2006) Pollination decays in biodiversity hotspots. Proc Natl Acad Sci U S A 103: 956–961.

26. Rathcke BJ, Jules ES (1993) Habitat fragmentation and plant pollinator interactions. Curr Sci 65: 273–277.

27. Murcia C (1996) Forest fragmentation and the pollination of neotropical plants. In: Schelhas J, Greenberg R, editors. Forest Patches in tropical landscapes. Washington, DC: Island Press. pp. 19–36.

28. Harris LF, Johnson SD (2004) The consequences of habitat fragmentation for plant-pollinator mutualism. Int J Trop Insect Sci 24: 29–43.

29. Aguilar R, Ashworth L, Galetto L, Aizen MA (2006) Plant reproductive susceptibility to habitat fragmentation: review and synthesis through a meta-analysis. Ecol Lett 9: 968–980.

30. Wilcock C, Neiland R (2002) Pollination failure in plants: why it happens and when it matters. Trends Plant Sci 7: 270–277.

31. Chambers JQ, Higuchi N, Schimel JP (1998) Ancient trees in Amazonia. Nature 391: 135–136.

32. Galindo-Leal C, Câmara IG (2005) Status do hotspot Mata Atlântica: uma síntese. In: Galindo-Leal C, Câmara IG, editors. Mata Atlântica: Biodiversidade, Ameaças e Perspectivas. Belo Horizonte: Fundação SOS Mata Atlântica & Conservação Internacional. pp. 3–11.

33. Ranta P, Blom T, Niemela J, Joensuu E, Siitonen M (1998) The fragmented Atlantic rain forest of Brazil: size, shape and distribution of forest fragments. Biodivers Conserv 7: 385–403.

34. Melo FPL, Lemire D, Tabarelli M (2007) Extirpation of large-seeded seedlings from the edge of a large Brazilian Atlantic forest fragment. Écoscience 14: 124–129.

35. Prance GT (1982) Forest refuges: evidence from woody angiosperms. In: Prance GT, editor. Biological diversification in the tropics. New York: Columbia University Press. pp. 137–158.

36. IBGE (1985) Atlas nacional do Brasil: Região Nordeste. Rio de Janeiro: Instituto Brasileiro de Geografia e Estatística.

37. Veloso HP, Rangel-Filho ALR, Lima JCA (1991) Classificação da vegetação brasileira adaptada a um sistema universal. IBGE.

38. Grillo A, Oliveira MA, Tabarelli M (2006) Árvores. In: Pôrto KC, Almeida-Cortez JS, Tabarelli M, editors. Diversidade biológica e conservação da Floresta Atlântica ao norte do Rio São Francisco. 1 ed. Brasília: Ministério do Meio Ambiente. pp. 191–216.

39. Grillo A (2005) As implicações da fragmentação e da perda de habitat sobre a assembléia de árvores na Floresta Atlântica ao norte do rio São Francisco [PhD Thesis]. Recife: Universidade Federal de Pernambuco.

40. Coimbra-Filho F, Câmara IG (1996) Os limites originais do Bioma Mata Atlântica na Região Nordeste do Brasil. Rio de Janeiro: Fundação Brasileira para Conservação da Natureza - FBCN.

41. Harper KA, Macdonald SE, Burton PJ, Chen JQ, Brosofske KD, et al. (2005) Edge influence on forest structure and composition in fragmented landscapes. Conserv Biol 19: 768–782.

42. Chiarello AG (1999) Effects of fragmentation of the Atlantic forest on mammal communities in south-eastern Brazil. Biol Conserv 89: 71–82.

43. Tabarelli M, Peres CA (2002) Abiotic and vertebrate seed dispersal in the Brazilian Atlantic forest: implications for forest regeneration. Biol Conserv 106: 165–176.

44. Bruna EM, Kress WJ (2002) Habitat fragmentation and the demographic structure of an Amazonian understory herb (Heliconia acuminata). Conserv Biol 16: 1256–1266.

45. Pôrto KC, Tabarelli M, Almeida-Cortez JS (2006) Diversidade biológica e conservação da Floresta Altântica ao norte do Rio São Francisco. Brasília: Ministério do Meio Ambiente.

46. Pimentel DS, Tabarelli M (2004) Seed dispersal of the palm Attalea oleifera in a remnant of the Brazilian Atlantic Forest. Biotropica 36: 74–84.

47. Ribeiro JELS, Hopkins MJG, Vicentini A, Sothers CA, Costa MAS, et al. (1999) Flora da Reserva Ducke: guia de identificação de uma floresta de terra-firme na Amazônia Central. Manaus: INPA-DFID.

48. Backes P, Irgang B (2004) Mata Atlântica: As árvores e a paisagem. Porto Alegre: Editora Paisagem do Sul.

49. Lorenzi H (2002) Árvores Brasileiras: manual de identificação e cultivo de plantas arbóreas nativas do Brasil (vol.1). Nova Odessa: Instituto Plantarum.

50. Lorenzi H (2002) Árvores Brasileiras: manual de identificação e cultivo de plantas arbóreas nativas do Brasil (vol.2). Nova Odessa: Instituto Plantarum.

51. Machado IC, Lopes AV (2002) A Polinização em ecossistemas de Pernambuco: uma revisão do estado atual do conhecimento. In: Tabarelli M, Silva JMC, editors. Diagnóstico da Biodiversidade de Pernambuco. Recife: Secretaria de Ciência Tecnologia e Meio-Ambiente, Fundação Joaquim Nabuco, Editora Massangana. pp. 583–596.

52. ter Steege H, Pitman NCA, Phillips OL, Chave J, Sabatier D, et al. (2006) Continental-scale patterns of canopy tree composition and function across Amazonia. Nature 443: 444–447.

53. Gorresen PM, Willig MR (2004) Landscape responses of bats to habitat fragmentation in Atlantic forest of Paraguay. J Mammal 85: 688–697.

54. Krebs C (1989) Ecological metodology. New York: Harper & Hall.

55. Petchey OL, Gaston KJ (2002) Extinction and the loss of functional diversity. Proc R Soc Lond Ser B-Biol Sci 269: 1721–1727.

56. Petchey OL, Hector A, Gaston KJ (2004) How do different measures of functional diversity perform? Ecology 85: 847–857.

57. Ricotta C (2005) A note on functional diversity measures. Basic Appl Ecol 6: 479–486.

58. Sokal RR, Rohlf FG (1995) Biometry. New York: W.H. Freeman and Company.

59. Clarke KR, Gorley RN (2001) PRIMER v5: User manual/tutorial. Playmouth: PRIMER-E Ltd.

60. Condit R, Pitman N, Leigh EG, Chave J, Terborgh J, et al. (2002) Beta-diversity in tropical forest trees. Science 295: 666–669.

61. Jones MM, Tuomisto H, Clark DB, Olivas P (2006) Effects of mesoscale environmental heterogeneity and dispersal limitation on floristic variation in rain forest ferns. J Ecol 94: 181–195.

62. Wilkinson L (1996) SYSTAT. Version 6.0. Chicago: SPSS.

63. Anon (2001) Primer 5 for Windows. Plymouth: PRIMER-E.

64. McCune B, Mefford MJ (1999) PC-ORD. Multivariate Analysis of Ecological Data. Version 4.36. Gleneden Beach: MjM Software.

65. Kang H, Bawa KS (2003) Effects of successional status, habit, sexual systems, and pollinators on flowering patterns in tropical rain forest trees. Am J Bot 90: 865–876.

66. Faegri K, Pijl L van der (1979) The principles of pollination ecology. Oxford: Pergamon Press.

67. Clark DB, Clark DA, Read JM (1998) Edaphic variation and the mesoscale distribution of tree species in a neotropical rain forest. J Ecol 86: 101–112.

68. Michalski F, Nishi I, Peres CA (2007) Disturbance-mediated drift in tree functional groups in Amazonian forest fragments. Biotropica. doi:10.1111/j.1744-7429.2007.00318.x.

69. Laurance WF, Lovejoy TE, Vasconcelos HL, Bruna EM, Didham RK, et al. (2002) Ecosystem decay of Amazonian forest fragments: A 22-year investigation. Conserv Biol 16: 605–618.

70. Laurance WF (2001) Fragmentation and plant communities: synthesis and implications for landscape management. In: Bierregaard RO Jr, Gascon C, Lovejoy TE, Mesquita RCG, editors. Lessons from Amazonia: the Ecology and Conservation of a Fragmented Forest. New Haven: New Haven: Yale University Press. pp. 158–168.

71. Laurance WF, Peres CA (2006) Emerging threats to tropical forests. Chicago: The University of Chicago Press.

72. Chazdon RL, Careaga S, Webb C, Vargas O (2003) Community and phylogenetic structure of reproductive traits of woody species in wet tropical forests. Ecol Monogr 73: 331–348.

73. Laurance WF, Delamonica P, Laurance SG, Vasconcelos HL, Lovejoy TE (2000) Conservation - Rainforest fragmentation kills big trees. Nature 404: 836–836.

74. Costa JBP (2007) Efeitos da fragmentação sobre a assembléia de plântulas em um trecho da floresta Atlântica Nordestina [Master thesis]. Recife: Universidade Federal de Pernambuco.

75. APG (= Angiosperm Phylogeny Group) II (2003) An update of the Angiosperm Phylogeny Group classification for the orders and families of flowering plants: APG II. Bot J Linnean Soc 141: 399–436.

76. Bawa KS (1990) Plant-pollinator interactions in tropical rain-forests. Annu Rev Ecol Syst 21: 399–422.

77. Bawa KS, Bullock SH, Perry DR, Coville RE, Grayum MH (1985) Reproductive biology of tropical lowland rain forest trees. II. Pollination systems. Am J Bot 72: 346–356.

78. Ghazoul J (2005) Pollen and seed dispersal among dispersed plants. Biol Rev 80: 413–443.

79. Kearns CA, Inouye DW, Waser NM (1998) Endangered mutualisms: The conservation of plant-pollinator interactions. Ann Rev Ecol Syst 29: 83–112.

80. Aizen MA, Feinsinger P (1994) Forest fragmentation, pollination, and plant reproduction in a chaco dry forest, Argentina. Ecology 75: 330–351.

81. Renner SS (1998) Effects of habitat fragmentation on plant pollinator interactions in the tropics. In: Newbery DM, Prins HHT, Brown ND, editors. Dynamics of Tropical Communities. London: Blackwell Scientific Publishers. pp. 339–360.

82. Hobbs RJ, Yates CJ (2003) Impacts of ecosystem fragmentation on plant populations: generalising the idiosyncratic. Aust J Bot 51: 471–488.

83. Sá-Neto RJ (2003) Efeito da fragmentação na comunidade de morcegos (Mammalia: Chiroptera) em remanescentes de floresta Atlântica, Usina Serra Grande - Alagoas [Master thesis]. Recife: Universidade Federal de Pernambuco.

84. Lopes AV, Medeiros PC, Aguiar AV, Machado IC (2006) Esfingídeos. In: Pôrto KC, Almeida-Cortez JS, Tabarelli M, editors. Diversidade biológica e conservação da floresta Atlântica ao norte do Rio São Francisco. Brasília: Ministério do Meio Ambiente. pp. 229–235.

85. Wirth R, Meyer ST, Leal IR, Tabarelli M (2007) Plant–herbivore interactions at the forest edge. Progr Bot 69: 423–448.

86. Endress PK (1994) Diversity and evolutionary biology of tropical flowers. Cambridge: Cambridge University Press.

87. Proctor M, Yeo P, Lack A (1996) The natural history of pollination. London: Harper Collins Publishers.

88. Machado IC, Lopes AV (2004) Floral traits and pollination systems in the Caatinga, a Brazilian tropical dry forest. Ann Bot 94: 365–376.

89. Richards AJ (1997) Plant breeding systems. London: Chapman & Hall.

90. Oliveira PE, Gibbs PE (2000) Reproductive biology of woody plants in a cerrado community of Central Brazil. Flora 195: 311–329.

CITATION

Originally published under the Creative Commons Attribution License. Girão LC, Lopes AV, Tarbelli M, Bruna EM. Changes in Tree Reproductive Traits Reduce Functional Diversity in a Fragmented Atlantic Forest Landscape. PLoS ONE. 2007; 2(9): e908. doi: 10.1371/journal.pone.0000908.

Genetic Subtraction Profiling Identifies Genes Essential for Arabidopsis Reproduction and Reveals Interaction Between the Female Gametophyte and the Maternal Sporophyte

Amal J. Johnston, Patrick Meier, Jacqueline Gheyselinck,
Samuel E. J. Wuest, Michael Federer, Edith Schlagenhauf,
Jörg D. Becker and Ueli Grossniklaus

ABSTRACT

Background

The embryo sac contains the haploid maternal cell types necessary for double fertilization and subsequent seed development in plants. Large-scale identification of genes expressed in the embryo sac remains cumbersome because of its inherent microscopic and inaccessible nature. We used genetic subtraction and comparative profiling by microarray between the Arabidopsis thaliana wild-type and a sporophytic mutant lacking an embryo sac in order to identify embryo sac expressed genes in this model organism. The influences of the embryo sac on the surrounding sporophytic tissues were previously thought to be negligible or nonexistent; we investigated the extent of these interactions by transcriptome analysis.

Results

We identified 1,260 genes as embryo sac expressed by analyzing both our dataset and a recently reported dataset, obtained by a similar approach, using three statistical procedures. Spatial expression of nine genes (for instance a central cell expressed trithorax-like gene, an egg cell expressed gene encoding a kinase, and a synergid expressed gene encoding a permease) validated our approach. We analyzed mutants in five of the newly identified genes that exhibited developmental anomalies during reproductive development. A total of 527 genes were identified for their expression in ovules of mutants lacking an embryo sac, at levels that were twofold higher than in the wild type.

Conclusion

Identification of embryo sac expressed genes establishes a basis for the functional dissection of embryo sac development and function. Sporophytic gain of expression in mutants lacking an embryo sac suggests that a substantial portion of the sporophytic transcriptome involved in carpel and ovule development is, unexpectedly, under the indirect influence of the embryo sac.

Background

The life cycle of plants alternates between diploid (sporophyte) and haploid (male and female gametophytes) generations. The multicellular gametophytes represent the haploid phase of the life cycle between meiosis and fertilization, during which the gametes are produced through mitotic divisions. Double fertilization is unique to flowering plants; the female gametes, namely the haploid egg cell and the homo-diploid central cell, are fertilized by one sperm cell each. Double fertilization produces a diploid embryo and a triploid endosperm, which are the two major constituents of the developing seed [1]. The egg, the central cell, and two accessory cell types (specifically, two synergid cells and three antipodal cells) are contained in the embryo sac, also known as the female gametophyte or megagametophyte, which is embedded within the maternal tissues of the ovule. As a carrier of maternal cell types required for fertilization, the embryo sac provides an interesting model in which to study a variety of developmental aspects relating to cell specification, cell polarity, signaling, cell differentiation, double fertilization, genomic imprinting, and apomixis [1-3].

Out of the 28,974 predicted open reading frames of Arabidopsis thaliana, a few thousand genes are predicted to be involved in embryo sac development [1,4]. These genes can be grouped into two major classes: genes that are necessary during female gametogenesis and genes that impose maternal effects through the female gametophyte, and thus play essential roles for seed development. To date, loss-of-function mutational analyses have identified just over 100 genes in Arabidopsis that belong to these two classes [5-14]. However, only a small number of genes have been characterized in depth. Cell cycle genes (for instance, PRO-LIFERA, APC2 [ANAPHASE PROMOTING COMPLEX 2], NOMEGA, and RBR1 [RETINOBLASTOMA RELATED 1]), transcription factors (for instance, MYB98 and AGL80 [AGAMOUS-LIKE-80]), and others (including CKI1 [CYTOKININ INDEPENDENT 1], GFA2 [GAMETOPHYTIC FACTOR 2], SWA1 [SLOW WALKER 1] and LPAT2 [LYSOPHOSPHATIDYL ACYL-TRANSFERASE 2]) are essential during embryo sac development [6,15-23]. Maternal effect genes include those of the FIS (FERTILIZATION INDEPEN-DENT SEED) class and many others that are less well characterized [9,13,24]. FIS genes are epigenetic regulators of the Polycomb group and control cell pro-liferation during endosperm development and embryogenesis [7,10,12,25,26]. Ultimately, the molecular components of cell specification and cell differentiation during megagametogenesis and double fertilization remain largely unknown, and alternate strategies are required for a high-throughput identification of candidate genes expressed during embryo sac development.

Although transcriptome profiling of Arabidopsis floral organs [27,28], whole flowers and seed [29], and male gametophytes [30-33] have been reported in previous studies, large-scale identification of genes expressed during female game-tophyte development remains cumbersome because of the microscopic nature of the embryo sac. Given the dearth of transcriptome data, we attempted to ex-plore the Arabidopsis embryo sac transcriptome using genetic subtraction and microarray-based comparative profiling between the wild type and a sporophytic mutant, coatlique (coa), which lacks an embryo sac. Using such a genetic subtrac-tion, genes whose transcripts were present in the wild type at levels higher than in coa could be regarded as embryo sac expressed candidate genes. While our work was in progress, Yu and coworkers [34] reported a similar genetic approach to reveal the identity of 204 genes expressed in mature embryo sacs. However, their analysis of the embryo sac transcriptome was not exhaustive because they used different statistical methodology in their data analysis. Thus, we combined their dataset with ours for statistical analyses using three statistical packages in order to explore the transcriptome more extensively. Here, we report the identity of 1,260 potentially embryo sac expressed genes, 8.6% of which were not found in tissue-specific sporophytic transcriptomes, suggesting selective expression in the embryo sac. Strong support for the predicted transcriptome was provided by the spatial expression pattern of 24 genes in embryo sac cells; 13 of them were previously identified as being expressed in the embryo sac by enhancer detectors or promoter-reporter gene fusions, and we could confirm the spatial expression of the corresponding transcripts by microarray analysis. In addition, we show embryo sac cell-specific expression for nine novel genes by in situ hybridization or reporter gene fusions. In order to elucidate the functional role of the identified genes, we sought to search for mutants affecting embryo sac and seed development by T-DNA mutagenesis. We describe the developmental anomalies evident in five mutants exhibiting lethality during female gametogenesis or seed development.

Genetic evidence suggests that the maternal sporophyte influences development of the embryo sac [1,35-37]. Because the carpel and sporophytic parts of the ovule develop normally in the absence of an embryo sac, it has been concluded that the female gametophyte does not influence gene expression in the surrounding tissue [2]. Our data clearly showed that 527 genes were over-expressed by at least twofold in the morphologically normal maternal sporophyte in two sporophytic mutants lacking an embryo sac. We confirm the gain of expression of 11 such genes in mutant ovules by reverse transcription polymerase chain reaction (RT-PCR). Spatial expression of five of these genes in carpel and ovule tissues of coa was confirmed by in situ hybridization, revealing that expression mainly in the carpel and ovule tissues is tightly correlated with the presence or absence of an embryo sac. In summary, our study provides two valuable datasets of the transcriptome of Arabidopsis gynoecia, comprising a total of 1,787 genes: genes that are expressed or enriched in the embryo sac and are likely function to control embryo sac and seed development; and a set of genes that are over-expressed in the maternal sporophyte in the absence of a functional embryo sac, revealing interactions between gametophytic and sporophytic tissues in the ovule and carpel.

Results

We intended to isolate genes that are expressed in the mature female gametophyte of A. thaliana, and are thus potentially involved in its development and function. To this end, the transcriptomes of the gynoecia from wild-type plants were compared with those of two sporophytic recessive mutants, namely coatlique (coa) and sporocyteless (spl), both of which lack a functional embryo sac. The coa mutant was isolated during transposon mutagenesis for its complete female sterility and partial male sterility in the homozygous state (Vielle-Calzada J-P, Moore JM, Grossniklaus U, unpublished data). Following tetrad formation three megaspores degenerated, producing one viable megaspore, but megagametogenesis was not initiated in coa. Despite the failure in embryo sac development, the integuments and endothelium in coa differentiated similar to wild-type ovules (Figure 1). In addition to our experiment with coa, we reanalyzed the dataset reported by Yu and coworkers [34], who used the spl mutant and corresponding wild type for a similar comparison. The spl mutant behaves both phenotypically and genetically very similar to coa [38]. The primary difference in the experimental set up between the present study and that conducted by Yu and coworkers [34] is that we did not dissect out the ovules from pistils, whereas Yu and coworkers extracted ovule samples by manual dissection from the carpel, which led to a lower dilution of 'contaminating' cells surrounding the embryo sac. However, our inclusion of intact pistils allowed us to elucidate the carpel-specific and ovule-specific effects controlled by the female gametophyte.

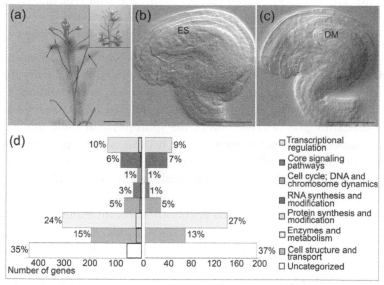

Figure 1. A genetic subtraction strategy for determination of the embryo sac transcriptome. (a) A branch of a coatlique (coa) showing undeveloped siliques. Arrows point to a small silique, which bears female sterile ovules inside the carpel (insert: wild-type Ler branch). (b) Morphology of a mature wild-type ovule bearing an embryo sac (ES) before anthesis. (c) A functional embryo sac is absent in coa (degenerated megaspores [DM]). Note that the ovule sporophyte is morphologically equivalent to that of the wild type. (d) Functional categories of genes identified by a microarray-based comparison of coa and sporocyteless (spl; based on data from Yu and coworkers [34]) with the wild type. The embryo sac expressed transcriptome is shown to the left. Embryo sac expressed genes were grouped as preferentially expressed in the embryo sac if they were not detected in previous sporophytic microarrays [28]. The size of the specific transcriptome in each class is marked on each bar by a dark outline. Functional categories of genes that were identified as over-expressed in the sporophyte of coa and spl are shown to the right. Scale bars: 1 cm in panel a (2 cm in the insert of panel a), and 50 μm in panels b and c.

Statistical Issues on the Microarray Data Analysis

To determine the embryo sac transcriptome, we used coa and wild-type pistil samples (late 11 to late 12 floral stages [39]) in three biologic replicates, and followed the Affymetrix standard procedures from cRNA synthesis to hybridization on the chip. Finally, raw microarray data from the coa and wild-type samples in triplicate were retrieved after scanning the Arabidopsis ATH1 'whole genome' chips, which represent 24,000 annotated genes, and they were subjected to statistical analyses. The normalized data were examined for their quality using cluster analysis [40]. There was strong positive correlation between samples within the three replicates of wild-type and coa (Pearson coefficients: r = 0.967 for for wild-type and r = 0.973 for coa). Therefore, the data were considered to be of good quality for further analyses. It was necessary to ensure that the arrays of both the wild type and coa did not differ in RNA quality and hybridization efficiency. The hybridization signal intensities of internal control gene probes were not significantly altered across the analysed arrays, hence assuring the reliability of the results (data not shown). The quality of data for the spl mutant and wild-type microarray was described previously [34]. Subsequently, differentially expressed genes were identified using three independent microarray data analysis software packages.

To identify genes that are expressed in the female gametophyte, we subtracted the transcriptomes of coa or spl from the corresponding wild type. Genes that were identified as being upregulated in wild-type gynoecia are candidates for female gametophytic expression, and genes highly expressed in coa and spl are probable candidates for gain-of-expression in the sporophyte of these mutants. However, this comparison was not straightforward because we were not in a position to compare the mere four cell types of the mature embryo sac with the same number of sporophytic cells. Whether using whole pistils or isolated ovules, a large excess of sporophytic cells surrounds the embryo sac. The contaminating cells originate from the ovule tissues such as endothelium, integuments and funiculus, or those surrounding the ovules such as stigma, style, transmitting tract, placenta, carpel wall and replum. Therefore, we anticipated that the transcript subtraction for embryo sac expression would suffer from high experimental noise. We examined the log transformed data points from the coa and spl datasets (with their corresponding wild-type data) in volcano plots. This procedure allows us to visualize the trade-offs between the fold change and the statistical significance. As we anticipated, the data points from the sporophytic gain outnumbered the embryo sac transcriptome data points on a high-stringency scale (data not shown). This problem of dilution in our data for embryo sac gene discovery was more pronounced in the coa dataset than that of spl, because we did not dissect out the ovules from the carpel. Therefore, we made the following decisions in analyzing the gametophytic data: to use advanced statistical packages that use different principles in their treatment of the data; and to set a lowest meaningful fold change in data comparison, in contrast to the usual twofold change as recommended in the literature.

In the recent past, many new pre-processing methods for Affymetrix GeneChip data have been developed, and there are conflicting reports about the performance of each algorithm [41-43]. Because there is no consensus about the most accurate analysis methods, contrasting methods can be combined for gene discovery [44]. We used the following three methods in data analyses: the microarray suite software (MAS; Affymetrix) and Genspring; the DNA Chip analyzer (dCHIP) package [45]; and GC robust multi-array average analysis (gcRMA) [46]. MAS uses a nonparametric statistical method in data analyses, whereas dCHIP uses an intensity modeling approach [47]. dCHIP removes outlier probe intensities, and reduces the between-replicate variation [48]. A more recent method, gcRMA uses a model-based background correction and a robust linear model to calculate signal intensities. Depending on the particular question to be addressed, one may wish to identify genes that are expressed in the embryo sac with the highest probability possible and to use a very stringent statistical treatment (for example, dCHIP), or one may wish to obtain the widest possible range of genes that are potentially expressed in the embryo sac and employ a less stringent method (for example, MAS). We did not wish to discriminate between the three methods in our analysis, and we provide data for all of them.

Although conventionally twofold change criteria have been followed in a number of microarray studies, it has been disputed whether fold change should be used at all to study differential gene expression (for review, see [49]). Based on studies correlating both microarray and quantitative RT-PCR data, it was suggested that genes exhibiting 1.4-fold change could be used reliably [50,51]. Tung and coworkers used a minimum fold change as low as 1.2 in order to identify differentially expressed genes in Arabidopsis pistils within specific cell types, and the results were spatially validated [52]. In order to make a decision on our fold change criterion in the data analysis, we examined the dataset for validation of embryo sac expressed genes that had previously been reported. We found that genes such as CyclinA2;4 (coa dataset) and ORC2 (spl dataset) were identified at a fold change of 1.28. In addition, out of the 43 predicted genes at 1.28-fold change from coa and spl datasets, 33% were present in triplicate datasets from laser captured central cells (Wuest S, Vijverberg K, Grossniklaus U, unpublished data), independently confirming their expression in at least one cell of the embryo sac. Therefore, the baseline cutoff for subtraction was set at 1.28-fold in the wild type, and a total of 1,260 genes were identified as putative candidates for expression in the female gametophyte.

However, it must be noted that lowering the fold change potentially increases the incidence of false-positive findings. By setting the baseline to 1.28, we could predict that false discovery rates (FDRs) would range between 0.05% and 3.00%, based on dCHIP and gcRMA analyses (data not shown). Convincingly, we we able to observe 24 essential genes and 17 embryo sac expressed genes at a fold change range between 1.28 and 1.6. Moreover, our data on homology of candidate genes to expressed sequence tags (ESTs) from monocot embryo sacs will facilitate careful manual omission of false-positive findings. The usefulness of this approach is also demonstrated by the observation that 84% of the essential genes and genes validated for embryo sac expression (n = 51) present in our datasets exhibited homology to the monocot embryo sac ESTs. Therefore, our practical strategy of using a low fold change cut-off probably helped in identifying low-abundance signals, which would otherwise be ignored or handled in an ad hoc manner.

In contrast to the embryo sac datasets, we applied a more stringent twofold higher expression as a baseline for comparison of the mutant sporophyte with the wild type. This is because we had large amounts of sporophytic cells available for comparison. In all, 527 genes were identified as candidate genes for gain of sporophytic expression in coa and spl mutant ovules. Because the transcriptome identified by three independent statistical methods and the resultant overlaps were rather different in size for both the gametophytic and sporophytic datasets, we report all the data across the three methods. This approach is validated by the fact that candidate genes found using only one statistical method can indeed be embryo sac expressed. Furthermore, only 8% of the validated genes (n = 51) were consistently identified by all three methods, demonstrating the need for independent statistical treatments. In short, our data analyses demonstrate the usefulness of employing different statistical treatments for microarray data.

Another practical consideration following our data analyses was the very limited overlap between coa and spl datasets. Although both mutants are genetically and phenotypically similar, the overlap is only 35 genes between the embryo sac datasets and 13 genes between the sporophytic datasets. In light of the validation in expression for 12 genes from the coa dataset, which were not identified from the spl dataset, we suggest that the limited overlap is not merely due to experimental errors. It is likely that the embryo sac transcriptome is substantial (several thousands of genes [2]), and two independent experiments identified different subsets of the same transcriptome. This is apparent from our validation of expression for several genes, which were exclusively found in only one microarray dataset. In terms of the sporophytic gene expression, we have shown that three sporophytic genes initially identified only in the spl microarray dataset were indeed over-expressed in coa tissues (discussed below). In short, despite the limited overlap between datasets, both the embryo sac and sporophytic datasets will be very useful in elucidating embryo sac development and its control of sporophytic gene expression.

Functional Classification of the Candidate Genes

The genes identified as embryo sac expressed or over-expressed sporophytic candidates were grouped into eight functional categories based on a classification system reported previously [53] (Figure 1). The gene annotations were improved based on the Gene Ontology annotations available from 'The Arabidopsis Information Resource' (TAIR). The largest group in both gene datasets consisted of genes with unknown function (35% of embryo sac expressed genes and 37% of over-expressed sporophytic candidate genes), and the next largest was the class of metabolic genes (24% and 27%; Figure 1). Overall, both the gametophytic and sporophytic datasets comprised similar percentages of genes within each functional category (Figure 1). In both datasets, we found genes that are predicted to be involved in transport facilitation and cell wall biogenesis (15% of embryo sac expressed genes and 13% of over-expressed sporophytic candidate genes), transcriptional regulation (10% and 9%), signaling (7% and 6%), translation and protein fate (5% each), RNA synthesis and modification (3% and 1%), and cell cycle and chromosome dynamics (1% each).

Validation of Expression for Known Embryo Sac-Expressed Genes

The efficacy of the comparative profiling approach used here was first confirmed by the presence of 18 genes that were previously identified as being expressed in the embryo sac. They included embryo sac expressed genes such as PROLIFERA, PAB2 and PAB5 (which encode poly-A binding proteins) and MEDEA, and genes with cell-specific expression such as central cell expressed FIS2 and FWA, synergid cell expressed MYB98, and antipodal cell expressed AT1G36340. Therefore, our comparative profiling approach potentially identified novel genes that could be expressed either throughout the embryo sac or in an expression pattern that is restricted to specific cell types.

In Situ Hybridization and Enhancer Detector Patterns Confirm Embryo Sac Expression of Candidate Genes

In order to validate the spatial expression of candidate genes in the wild-type embryo sac, the six following genes were chosen for mRNA in situ hybridization on paraffin-embedded pistils: AT5G40260 (encoding nodulin; 1.99-fold) and AT4G30590 (encoding plastocyanin; 1.88-fold); AT5G60270 (encoding a receptor-like kinase; 1.56-fold) and AT3G61740 (encoding TRITHORAX-LIKE 3 [ATX3]; 1.47-fold); and AT5G50915 (encoding a TCP transcription factor; 1.36-fold) and AT1G78940 (encoding a protein kinase; 1.35-fold). Broad expression in all cells of the mature embryo sac was observed for genes AT5G40260, AT4G30590, AT5G60270, and AT4G01970 (Figure 2). The trithorax group gene ATX3 and AT5G50915 were predominantly expressed in the egg and the central cell, and the expression of the receptor-like kinase gene AT5G60270 was found to be restricted to the egg cell alone (Figure 2). In addition to the in situ hybridization experiments, we examined the expression of transgenes where specific promoters drive the expression of the bacterial uidA gene encoding β-galacturonidase (GUS) or in enhancer detector lines. We show that CYCLIN A2;4 (1.28-fold) and AT4G01970 (encoding a galactosyl-transferase; about 1.51-fold) were broadly expressed in the embryo sac, and that PUP3 (encoding a purine permease; 1.3-fold) was specifically expressed in the synergids (Figure 2). CYCLIN A2;4 appears to be expressed also in the endothelial layer surrounding the embryo sac (Figure 2e). Diffusion of GUS activity did not permit us to distinguish unambiguously embryo sac expression from endothelial expression. In short, both broader and cell type specific expression patterns in the embryo sac were observed for the nine candidate genes. Hence, we could validate the minimal fold change cut-off of 1.28 and the statistical methods employed in this study.

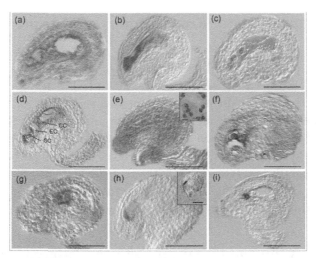

Figure 2. Confirmation of embryo sac expression for selected genes.

Embryo sac expression of nine candidate genes is shown by in situ hybridization (panels a, c, d, f, g, and i) or histochemical reporter gene (GUS) analysis (b, e, and h). Illustrated is the in situ expression of broadly expressed genes: (a) AT1G78940 (encoding a protein kinase that is involved in regulation of cell cycle progression), (c) AT5G40260 (encoding a nodulin), and (d) AT4G30590 (encoding a plastocyanin). Also shown is the restricted expression of (f) AT3G61740 (encoding the trithorax-like protein ATX3), (g) AT5G50915 (encoding a TCP transcription factor), and (i) AT5G60270 (encoding a protein kinase). The corresponding sense control for panels a, b, c, d, f, g, and i did not show any detectable signal (data not shown). GUS staining: (b) an enhancer-trap line for AT4G01970 (encoding a galactosyltransferase) shows embryo sac expression, (e) a promoter-GUS line for AT1G80370 (encoding CYCLIN A2;4) shows a strong and specific expression in the embryo sac and endothelium (insert: shows several ovules at lower magnification), and (h) a promoter-GUS line for AT1G28220 (encoding the purine permease PUP3) shows synergid specific expression (insert; note the pollen-specific expression of PUP3-GUS when used as a pollen donor on a wild-type pistil). CC, central cell; EC, egg cell; SC, synergids. Scale bars: 50 μm in panels a to i; and 100 μm and 50 μm in the inserts of panels e and h, respectively.

Embryo Sac Enriched Genes

Our strategic approach to exploring the embryo sac transcriptome was twofold: we aimed first to identify embryo sac expressed genes; second to describe the gametophyte enriched (male and female) transcriptome; and finally to define the embryo sac enriched (female only) transcriptome. Although the first category does not consider whether an embryo sac expressed gene is also expressed in the sporophyte, the second class of genes are grouped for their enriched expression in the male (pollen) and female gametophyte, but not in the sporophyte. The embryo sac enriched transcriptome is a subset of the gametophyte enriched transcriptome, wherein male gametophyte expressed genes are omitted. Of the embryo sac expressed genes, 32% were also present in the mature pollen transcriptome, and the vast majority (77%) were expressed in immature siliques as expected. Because large-scale female gametophytic cell expressed transcriptome data of Arabidopsis based on microarray or EST analyses are not yet available, we compared our data with the publicly available cell specific ESTs from maize and wheat by basic local alignment search tool (BLAST) analysis. Large-scale monocot ESTs are available only for the embryo sac and egg cells but not for the central cells (only 30 central cell derived ESTs from [54]). Therefore, we included the ESTs from immature endosperm cells at 6 days after pollination in the data comparison. Of our candidate genes, 38% were similar to the monocot embryo sac ESTs, 33% to the egg ESTs, and 53% to the central cell and endosperm ESTs.

Genes that were enriched in both the male and female gametophytes, or only in the embryo sac, were identified by subtracting these transcriptomes from a vast array of plant sporophytic transcriptomes of leaves, roots, whole seedlings, floral organs, pollen, and so on. The transcriptomes of the immature siliques were omitted in this subtraction scheme because often the gametophyte enriched genes are also present in the developing embryo and endosperm. We found 129 gametophyte enriched and 108 embryo sac enriched genes, accounting for 10% and 8.6%, respectively, of the embryo sac expressed genes (Table 1). Among the embryo sac enriched genes, 52% are uncategorized, 17% are enzymes or genes that are involved in metabolism, 15% are involved in cell structure and transport, 8% are transcriptional regulators, 4% are involved in translational initiation and modification, 3% are predicted to be involved in RNA synthesis and modification, and 2% in signaling (Figure 1 and Table 1). Of the embryo sac enriched transcripts, 31% were present in the immature siliques, suggesting their expression in the embryo and endosperm (Table 1). Furthemore, 26% of the embryo sac enriched genes were similar to monocot ESTs from the embryo sac or egg, and 41% were similar to central cell and endosperm ESTs (Table 1).

Table 1. Enriched expression of genes in the embryo sac cells was distinguished by their absence of detectable expression in sporophytic and pollen transcriptomes

Gene ID	Gene description	Study[a]	Homology to At siliques transcriptome[b]	Orthologous Zm EST[c]		
				ES	Egg	CC and EN
Transcriptional Regulation						
At5g06070	Zinc Finger (C2H2 Type) Family Protein (RBE)	2	0	0	0	0
At1g75430	Homeodomain Protein	1	0	0	0	0
At2g01500	Homeodomain-Leucine Zipper (WOX6, PFA2)	1	0	0	0	1
At2g24840	MADS-Box Protein Type I (AGL61)	2	0	1	1	1
At1g02580	MEDEA (MEA)	2	0	0	1	1
At5g11050	MYB Transcription Factor (MYB64)	1, 2	0	0	0	1
At3g29020	MYB Transcription Factor (MYB110)	2	0	0	0	1
At5g35550	MYB Transcription Factor (MYB123) (TT2)	2	1	0	0	1
At4g18770	MYB Transcription Factor (MYB98)	2	0	0	0	1
Core Signaling Pathways						
At5g12380	Annexin	2	0	1	1	1
At2g20660	Rapid Alkalinization Factor (RALF)	2	0	0	0	0
RNA Synthesis And Modification						
At1g63070	PPR Repeat-Containing Protein	1	0	0	0	1
At2g20720	PPR Repeat-Containing Protein	1	0	0	0	0
At3g54490	RPB5 RNA Polymerase Subunit	2	0	1	1	1
Protein Synthesis And Modification						
At5g11360	Protein Involved in Amino Acid Phosphorylation	2	1	1	0	0
At4g15040	Subtilase Family Protein, Proteolysis	2	0	1	1	1
At5g58830	Subtilase Family Protein, Proteolysis	2	1	0	1	1
At1g36340	Ubiquitin-Conjugating Enzyme	2	1	1	1	1
Enzymes And Metabolism						
At3g30540	(1-4)-Beta-Mannan Endohydrolase Family	2	0	0	0	0
At1g47780	Acyl-Protein Thioesterase-Related	2	0	1	0	0
At1g31450	Aspartyl Protease Family Protein	2	0	1	1	1
At1g69100	Aspartyl Protease Family Protein	2	0	0	1	1
At2g28010	Aspartyl Protease Family Protein	2	0	0	1	1
At2g34890	CTP Synthase, UTP-Ammonia Ligase	2	1	1	0	1
At4g39650	Gamma-Glutamyltransferase	2	1	0	0	0
At4g30540	Glutamine Amidotransferase	2	0	0	0	1
At3g48950	Glycoside Hydrolase Family 28 Protein	2	1	0	1	1
At2g42930	Glycosyl Hydrolase Family 17 Protein	2	0	1	1	1
At4g09090	Glycosyl Hydrolase Family 17 Protein	2	1	1	1	1
At1g56530	Hydroxyproline-Rich Glycoprotein	1	0	0	0	0
At1g06020	Pfkb-Type Carbohydrate Kinase	2	1	0	1	1
At2g43860	Polygalacturonase	2	0	0	0	1
At1g78400	Glycoside Hydrolase Family 28 Protein	1	0	0	0	1
At5g22960	Serine Carboxypeptidase A10 Family Protein	1	0	0	1	1
At4g21630	Subtilase Family Protein	2	1	1	1	1
At4g26280	Sulfotransferase Family Protein	2	0	0	0	0
Cell Structure And Transport						
At1g10010	Amino Acid Permease Involved In Transport	2	1	1	0	0
At4g20800	FAD-Binding Domain-Containing Protein	2	0	0	1	0
At1g34575	FAD-Binding Domain-Containing Protein	1	1	0	1	0

Table 1 (*Continued*)

Gene	Description					
At1g48010	Invertase/Pectin Methylesterase Inhibitor Family Protein	2	0	0	0	0
At3g17150	Invertase/Pectin Methylesterase Inhibitor Family Protein	2	I	0	0	0
At3g55680	Invertase/Pectin Methylesterase Inhibitor Family Protein	I	0	0	0	0
At2g47280	Pectinesterase Family Protein	2	0	0	0	0
At4g00190	Pectinesterase Family Protein	2	0	0	0	0
At5g18990	Pectinesterase Family Protein	2	0	0	0	0
At1g56620	Pectinesterase Inhibitor	2	0	0	0	0
At2g23990	Plastocyanin-Like	2	I	0	0	I
At1g73560	Lipid Transfer Protein (LTP) Family Protein	2	I	0	0	0
At5g56480	Lipid Transfer Protein (LTP) Family Protein	2	0	0	0	I
At1g63950	Lipid Transfer Protein (LTP) Family Protein	2	I	0	0	0
At3g05460	Sporozoite Surface Protein-Related	2	I	0	0	0
At5g06170	Sucrose Transporter	2	I	0	0	I
Uncategorized						
At1g24000	Bet V I Allergen Family Protein	2	I	0	0	0
At3g42130	Glycine-Rich Protein	I	0	0	0	0
At3g17140	Invertase Inhibitor-Related	2	0	0	0	0
At5g09360	Laccase-Like Protein Laccase	2	0	I	I	I
At1g79960	Ovate Protein-Related	2	0	0	0	0
At3g59260	Pirin	2	0	0	0	0
At4g30070	Plant Defensin-Fusion Protein	2	I	0	0	0
At5g38330	Plant Defensin-Fusion Protein	2	0	0	0	0
At2g01240	Reticulon Family Protein (RTNLB15)	2	0	I	0	0
At3g17080	Self-Incompatibility Protein-Related	2	0	0	0	0
At5g12060	Self-Incompatibility Protein-Related	2	I	0	0	0
At3g28020	Unknown	I	0	0	0	0
At3g19780	Unknown	I	0	0	0	0
At5g30520	Unknown	I	0	0	0	0
At3g45380	Unknown	I	0	0	0	0
At4g23780	Unknown	I	0	0	0	0
At1g54926	Unknown	I	0	0	0	0
At3g23720	Unknown	I	0	0	0	0
At1g47470	Unknown	I, 2	0	0	0	0
At1g32680	Unknown	I	0	0	0	0
At1g11690	Unknown	I	0	0	0	0
At2g04870	Unknown	I	0	0	0	0
At2g06630	Unknown	I	0	0	0	0
At4g09400	Unknown	I	0	0	0	0
At1g21950	Unknown	2	I	0	0	0
At4g11510	Unknown	2	I	0	0	0
At5g25960	Unknown	2	0	0	0	I
At1g60985	Unknown	2	0	0	0	0
At1g63960	Unknown	2	0	0	0	0
At1g78710	Unknown	2	I	I	0	I
At2g02515	Unknown	2	I	0	0	0
At2g20070	Unknown	2	0	0	0	0
At2g21740	Unknown	2	0	0	I	0
At2g30900	Unknown	2	0	I	0	I
At3g04540	Unknown	2	0	0	0	0

Table 1 (*Continued*)

At3g13630	Unknown	2	I	0	0	0
At3g43500	Unknown	2	0	0	0	0
At3g57850	Unknown	2	0	0	0	0
At4g07515	Unknown	2	0	0	0	0
At4g10220	Unknown	2	I	0	0	0
At4g17505	Unknown	2	0	0	0	I
At5g17130	Unknown	2	0	0	0	0
At5g25950	Unknown	2	I	0	0	I
At5g46300	Unknown	2	I	0	0	0
At5g64720	Unknown	2	0	0	I	0
At1g52970	Unknown	2	0	0	0	0
At5g42955	Unknown	2	0	0	0	0
At2g21655	Unknown	2	0	0	0	0
At2g20595	Unknown	2	0	0	0	0
At1g45190	Unknown	2	0	0	0	0
At5g43510	Unknown	2	I	0	0	0
At2g15780	Unknown, Blue Copper-Binding Protein	I	0	0	0	0
At1g24851	Unknown	I	0	0	0	0
At1g30030	Non-LTR retrotransposon family (LINE)	I	0	ND	ND	ND
At2g34130	CACTA-like transposase family	2	0	ND	ND	ND
At3g42930	CACTA-like transposase family	I	0	ND	ND	ND

Targeted Reverse Genetic Approaches Identified Female Gametophytic and Zygotic Mutants

Initial examination of our dataset for previously characterized genes revealed that the dataset contained 33 genes that were reported to be essential for female gametophyte or seed development (Figure 3). Given the availability of T-DNA mutants from the Arabidopsis stock centers, we wished to examine T-DNA knockout lines of some selected embryo sac expressed genes for ovule or seed abortion. During the first phase of our screen using 90 knockout lines, we identified eight semi-sterile mutants with about 50% infertile ovules indicating gametophytic lethality, and four mutants with about 25% seed abortion suggesting zygotic lethality (Table 2). When we examined the mutant ovules of gametophytic mutants, we found that seven mutants exhibited a very similar terminal phenotype: an arrested one-nucleate embryo sac. Co-segregation analysis by phenotyping and genotyping of one such mutant, namely frigg (fig-1) demonstrated that the mutant was not tagged, and the phenotype caused by a possible reciprocal translocation that may have arisen during T-DNA mutagenesis (Table 2). Preliminary data suggested that the six other mutants with a similar phenotype were not linked to the gene disruption either. Although not conclusively shown, it is likely that these mutants carry a similar translocation and, therefore, we did not analyze them further. These findings demonstrate that among the T-DNA insertion lines available, a rather high percentage (7/90 [8%]) exhibit a semisterile phenotype that is not due to the insertion. Therefore, caution must be exercised in screens for gametophytic mutants among these lines.

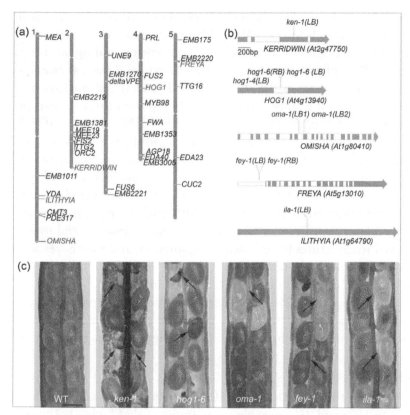

Figure 3. Genes essential for female gametogenesis, fertilization, and seed development are present in the embryo sac transcriptome datasets.

(a) Chromosomal locations of 35 essential genes. Five genes that are described in the current work are shown in blue. (b) Five genes and the locations of corresponding mutant alleles described in this work. Exons are shaded in orange. The genes were named after the following Goddesses: KERRIDWIN, the Welsh triple Goddess of trinity known for nurturing children; OMISHA, Indian Goddess of birth and death; FREYA, the Norse Goddess of fertility; and ILITHYIA, the Greek Goddess of childbirth. HOG1, HOMOLOGY DEPENDENT GENE SILENCING 1; LB and RB, left and right borders of the T-DNA. (c) Mutants were identified based on infertile ovules (ken-1) or seed abortion (hog1-6, oma-1, fey-1, and ila-1). The arrows identify the defective ovules. Scale bar: 100 μm in panel c.

Table 2. Genetics of mutant alleles affecting the female gametophyte and seed development

Mutant[a,b]	Segregation ratio[c,d]	χ^2 (segregation ratio)[e]	Seed abortion[f]	χ^2 (seed abortion)[g]	Mutant embryo sac phenotype
ken-1	0.97 (n = 290)	0.06**	54% (n = 327)	2.23*	54% unfused polar nuclei (n = 327)
fig-1[h]	ND	ND	53% (n = 258)	0.99*	53% arrested one-nucleate embryo sac (n = 258)
hog1-4	1.96 (n = 548)	0.04**	26% (n = 552)	0.24*	ND
hog1-6	2.11 (n = 351)	0.21**	24% (n = 420)	0.11*	22% aberrant early endosperm mitosis and zygote (n = 318)
oma-1	2.10 (n = 251)	0.13**	18% (n = 514)	13.1#	17% arrested, arrested mid-globular embryo (n = 269)
fey-1	1.99 (n = 425)	0.00**	21% (n = 414)	4.41**	19% arrested, arrested late-globular embryo (n = 243)
ila-1	1.92 (n = 038)	0.01**	23% (n = 352)	0.74*	20% and 3% arrested torpedo and late heart embryo (n = 352)

In about 54% of the ovules, the polar nuclei failed to fuse in kerridwin (ken-1), a mutant allele of AT2G47750, which encodes an auxin-responsive GH3 family protein (Figure 4 and Table 2). The corresponding wild-type pistils exhibited 9% unfused polar nuclei when examined 2 days after emasculation, and the remaining ovules had one fused central cell nucleus (n = 275). The hog1-6 mutant is allelic to the recently reported hog1-4, disrupting the HOMOLOGY DEPENDENT GENE SILENCING 1 gene (HOG1; AT4G13940), and they both were zygotic lethal, producing 24% to 26% aborted seeds (Table 2) [55]. Both these mutants exhibit anomalies during early endosperm division and zygote development (Figure 4i-l). In wild-type seeds, the endosperm remains in a free-nuclear state before cellularization around 48 to 60 hours after fertilization (HAP), and the embryo is at the globular stage (Figure 4f). In hog1-6, at about the same time the endosperm nuclei displayed irregularities in size, shape and number, and they never were uniformly spread throughout the seed (Figure 4i-l; n = 318). The irregular mitotic nuclei were clustered into two to four domains. The zygote remained at the single-cell stage, and in 2% of the cases it went on to the two-cell stage. In very rare instances (five observations), two large endosperm nuclei were observed while the embryo remained arrested at single-cell stage in hog1-4 (Figure 4k).

In omisha (oma-1) and freya (fey-1), the T-DNA disrupted AT1G80410 (encoding an acetyl-transferase) and AT5G13010 (encoding an RNA helicase), leading to 18% and 21% seed abortion, respectively (Table 2). The embryo arrested around the globular stage in both mutants (Figure 5f-i). The arrested mid-globular embryo cells (17%; n = 269) were larger in size in oma-1, whereas the corresponding wild type progressed to late-heart and torpedo stages with cellularized endosperm (Figure 5g). In the aborted fey-1 seeds, the cells of late-globular embryos (19%; n = 243) were much larger and irregular in shape than in the wild type, but no endosperm phenotype was discernible (Figure 5i). In most cases, giant suspensor cells were seen in fey-1, and there were more cells in the mutant suspensor than in that of the wild type (Figure 5i). ILITHYIA disrupts AT1G64790 encoding a translational activator, and the ila-1 embryos arrested when they reached the torpedo stage (Figure 4j and Table 2; n = 352). A small proportion of ila-1 embryos arrested at a late heart stage (11 observations). The results from the first phase of our targeted reverse genetic approach showed that there are mutant phenotypes for embryo sac expressed candidate genes, and that these gene disruptions lead to lethality during female gametophyte or seed development.

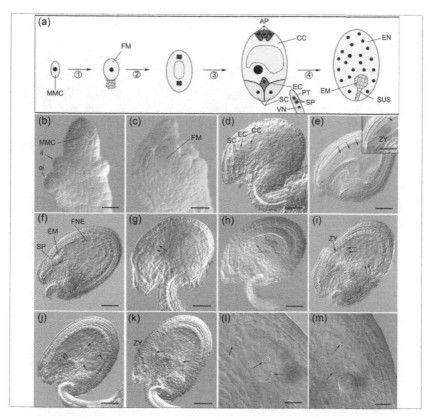

Figure 4. Female gametophytic and early zygotic mutant phenotypes highlight the essential role of corresponding genes for reproductive development.

(a) A cartoon showing the ontogeny of the wild-type female gametophyte in Arabidopsis and the early transition to seed development. A haploid functional megaspore (FM) develops from a diploid megaspore mother cell (MMC) upon two meiotic divisions (1). Three syncitial mitotic divisions (2) convert the FM into an eight-nuclear cell. Upon nuclear migration, cellularization, nuclear fusion and differentiation (3), a cellularized seven-celled embryo sac forms. It contains an egg cell (EC) and two synergid cells (SC) at the micropylar pole, three antipodals (AP) at the chalazal pole, and one vacuolated homo-diploid central cell (CC) in the middle. Subsequently, the AP cells degenerate. Degeneration of one SC precedes the entry of one pollen tube (PT), and two sperm cells (SP) independently fertilize the egg and central cell, leading to the development of a diploid embryo (EM) and triploid endosperm (EN) respectively. SUS, suspensor, VN, vegetative nucleus. (b-f) Morphology of wild-type ovules corresponding to representative events described above is depicted (ii indicates inner integuments, and oi indicates outer integuments). Both synchronous and asynchronous free nuclear mitotic divisions (as shown in panel e; arrows) lead to development of the free nuclear endosperm (FNE) as shown in panel f. The insert in panel e depicts a developing zygote (ZY). (g) In kerridwin (ken-1), two polar nuclei in the central cell fail to fuse. (h) Female gametophyte development did not initiate beyond the one-nucleate embryo sac stage (arrows) in frigg (fig-1). (i-l) Anomalies in early endosperm and zygotic development in hog1 (homology dependent gene silencing 1) mutants. The zygote did not develop beyond single cell stage, and subsequent divisions and cytokinesis did not occur (panel i, j, and k). The arrows in panels i and j identify the irregular nature of free nuclear mitotic divisions in hog-1 endosperm. The endosperm nuclei were irregular in size and they were often clustered. Compare the large and small irregular endosperm nuclei in hog1-6 (panel l) with the regular free nuclear endosperm nuclei in (m) the wild type. Scale bars: 20 μm for panels d to k, and the insert of panel e; and 50 μm in panels b, c, l, and m.

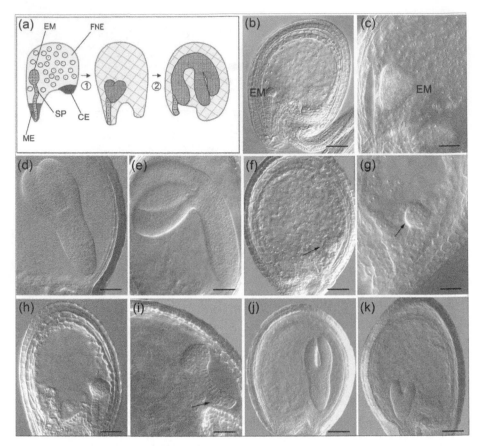

Figure 5. Mutants arrested late in seed development.

(a) Shown is a scheme of seed development in Arabidopsis. A globular embryo (EM) develops into heart stage (1). Note that the peripheral endosperm nuclei surrounding the globular embryo are organized into three distinct domains: micropylar endosperm (ME), chalazal endosperm (CE), and free nuclear endosperm (FNE). Following rapid cellularization of endosperm, a torpedo stage embryo and then an upturned-U stage embryo is formed (2). (b-e) Morphology of wild-type seed development corresponding to representative events described above. (f) In oma-1 the embryo arrested at the mid-globular stage. The size of cells in embryo and endosperm were larger than that in (g) the wild type. (h,i) In fey-1 the embryo arrested at around the late globular stage. Note that the cells of the embryo and suspensor were large, and the suspensor displays a bend due to the irregularly bulged cells (panel i, arrow). (j) The majority of the ila-1 embryos arrested when they were at upturned U stage. (k) A small fraction of late-heart ila-1 embryos could also be observed. Scale bars: 10 μm for panels b, f, h, j, and k; and 20 μm for panels c, d, e, g, and i.

Transcription Factors, Homeotic Genes, and Signaling Proteins are Over-Expressed in the Absence of an Embryo Sac

Even though the two mutants we used in this study exhibit morphologically normal carpels and ovules in the absence of an embryo sac, we considered whether the gene expression program within the sporophyte is altered. The genes exhibiting higher levels of expression in the coa and spl mutants could be regarded as candidate genes that were deregulated in the maternal sporophyte because of the absence of a functional embryo sac in these mutants. Of the 527 genes identified for their maternal-gain-of-expression in coa and spl, about 9% were predicted to be involved in transcriptional regulation and 7% were signaling proteins (Figure 1). Among the genes encoding transcription factors, there were eight MYB class protein genes, seven zinc-finger protein genes including SUPERMAN and NUBBIN, five homeo box genes including SHOOT MERISTEMLESS (STM), five genes each encoding basic helix-loop-helix (bHLH) and SQUAMOSA-binding proteins, three genes encoding basic leucin zipper (bZIP) proteins, and two genes each encoding APETALA2-domain and NAC-domain transcription factors. No MADS box genes were represented. The genes encoding signaling proteins included the auxin-responsive genes AUXIN RESISTANT 2/3 (AXR2 and AXR3), three genes encoding DC1-domain-containing proteins, ten genes encoding kinases and related proteins, two genes encoding phosphatases, four LRR-protein genes, five auxin response regulator genes, and the two zinc-finger protein genes SHORT INTERNODES (SHI) and STYLISH2 (STY2). When we examined the whole dataset for genes encoding secreted proteins, 87 predicted proteins fulfilled the criteria; 24% were below 20 kDa in size, which included a peptidase and two lipid transfer proteins (data not shown).

The Carpel is the Major Target Tissue for Over-Expression Caused by the Lack of an Embryo Sac

In order to confirm that the genes we identified truly reflect a gain of expression in the maternal sporophyte of the mutant, we examined the expression levels and patterns of 11 candidate genes in coa and wild-type gynoecium by RT-PCR or in situ hybridization. Figure 6a shows an RT-PCR panel confirming that eight genes from the coa dataset and three genes from the spl dataset were more highly expressed in coa than in wild-type pistils. We present evidence that the genes we identified for their gain of expression in spl were indeed over-expressed in coa as well, suggesting that the genes are generally over-expressed in the absence of an embryo sac, regardless of the mutation (Figure 6a). Figure 6 shows the expression of the following genes in the coa gynoecium as detected by in situ hybridization: AT4G12410 (a SAUR [auxin-responsive Small Auxin Up RNA] gene; Figure 6b), AT1G75580 (an auxin-responsive gene; Figure 6c), AT5G03200 (encoding C3HC4-type RING finger protein; Figure 6d), AT5G15980 (encoding PPR repeat-containing protein; Figure 6e), and STM (a homeo box gene; Figure 6g). Surprisingly, all of the five genes exhibited similar expression patterns: strong expression in the carpel wall and septum, and relatively low expression in the sporophytic tissues of the ovules surrounding the embryo sac. In case of AT4G12410, we did not detect expression in the wild-type pistils. For the other four genes, the spatial expression patterns in the wild-type ovule and carpel tissues were comparable to that in coa, but the expression levels were far lower than in the mutant (data not shown). In summary, we provide evidence that a significant fraction of the sporophytic transcriptome can be modulated by the presence or absence of an embryo sac.

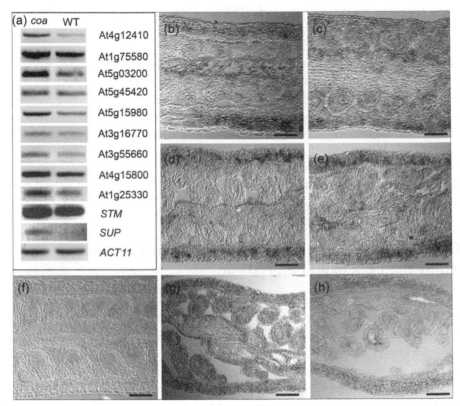

Figure 6. Gain of expression in the sporophyte in the absence of a functional embryo sac: expression analysis in the coatlique (coa) mutant.

(a) RT-PCR for 11 genes in coa and wild-type (WT) pistils. Equal loading of both coa and WT cDNA templates in PCR was monitored by expression of ACT11. SUP, SUPERMAN. Also shown are in situ expression patterns of the following genes in coa pistil tissues: (b) AT4G12410, encoding an auxin-responsive Small Auxin Up RNA (SAUR) protein; (c) AT1G75580, encoding an auxin-responsive protein; (d) AT5G03200, encoding a C3HC4-type RING finger protein; and (e) at5g15980, encoding a PPR repeat containing protein. The corresponding sense control probes did not show any expression (data not shown). (f) AT4G12410 did not show any detectable expression pattern in wild-type pistils. The other four genes exhibited spatial expression patterns in the wild-type ovule and carpel tissues comparable to that of coa, but their wild-type expression levels were much lower than in coa (data not shown). (g) We initially identified the over-expression of STM in the ovule tissues of spl (sensu microarray data), and confirmed that this gene is over-expressed in the carpel and ovules of coa as well (panels a and g). (h) A comparable but less intense spatial expression pattern of STM was seen in wild-type pistils. Scale bars: 100 μm in panels b to h.

Discussion

A Comparative Genetic Subtraction Approach Identifies Embryo Sac Expressed Candidate Genes

The female gametophyte or the embryo sac develops from a single functional megaspore cell through a series of highly choreographed free-nuclear mitotic divisions [1,2]. Understanding the molecular pathways that govern embryo sac development and function, as well as subsequent seed development, has important implications for both basic plant developmental biology and plant breeding. Despite the possible involvement of a few thousands of genes in this essential developmental pathway, only a few more than 100 genes have been identified by loss-of-function mutations, and most of them have not been studied in detail [14]. In the present study we provide an alternative strategy to identify genes that are expressed in the embryo sac of A. thaliana, namely comparative whole-genome transcriptional profiling by microarray, which led to a candidate dataset of 1,260 genes.

Our approach, similar to that employed by Yu and coworkers [34], is different from that used in previously reported whole-genome transcriptional profiling experiments (for example, pollen transcriptome [33] and whole flower and silique transcriptome [29]) in that we deduced the transcriptome of the few-celled female gametophyte by simple genetic subtraction using a mutant that lacks an embryo sac. Putative embryo sac expressed candidates included a significant number of genes that are involved in transcriptional regulation, signaling, translational regulation, protein degradation, transport and metabolism, and a majority of genes that were not identified in previous studies. Similar to previous transcriptional profiling reports, the largest functional category of embryo sac expressed genes was plant metabolism [29,52,56]. Percentages of genes classified into transcriptional regulation and signaling were comparable across embryo sac and pollen expressed transcriptomes (about 6% to 10%) and, interestingly, these categories are larger in both gametophytic transcriptomes than the general sporophytic transcriptomes such as leaf, stem, and root [28]. In a much larger dataset of pollen samples, Pina and colleagues [33] reported a little over 16% of pollen expressed genes as part of the signaling category. It is possible that the mature pollen transcriptome is more active in terms of signal transduction processes than that of the embryo sac, given its role during polarized tip growth through the female reproductive tract, and the gametic interaction at fertilization (for review [57]). We could not compare other functional categories across other organ-specific transcriptome datasets because the methods employed for functional classification were very different. Briefly, our work provides novel data for organ specific expression in Arabidopsis and, in particular, it illustrates the similarities and dissimilarities between male and female gametophytic expression.

Interesting insights can be gained from the subset of embryo sac expressed genes (8.6%) that was subtracted for their enriched expression only in the embryo sac. It was recently reported that 10% to 11% of the pollen transcriptome was selectively expressed in the pollen, as evident from their absence of expression in the sporophytic transcriptomes (n = 1,584 in [30] and n = 6,587 in [33]). In a very similar study [32], it was reported that 9.7% of the 13,977 male gameto-phytically expressed genes were specific for the male gametophyte. Even though the complete embryo sac transcriptome is yet to be determined, it appears that the enriched transcriptome of the embryo sac we report here is similar in size to that of pollen. Male and female gametophyte enriched transcriptomes appear to be much larger than the specific transcriptomes of vegetative organs such as leaf and entire seedlings, which accounted for 2% to 4% of their corresponding complete transcriptomes [33]. When we compared the genes with enriched expression in the embryo sac or pollen, the embryo sac appears to harbor more transcriptional regulators than pollen (8% versus 3%) [30]. However, the pollen transcriptome exhibited a greater abundance of signaling proteins than the embryo sac (23% versus 2%). This implies that either the pollen is more active in signaling than the female gametophyte at the time around fertilization, or that the sensitivity of detecting signaling genes in the embryo sac will have to be improved in the future studies. The promise of our approach to deducing genes with enriched expression was supported by the presence of essential genes that are female gametophyte spe-cific, such as MEDEA and MYB98 in our dataset [12,22]. Furthermore, temporal and spatial expression of nine transcripts in this study, and 18 other genes from previous studies, suggests that the whole dataset of embryo sac expressed genes may comprise genes that are expressed either in the entire embryo sac or restricted to a few or single cell types.

A significant fraction of genes were probably undetected by this experiment for two reasons: relatively similar or higher expression in the maternal sporophytic tissues; and low level of expression in the embryo sac, similar to most of the known female gametophytic genes. For example, cell cycle genes are barely repre-sented among our candidate genes. In contrast, the pollen transcriptome has been reported to be enriched with several core cell cycle transcripts [33]. Although our comparative approach is very different from that reported by Pina and coworkers [33], there could be a large number of cell cycle regulators that are expected to be expressed during embryo sac development, suggesting a need for improvements in embryo sac isolation and subsequent transcriptome analysis. Unlike the relative ease in isolating some embryo sac cell types in maize and wheat, large-scale isola-tion of the embryo sac cells is not possible in Arabidopsis [58,59]. Following the work conducted by Yu and coworkers [34], we present here a large-scale study to explore embryo sac expressed genes in Arabidopsis. If the scale of gene discovery is to be improved much further, then methods to isolate embryo sac cells using methods such as florescence-activated cell sorting, targeted genetic ablation by expression of a cell-autonomous cytotoxin, or laser-assisted microdissection must be developed [51,60-62].

The Embryo Sac Expressed Candidate Genes May Be Essential for Female Gametophyte and Seed Development

Once we had validated the expression of the embryo sac expressed genes, we considered whether these genes could play essential roles during embryo sac and seed development. It is apparent from our work on five mutants, and mutant data from the literature, that the embryo sac expressed genes that we report here may play a crucial role during the embryo sac development or later during seed formation. HOG1 is of special interest because we have provided evidence for allelic phenotypic complementation by two mutant alleles. HOG1 is proposed to act upstream of METHYL TRANSFERASE 1 (MET1) and CMT3 among other methylases, and mutants for HOG1 have high levels of global hypomethylation [54]. It has become clear that DNA hypomethylation plays a crucial role during gametogenesis, and that mutations affecting the genes in this pathway such as HOG1, MET1, and CMT3 affect embryo and endosperm development [55,63,64]. It is interesting to note that we identified CMT3, MEA, and FIS2 that are associated with pathways involving DNA and histone methylation [63,65-68].

We have shown that our dataset will be a resource for targeted reverse genetic approaches. The extensive reverse genetic tools available for Arabidopsis researchers make such a large-scale functional study possible [69]. While screening for female gametophytic mutants through T-DNA mutagenesis, we unexpectedly observed a number of female gametophytic mutants that had a very similar phenotype: a complete arrest of female gametogenesis at the one-nuclear stage. These, however, were not linked to the gene disruption. Agrobacterium-mediated Arabidopsis T-DNA mutagenesis has been facilitated by floral dipping, which involves integration of the T-DNA through the ovule, and the chromosomes of the female gametophyte are the main target for T-DNA insertion [70]. Based on our results from this study, and other independent observations (Johnston AJ, Grossniklaus U, unpublished data), we believe that these unlinked gametophytic lethal events arose because of translocations and other rearrangements of maternal chromosomes during the integration of the T-DNA, and we advise due caution in mutant screening.

Communication Between the Embryo Sac and the Surrounding Sporophyte May Be Important for Reproductive Development

In Arabidopsis, the sporophytic and gametophytic tissues are intimately positioned next to each other within the ovule. Independent studies on Arabidopsis ovule mutants suggest that the development of the female gametophyte might require highly synchronized morphogenesis of the maternal sporophyte surrounding the gametophyte [1,35,37]. This notion is exemplified by the fact that megagametogenesis is largely perturbed in most of the known sporophytic ovule development mutants. For example, in short integument 1 (sin1) the ovules display uncoordinated growth patterns of integuments and the nucellus, and embryo sac development is not initiated [35,71]. In bell1 and aintegumenta mutants, in which integument morphogenesis and identity are disrupted, embryo sac development is arrested [35,37,72,73]. Therefore, early acting sporophytic genes in the ovule also affect female gametophyte development. On the contrary, in several mutations where female gametogenesis is completely or partially blocked, the ovule sporophyte appears morphologically normal. In coa and spl, or female gametophytic mutations such as hadad and nomega, embryo sac development is blocked either at the onset or during megagametogenesis, but ovule morphogenesis continues normally until anthesis [8,18,38]. It was therefore thought that the embryo sac does not influence the development of the sporophytic parts of the surrounding ovule and carpel tissues [2].

Our data clearly demonstrate that in the absence of an embryo sac there was a predominant transcriptional upregulation of transcription factors, and signaling molecules in the carpel and the ovule. It is interesting to note that we identified genes that were previously implicated in gynoecium patterning such as NUBBIN, SHI and STY2 for their gain of expression in the sporophyte [74-77]. Based on the proposed functionalities of these and other genes in our dataset, we suggest that signaling pathways involving auxin and gibberellic acid could possibly be triggered in the carpel and ovule sporophyte, in the absence of an embryo sac. We anticipate that sporophytic patterning genes and signaling molecules are under indirect repressive control by the female gametophyte. Impairment of this signaling cascade leads to deregulation of the sporophytic transcriptome.

Conclusion

Understanding gene expression and regulation during embryo sac development demands large-scale experimental strategies that subtract the miniature haploid embryo sac cells from the thousands of surrounding sporophytic cells. We used a simple genetic subtraction strategy, which successfully identified a large number of candidate genes that are expressed in the cell types of the embryo sac. The wealth of data reported here lays the foundation to elucidate the regulatory networks of transcriptional regulation, signaling, transport, and metabolism that operate in these unique cell types of the haploid phase of the life cycle. Given that many of the genes in our expression dataset are essential to female gametophyte and seed development, targeted functional studies with further candidate genes promise to yield novel insights into the development and function of the embryo sac. Another major finding of this work is the identification of 108 genes that are enriched for embryo sac expression and thus probably play important roles for the differentiation and function of these specific cell types. The surprising finding that many genes are deregulated in sporophytic tissues in the absence of an embryo sac suggests a much more complex interplay of the haploid gametophytic with the diploid sporophytic tissues than was previously anticipated. Understanding the sporophytic regulatory network governed by the embryo sac will be of key interest for future studies.

Materials and Methods

Plant Material and Growth Conditions

The coatlique (coa) mutant was identified in Arabidopsis var Landsberg (erecta mutant; Ler) background and Ler was used as a wild-type control in the microarray and in situ hybridization experiments. Before transplanting, seeds were sown on Murashige and Skoog media (1% sucrose and 0.9% agar; pH 5.7) supplemented with appropriate selection markers and stratified for two days at 4°C (see Table 1 for description of mutants plants and selection markers). The seeds were germinated and grown for up to 15 days under 16-hour light/8-hour dark cycles at 22°C. Plants were then transplanted into ED73 soil (Einheitserde, Schopfheim, Germany) and grown in greenhouse conditions under a 16-hour photo-period at 22°C and 60% to 70% relative humidity.

Histological Analysis

For phenotypic characterization, the gynoecia of Arabidopsis wild-type, coa and gametophytic mutants, and siliques of the zygotic mutants were cleared in accordance with a protocol described in the report by Yadegari and co-workers [78]. Samples were observed using a Leica DMR microscope (Leica Microsystems, Mannheim, Germany) under differential interference contrast (DIC) optics.

Transcriptional Profiling by Oligonucleotide Array

Transcriptional profiling by Affymetrix microarray was done using coa and wild-type pistils. In particular, emphasis was given to the low-level analysis of the microarray data, because the low fold change cut-off used for the embryo sac dataset could potentially introduce a large number of false positives. We chose to use three independent statistical packages (dCHIP, gcRMA and Gene Spring), with the most and least stringent being dCHIP and Gene Spring analysis, respectively. For dCHIP analysis, only those genes within replicate arrays called 'present' within a variation of $0 <$ median (standard deviation/mean) < 0.5 were retained for downstream analysis. By setting P to < 0.1 and differential fold change expression cut-off to 1.28-fold, we could predict that the median FDR ranges from 1% (spl dataset) to 3% (coa dataset) in the dCHIP analysis. The dilution of gametophytic cells in an excess of sporophytic tissues was higher in coa samples than in spl samples (discussed in Results, above), which may be the reason for the increase in the FDR. In such cases, standard error values of the signal averages provide an indication for manual omission of false positives. In the analysis using gcRMA, pre-processed signal values were statistically analyzed using an empirical bayesian approach and the FDR was calculated for each gene using the options implemented in the Bioconductor software version 2.3.0 [79]. Only those genes with a FDR below 0.05 were considered to be differentially expressed. Manual omission of false-positive findings is possible in this type of analysis, if the standard error estimates of the mean RMA values (signal) and the absolute FDR values are to be used as indicators of false discovery. The sporophytic datasets did not impose such problems because the fold change cut-off was set to twofold as a stringent baseline, in addition to the analysis using three statistical methods.

Bioinformatics Analyses

The candidate genes were functionally classified according to the Gene Ontology data from TAIR or published evidence where appropriate. Annotations were improved mainly for the transcription factors from the Arabidopsis Gene Regulatory Information Server [80]. The secreted proteins were chosen based on the protein sequence analysis using TargetP with the top two reliability scores out of five [81]. A total of 32,349 maize and wheat EST sequences extracted from libraries specific for the embryo sac, egg, central cell, and early endosperm were obtained from various sources. The pools of EST sequences were converted to local BLASTable databases using NCBI software [82]. A PERL script was written to perform the mapping of A. thaliana female gametophyte transcriptome data to the EST datasets. An EST sequence is considered similar to an Arabidopsis protein if it matches at an e-value cutoff threshold of 10^{-8} by TBLASTN [81]. For comparisons with sporophytic transcriptomes, the highly standardized experiment conducted by Schmid and coworkers [28] was chosen. Presence/absence calls calculated from the microarray analyses were downloaded for selected tissues from the TAIR website [83]. A gene was declared to be expressed in a tissue when a presence call was assigned to it in at least two out of three replicates.

In Situ Hybridization

Inflorescences and emasculated pistils were paraplast embedded using the protocol of Kerk and colleagues [84] with minor modifications. Unique gene-specific probes of about 200 to 300 base pairs were cloned into pDRIVE (Qiagen, Basel, Switzerland) and used as templates for generating digoxygenin-UTP-labeled riboprobes by run-off transcription using T7 RNA polymerase, in accordance with the manufacturer's protocol (Roche Diagnostics, Basel, Switzerland). In situ hybridization was performed on 8 to 10 μm semi-thin paraffin sections, as described by Vielle-Calzada and coworkers [85] with minor modifications.

Histochemical GUS Expression

Embryo sac expression of the GUS reporter gene (β-glucuronidase) in the promoter-GUS lines and transposants was detected as described by Vielle-Calzada and coworkers [86].

Image Processing

All of the images were recorded using a digital Magnafire camera (Optronics, Goleta, CA, USA), and they were edited for picture quality using Adobe Photoshop version CS (Adobe Systems Inc., San Jose, CA, USA).

Abbreviations

BLAST, basic local alignment search tool; coa, coatlique mutant; dCHIP, DNA-Chip Analyzer; FDR, false discovery rate; gcRMA, GC robust multi-array average; MAS, MicroArray Suite (Affymetrix); RT-PCR, reverse transcription polymerase chain reaction; spl, sporocyteless mutant; TAIR, The Arabidopsis Information Resource.

Authors' Contributions

UG conceived of and supervised the project. AJJ, PM and UG designed and interpreted the experiments. AJJ, PM, JG and MF performed the experiments. AJJ, SEJW, ES and JDB contributed statistical and bioinformatics analyses. UG contributed reagents and materials. AJJ and UG wrote the paper.

Acknowledgements

The microarray experiment was carried out in the microarray facility at the Functional Genomics Centre, Univeristy of Zürich. We are indebted to Andrea Patrignani, Ulrich Wagner, and Kathrin Michel (University of Zürich) for help during microarray experiments and data analyses. We acknowledge Venkatesan Sundaresan (University of California, Davis) for provision of spl microarray data. We thank the Arabidopsis Stock Centres in Nottingham (NASC) and Ohio (ABRC) for providing seeds of SALK, SAIL (Syngenta), and Spm (JIC) insertional mutants; Jean-Philippe Vielle-Clazada (CINVESTAV-Irapuato) for the initial isolation of coa; and Arturo Bolaños (University of Zürich) for help with plant care. Special thanks are due to Ian Furner (University of Cambridge), Wolf B Frommer (Carnegie Institute), and Takashi Aoyama (Kyoto University) for provision of the seeds of GT1724 (hog1-4), PUP3-GUS and CyclinA2;4-GUS, respectively. We are grateful to Sharon Kessler (University of Zürich) for critical reading of this manuscript. JDB was supported by a fellowship SFRH/BPD/3619/2000 from Fundacáo para a Ciência e a Tecnologia, Portugal. This project was supported by the University of Zürich, and grants of the Swiss National Science Foundation and the 'Stiftung für wissenschaftliche Forschung' to UG.

References

1. Grossniklaus U, Schneitz K: The molecular and genetic basis of ovule and megagametophyte development. Semin Cell Dev Biol 1998, 9:227–238.

2. Drews GN, Yadegari R: Development and function of the angiosperm female gametophyte. Annu Rev Genet 2002, 36:99–124.

3. Koltunow AM, Grossniklaus U: Apomixis: a developmental perspective. Annu Rev Biol 2003, 54:547–574.

4. Rhee SY, Beavis W, Berardini TZ, Chen G, Dixon D, Doyle A, Garcia-Hernandez M, Huala E, Lander G, Montoya M, et al.: The Arabidopsis Information Resource (TAIR): a model organism database providing a centralized, curated

gateway to Arabidopsis biology, research materials and community. Nucleic Acids Res 2003, 31:224–228.

5. Christensen CA, Subramanian S, Drews GN: Identification of gametophytic mutations affecting female gametophyte development in Arabidopsis. Dev Biol 1998, 202:136–151.

6. Christensen CA, Gorsich SW, Brown RH, Jones LG, Brown J, Shaw JM, Drews GN: Mitochondrial GFA2 is required for synergid cell death in Arabidopsis. Plant Cell 2002, 14:2215–2232.

7. Guitton AE, Page DR, Chambrier P, Lionnet C, Faure JE, Grossniklaus U, Berger F: Identification of new members of Fertilization Independent Seed Polycomb Group pathway involved in the control of seed development in Arabidopsis thaliana. Development 2004, 131:2971–2981.

8. Moore JM, Calzada JP, Gagliano W, Grossniklaus U: Genetic characterization of hadad, a mutant disrupting female gametogenesis in Arabidopsis thaliana. Cold Spring Harb Symp Quant Biol 1997, 62:35–47.

9. Moore JM: Isolation and characterization of gametophytic mutants in Arabidopsis thaliana. PhD thesis. State University of New York at Stony Brook, Graduate Program in Genetics; 2002.

10. Ohad N, Yadegari R, Margossian L, Hannon M, Michaeli D, Harada JJ, Goldberg RB, Fischer RL: Mutations in FIE, a WD Polycomb group gene, allow endosperm development without fertilization. Plant Cell 1999, 11:407–416.

11. Chaudhury AM, Ming L, Miller C, Craig S, Dennis ES, Peacock WJ: Fertilization-independent seed development in Arabidopsis thaliana. Proc Natl Acad Sci USA 1997, 94:4223–4228.

12. Grossniklaus U, Vielle-Calzada JP, Hoeppner MA, Gagliano WB: Maternal control of embryogenesis by MEDEA, a Polycomb group gene in Arabidopsis. Science 1998, 280:446–450.

13. Pagnussat GC, Yu HJ, Ngo QA, Rajani S, Mayalagu S, Johnson CS, Capron A, Xie LF, Ye D, Sundaresan V: Genetic and molecular identification of genes required for female gametophyte development and function in Arabidopsis. Development 2005, 132:603–614.

14. Brukhin V, Curtis MD, Grossniklaus U: The angiosperm female gametophyte: No longer the forgotten generation. Curr Sci 2005, 89:1844–1852.

15. Springer PS, McCombie WR, Sundaresan V, Martienssen RA: Gene trap tagging of PROLIFERA, an essential MCM2-3-5-like gene in Arabidopsis. Science 1995, 268:877–880.

16. Capron A, Serralbo O, Fulop K, Frugier F, Parmentier Y, Dong A, Lecureuil A, Guerche P, Kondorosi E, Scheres B, Genschik P: The Arabidopsis anaphase-promoting complex or cyclosome: molecular and genetic characterization of the APC2 subunit. Plant Cell 2003, 15:2370–2382.

17. Hejatko J, Pernisova M, Eneva T, Palme K, Brzobohaty B: The putative sensor histidine kinase CKI1 is involved in female gametophyte development in Arabidopsis. Mol Genet Genomics 2003, 269:443–453.

18. Kwee HS, Sundaresan V: The NOMEGA gene required for female gametophyte development encodes the putative APC6/CDC16 component of the anaphase promoting complex in Arabidopsis. Plant J 2003, 36:853–866.

19. Ebel C, Mariconti L, Gruissem W: Plant retinoblastoma homologues control nuclear proliferation in the female gametophyte. Nature 2004, 429:776–780.

20. Kim HU, Li Y, Huang AH: Ubiquitous and endoplasmic reticulum-located lysophosphatidyl acyltransferase, LPAT2, is essential for female but not male gametophyte development in Arabidopsis. Plant Cell 2005, 17:1073–1089.

21. Shi DQ, Liu J, Xiang YH, Ye D, Sundaresan V, Yang WC: SLOW WALKER1, essential for gametogenesis in Arabidopsis, encodes a WD40 protein involved in 18S ribosomal RNA biogenesis. Plant Cell 2005, 17:2340–2354.

22. Kasahara RD, Portereiko MF, Sandaklie-Nikolova L, Rabiger DS, Drews GN: MYB98 is required for pollen tube guidance and synergid cell differentiation in Arabidopsis. Plant Cell 2005, 17:2981–2992.

23. Portereiko MF, Lloyd A, Steffen JG, Punwani JA, Otsuga D, Drews GN: AGL80 is required for central cell and endosperm development in Arabidopsis. Plant Cell 2006, 18:1862–1872.

24. Grini PE, Jürgens G, Hülskamp M: Embryo and endosperm development are disrupted in the female gametophytic capulet mutants of Arabidopsis. Genetics 2002, 162:1911–1925.

25. Luo M, Bilodeau P, Koltunow A, Dennis ES, Peacock WJ, Chaudhury AM: Genes controlling fertilization-independent seed development in Arabidopsis thaliana. Proc Natl Acad Sci USA 1999, 96:296–301.

26. Köhler C, Hennig L, Bouveret R, Gheyselinck J, Grossniklaus U, Gruissem W: Arabidopsis MSI1 is a component of the MEA/FIE Polycomb group complex and required for seed development. EMBO J 2003, 22:4804–4814.

27. Wellmer F, Riechmann JL, Alves-Ferreira M, Meyerowitz EM: Genome-wide analysis of spatial gene expression in Arabidopsis flowers. Plant Cell 2004, 16:1314–1326.

28. Schmid M, Davison TS, Henz SR, Pape UJ, Demar M, Vingron M, Scholkopf B, Weigel D, Lohmann JU: A gene expression map of Arabidopsis thaliana development. Nat Genet 2005, 37:501–506.

29. Hennig L, Gruissem W, Grossniklaus U, Köhler C: Transcriptional programs of early reproductive stages in Arabidopsis. Plant Physiol 2004, 135:1765–1775.

30. Becker JD, Boavida LC, Carneiro J, Haury M, Feijo JA: Transcriptional profiling of Arabidopsis tissues reveals the unique characteristics of the pollen transcriptome. Plant Physiol 2003, 133:713–725.

31. Honys D, Twell D: Comparative analysis of the Arabidopsis pollen transcriptome. Plant Physiol 2003, 132:640–652.

32. Honys D, Twell D: Transcriptome analysis of haploid male gametophyte development in Arabidopsis. Genome Biol 2004, 5:R85.

33. Pina C, Pinto F, Feijo JA, Becker JD: Gene family analysis of the Arabidopsis pollen transcriptome reveals biological implications for cell growth, division control, and gene expression regulation. Plant Physiol 2005, 138:744–756.

34. Yu HJ, Hogan P, Sundaresan V: Analysis of the female gametophyte transcriptome of Arabidopsis by comparative expression profiling. Plant Physiol 2005, 139:1853–1869.

35. Robinson-Beers K, Pruitt RE, Gasser CS: Ovule development in wild-type Arabidopsis and two female-sterile mutants. Plant Cell 1992, 4:1237–1249.

36. Reiser L, Modrusan Z, Margossian L, Samach A, Ohad N, Haughn GW, Fischer RL: The BELL1 gene encodes a homeodomain protein involved in pattern formation in the Arabidopsis ovule primordium. Cell 1995, 83:735–742.

37. Ray S, Golden T, Ray A: Maternal effects of the short integument mutation on embryo development in Arabidopsis. Dev Biol 1996, 180:365–369.

38. Yang WC, Ye D, Xu J, Sundaresan V: The SPOROCYTELESS gene of Arabidopsis is required for initiation of sporogenesis and encodes a novel nuclear protein. Genes Dev 1999, 13:2108–2117.

39. Smyth DR, Bowman JL, Meyerowitz EM: Early flower development in Arabidopsis. Plant Cell 1990, 2:755–767.

40. Eisen MB, Spellman PT, Brown PO, Botstein D: Cluster analysis and display of genome-wide expression patterns. Proc Natl Acad Sci USA 1998, 95:14863–14868.

41. Shedden K, Chen W, Kuick R, Ghosh D, Macdonald J, Cho KR, Giordano TJ, Gruber SB, Fearon ER, Taylor JM, Hanash S: Comparison of seven methods for producing Affymetrix expression scores based on False Discovery Rates in disease profiling data. BMC Bioinformatics 2005, 6:26.

42. Harr B, Schlotterer C: Comparison of algorithms for the analysis of Affymetrix microarray data as evaluated by co-expression of genes in known operons. Nucleic Acids Res 2006, 34:e8.

43. Irizarry RA, Wu Z, Jaffee HA: Comparison of Affymetrix GeneChip expression measures. Bioinformatics 2006, 22:789–794.

44. Millenaar FF, Okyere J, May ST, van Zanten M, Voesenek LA, Peeters AJ: How to decide? Different methods of calculating gene expression form short oligonucleotide array data will give different results. BMC Bioinformatics 2006, 7:137.

45. Li C, Wong WH: Model-based analysis of oligonucleotide arrays: expression index computation and outlier detection. Proc Natl Acad Sci USA 2001, 98:31–36.

46. Wu HM, Wong E, Ogdahl J, Cheung AY: A pollen tube growth-promoting arabinogalactan protein from Nicotiana alata is similar to the tobacco TTS protein. Plant J 2000, 22:165–176.

47. Rajagopalan D: A comparison of statistical methods for analysis of high density oligonucleotide array data. Bioinformatics 2003, 19:1469–1476.

48. Barash Y, Dehan E, Krupsky M, Franklin W, Geraci M, Friedman N, Kaminski N: Comparative analysis of algorithms for signal quantitation from oligonucleotide microarrays. Bioinformatics 2004, 20:839–846.

49. Cui X, Churchill GA: Statistical tests for differential expression in cDNA microarray experiments. Genome Biol 2003, 4:210.

50. Wurmbach E, Yuen T, Sealfon SC: Focused microarray analysis. Methods 2003, 31:306–316.

51. Morey JS, Ryan JC, Van Dolahl FM: Microarray validation: factors influencing correlation between oligonucleotide microarrays and real-time PCR. Biol Proceed Online 2006, 8:175–193.

52. Tung CW, Dwyer KG, Nasrallah ME, Nasrallah JB: Genome-wide identification of genes expressed in Arabidopsis pistils specifically along the path of pollen tube growth. Plant Physiol 2005, 138:977–989.

53. Tzafrir I, Pena-Muralla R, Dickerman A, Berg M, Rogers R, Hutchens S, Sweeney TC, McElver J, Aux G, Patton D, et al.: Identification of genes required for embryo development in Arabidopsis. Plant Physiol 2004, 135:1206–1220.

54. Le Q, Gutierrez-Marcos JF, Costa LM, Meyer S, Dickinson HG, Lorz H, Kranz E, Scholten S: Construction and screening of substracted cDNA libraries form limited populations of plant cells: a comparative analysis of gene expression between maize egg cells and central cells. Plant J 2005, 44:167–178.

55. Rocha PS, Sheikh M, Melchiorre R, Fagard M, Boutet S, Loach R, Moffatt B, Wagner C, Vaucheret H, Furner I: The Arabidopsis HOMOLOGY-DEPEN-DENT GENE SILENCING1 gene codes for an S-adenosyl-L-homocysteine hydrolase required for DNA methylation-dependent gene silencing. Plant Cell 2005, 17:404–417.

56. Zhu JK: Cell signaling under salt, water and cold stresses. Curr Opin Plant Biol 2001, 4:401–406.

57. Weterings K, Russell SD: Experimental analysis of the fertilization process. Plant Cell 2004, 16(Suppl):S107-S118.

58. Sprunck S, Baumann U, Edwards K, Langridge P, Dresselhaus T: The transcript composition of egg cells changes significantly following fertilization in wheat (Triticum aestivum L.). Plant J 2005, 41:660–672.

59. Yang H, Kaur N, Kiriakopolos S, McCormick S: EST generation and analyses towards identifying female gametophyte-specific genes in Zea mays L. Planta 2006, 224:1004–1014.

60. Birnbaum K, Sasha DE, Wang JY, Jung JW, Lambert GM, Galbraith DW, Benfey PN: A gene expression map of the Arabidopsis root. Science 2003, 302:1956–1960.

61. Engel ML, Chaboud A, Dumas C, McCormick S: Sperm cells of Zea mays have a complex complement of mRNAs. Plant J 2003, 34:697–707.

62. Day RC, Grossniklaus U, Macknight RC: Be more specific! Laser-assisted microdissection of plant cells. Trends Plant Sci 2005, 10:397–406.

63. Xiao W, Custard KD, Brown RC, Lemmon BE, Harada JJ, Goldberg RB, Fischer RL: DNA methylation is critical for Arabidopsis embryogenesis and seed viability. Plant Cell 2006, 18:805–814.

64. Takeda S, Paszkowski J: DNA methylation and epigenetic inheritance during plant gametogenesis. Chromosoma 2006, 115:27–35.

65. Köhler C, Hennig L, Spillane C, Pien S, Gruissem W, Grossniklaus U: The Polycomb group protein MEDEA regulates seed development by controlling expression of the MADS-box gene PHERES1. Genes Dev 2003, 17:1540–1553.

66. Lindroth AM, Cao X, Jackson JP, Zilberman D, McCallum CM, Henikoff S, Jacobsen SE: Requirement of CHROMOMETHYLASE3 for maintenance of CpXpG methylation. Science 2001, 292:2077–2080.

67. Jullien PE, Kinoshita T, Ohad N, Berger F: Maintenance of DNA methylation during the Arabidopsis life cycle is essential for parental imprinting. Plant Cell 2006, 18:1360–1372.

68. Makarevich G, Leroy O, Akinci U, Schubert D, Clarenz O, Goodrich J, Grossniklaus U, Köhler C: Different Polycomb group complexes regulate common target genes in Arabidopsis. EMBO Rep 2006, 7:947–952.

69. Alonso JM, Stepanova AN, Leisse TJ, Kim CJ, Chen H, Shinn P, Stevenson DK, Zimmerman J, Barajas P, Cheuk R, et al.: Genome-wide insertional mutagenesis of Arabidopsis thaliana. Science 2003, 301:653–657.

70. Bechtold N, Jaudeau B, Jolivet S, Maba B, Vezon D, Voisin R, Pelletier G: The maternal chromosome set is the target of the T-DNA in the in planta transformation of Arabidopsis thaliana. Genetics 2000, 155:1875–1887.

71. Lang JD, Ray S, Ray A: sin1, a mutation affecting female fertility in Arabidopsis, interacts with mod1, its recessive modifier. Genetics 1994, 137:1101–1110.

72. Modrusan Z, Reiser L, Feldmann KA, Fischer RL, Haughn GW: Homeotic transformation of ovules into carpel-like structures in Arabidopsis. Plant Cell 1994, 6:333–349.

73. Klucher KM, Chow H, Reiser L, Fischer RL: The AINTEGUMENTA gene of Arabidopsis required for ovule and female gametophyte development is related to the floral homeotic gene APETALA2. Plant Cell 1996, 8:137–153.

74. Fridborg I, Kuusk S, Moritz T, Sundberg E: The Arabidopsis dwarf mutant shi exhibits reduced gibberellin responses conferred by overexpression of a new putative zinc finger protein. Plant Cell 1999, 11:1019–1032.

75. Kuusk S, Sohlberg JJ, Long JA, Fridborg I, Sundberg E: STY1 and STY2 promote the formation of apical tissues during Arabidopsis gynoecium development. Development 2002, 129:4707–4717.

76. Kuusk S, Sohlberg JJ, Magnus Eklund D, Sundberg E: Functionally redundant SHI family genes regulate Arabidopsis gynoecium development in a dose-dependent manner. Plant J 2006, 47:99–111.

77. Dinneny JR, Weigel D, Yanofsky MF: NUBBIN and JAGGED define stamen and carpel shape in Arabidopsis. Development 2006, 133:1645–1655.

78. Yadegari R, Paiva G, Laux T, Koltunow AM, Apuya N, Zimmerman JL, Fischer RL, Harada JJ, Goldberg RB: Cell differentiation and morphogenesis are uncoupled in Arabidopsis raspberry embryos. Plant Cell 1994, 6:1713–1729.

79. Bioconductor [http://www.bioconductor.org].

80. Arabidopsis Gene Regulatory Information Server [http://arabidopsis.med.ohio-state.edu/RGNet].

81. Nielsen H, Engelbrecht J, Brunak S, von Heijne G: Identification of prokaryotic and eukaryotic signal peptides and prediction of their cleavage sites. Protein Eng 1997, 10:1–6.

82. Altschul SF, Madden TL, Schaffer AA, Zhang J, Zhang Z, Miller W, Lipman DJ: Gapped BLAST and PSI-BLAST: A new generation of protein database search programs. Nucleic Acids Res 1997, 25:3389–3402.

83. TAIR [http://www.arabidopsis.org].

84. Kerk NM, Ceserani T, Tausta SL, Sussex IM, Nelson TM: Laser capture microdissection of cells from plant tissues. Plant Physiol 2003, 132:27–35.

85. Vielle-Calzada JP, Thomas J, Spillane C, Coluccio A, Hoeppner MA, Grossniklaus U: Maintenance of genomic imprinting at the Arabidopsis medea locus requires zygotic DDM1 activity. Genes Dev 1999, 13:2971–2982.

86. Vielle-Calzada JP, Baskar R, Grossniklaus U: Delayed activation of the paternal genome during seed development. Nature 2000, 404:91–94.

87. Acosta-Garcia G, Vielle-Calzada JP: A classical arabinogalactan protein is essential for the initiation of female gametogenesis in Arabidopsis. Plant Cell 2004, 16:2614–2628.

88. Palanivelu R, Belostotsky DA, Meagher RB: Conserved expression of Arabidopsis thaliana poly(A) binding protein 2 (PAB2) in distinct vegetative and reproductive tissues. Plant J 2000, 22:199–210.

89. Belostotsky DA, Meagher RB: A pollen-, ovule-, and early embryo-specific poly(A) binding protein from Arabidopsis complements essential functions in yeast. Plant Cell 1996, 8:1261–1275.

90. Suzuki M, Kato A, Komeda Y: An RNA binding protein, AtRBP1, is expressed in actively proliferative regions in Arabidopsis thaliana. Plant Cell Physiol 2000, 41:282–288.

91. Mandel MA, Yanofsky MF: The Arabidopsis AGL9 MADS box gene is expressed in young flower primordia. Sex Plant Reprod 1998, 11:22–28.

92. Kinoshita T, Miura A, Choi Y, Kinoshita Y, Cao X, Jacobsen SE, Fischer RL, Kakutani T: One-way control of FWA imprinting in Arabidopsis endosperm by DNA methylation. Science 2004, 303:521–523.

93. Collinge MA, Spillane C, Köhler C, Gheyselinck J, Grossniklaus U: Genetic interaction of an origin recognition complex subunit and the Polycomb group gene MEDEA during seed development. Plant Cell 2004, 16:1035–1046.

94. Lefebvre V, North H, Frey A, Sotta B, Seo M, Okamoto M, Nambara E, Marion-Poll A: Functional analysis of Arabidopsis NCED6 and NCED9 genes indicates that ABA synthesized in the endosperm is involved in the induction of seed dormancy. Plant J 2006, 45:309–319.

95. Seedgenes [http://www.seedgenes.org].

96. Pepper A, Delaney T, Washburn T, Pool D, Chory J: DET1, a negative regulator of light-mediated development and gene expression in Arabidopsis encodes a novel nuclear-localized protein. Cell 1994, 78:109–116.

97. Castle LA, Meinke DW: A FUSCA gene of Arabidopsis encodes a novel protein essential for plant development. Plant Cell 1994, 6:25–41.

98. Nesi N, Debeaujon I, Jond C, Stewart AJ, Jenkins GI, Caboche M, Lepiniec L: The TRANSPARENT TESTA16 locus encodes the ARABIDOPSIS BSISTER MADS domain protein and is required for proper development and pigmentation of the seed coat. Plant Cell 2002, 14:2463–2479.

99. Johnson CS, Kolevski B, Smyth DR: TRANSPARENT TESTA GLABRA2, a trichome and seed coat development gene of Arabidopsis, encodes a WRKY transcription factor. Plant Cell 2002, 14:1359–1375.

100. Aida M, Ishida T, Tasaka M: Shoot apical meristem and cotyledon formation during Arabidopsis embryogenesis: interaction among the CUP-SHAPED COTYLEDON and SHOOT MERISTEMLESS genes. Development 1999, 126:1563–1570.

101. Nakaune S, Yamada K, Kondo M, Kato T, Tabata S, Nishimura M, Hara-Nishimura I: A vacuolar processing enzyme, deltaVPE, is involved in seed coat formation at the early stage of seed development. Plant Cell 2005, 17:876–887.

102. Lukowitz W, Roede A, Parmenter D, Somerville C: A MAPKK kinase gene regulates extra-embryonic cell fate in Arabidopsis. Cell 2004, 116:109–119.

103. Cushing DA, Forsthoefel NR, Gestaut DR, Vernon DM: Arabidopsis emb175 and other ppr knockout mutants reveal essential roles for pentatricopeptide repeat (PPR) proteins in plant embryogenesis. Planta 2005, 221:424–436.

104. Lai J, Dey N, Kim CS, Bharti AK, Rudd S, Mayer KF, Larkins BA, Becraft P, Messing J: Characterization of the maize endosperm transcriptome and its comparison to the rice genome. Genome Res 2004, 14:1932–1937.

105. NASC The European Arabidopsis Stock Centre [http://www.arabidopsis.info].

106. Array Express [http://www.ebi.ac.uk/arrayexpress].

107. dChip [http://www.dchip.org].

108. Li C, Hung Wong W: Model-based analysis of oligonucleotide arrays: model validation, design issues and standard error application. Genome Biol 2001, 2:Research0032.1-0032.11.

109. Wu Z, Irizarry RA, Gentleman R, Murillo FM, Spencer F: A model based background adjustment for oligonucleotide expression arrays. Technical

Report. In Johns Hopkins University Department of Biostatistics Working Papers. Baltimore, MD: Johns Hopkins University; 2003.

110. Smyth GK: Linear models and empirical bayes methods for assessing differential expression in microarray experiments. Stat Appl Genet Mol Biol 2004, 3:Article3. Article3 Return to text

111. Storey JD, Tibshirani R: Statistical significance for genomewide studies. Proc Natl Acad Sci USA 2003, 100:9440–9445.

CITATION

Originally published under the Creative Commons Attribution License. Johnston AJ, Meier P, Gheyselinck SEJ, Federer N, Schlagenhauf E, Becker JD, Grossniklaus U. Genetic Subtraction Profiling Identifies Genes Essential for Arabidopsis Reproduction and Reveals Interaction Between the Female Gametophyte and the Maternal Sporophyte. Genome Biol. 2007;8(10):R204. doi:10.1186/gb-2007-8-10-r204.

Arabidopsis WRKY2 Transcription Factor Mediates Seed Germination and Postgermination Arrest of Development by Abscisic Acid

Wenbo Jiang and Diqiu Yu

ABSTRACT

Background

Plant WRKY DNA-binding transcription factors are key regulators in certain developmental programs. A number of studies have suggested that WRKY genes may mediate seed germination and postgermination growth. However, it is unclear whether WRKY genes mediate ABA-dependent seed germination and postgermination growth arrest.

Results

To determine directly the role of Arabidopsis WRKY2 transcription factor during ABA-dependent seed germination and postgermination growth arrest, we isolated T-DNA insertion mutants. Two independent T-DNA insertion mutants for WRKY2 were hypersensitive to ABA responses only during seed germination and postgermination early growth. wrky2 mutants displayed delayed or decreased expression of ABI5 and ABI3, but increased or prolonged expression of Em1 and Em6. wrky2 mutants and wild type showed similar levels of expression for miR159 and its target genes MYB33 and MYB101. Analysis of WRKY2 expression level in ABA-insensitive and ABA-deficient mutants abi5-1, abi3-1, aba2-3 and aba3-1 further indicated that ABA-induced WRKY2 accumulation during germination and postgermination early growth requires ABI5, ABI3, ABA2 and ABA3.

Conclusion

ABA hypersensitivity of the wrky2 mutants during seed germination and postgermination early seedling establishment is attributable to elevated mRNA levels of ABI5, ABI3 and ABI5-induced Em1 and Em6 in the mutants. WRKY2-mediated ABA responses are independent of miR159 and its target genes MYB33 and MYB101. ABI5, ABI3, ABA2 and ABA3 are important regulators of the transcripts of WRKY2 by ABA treatment. Our results suggest that WRKY2 transcription factor mediates seed germination and postgermination developmental arrest by ABA.

Background

Abscisic acid (ABA) is a phytohormone regulating plant responses to a variety of environmental stress, particularly water deprivation, notably by regulating stomatal aperture [1-4]. It also plays an essential role in mediating the initiation and maintenance of seed dormancy [5]. Late in seed maturation, the embryo develops and enters a dormant state that is triggered by an increase in the ABA concentration. This leads to the cessation of cell division and activation of genes encoding seed storage proteins and proteins required to establish desiccation tolerance [6]. Exposure of seeds to ABA during germination leads to rapid but reversible arrest in development. ABA-mediated postgermination arrest allows germinating seedlings to survive early water stress [5]. Based on ABA inhibition of seed germination, mutants with altered ABA sensitivity have been identified. These screens have led to the identification of several ABA-insensitive genes [7-13]. The transcription factors ABI3 and ABI5 are known to be important regulators of ABA-dependent growth arrest during germination [14,15]. ABI5, an ABA-insensitive gene, encodes a basic leucine zipper transcription factor. Expression of ABI5 defines a narrow developmental checkpoint following germination, during which Arabidopsis plants sense the water status in the environment. ABI5 is a rate-limiting factor conferring ABA-mediated postgermination developmental growth arrest [5].ABI3 is also reactivated by ABA during a short development window. Like ABI5, ABI3 is also required for the ABA-dependent postgermination growth arrest [15]. However, ABI3 acts upstream of ABI5 and is essential for ABI5 gene expression [15]. In arrested, germinated embryos, ABA can activate de novo late embryogenesis programs to confer osmotic tolerance. During a short development window, ABI3, ABI5 and late embryogenesis genes are reactivated by ABA. ABA can activate ABI5 occupancy on the promoter of several late embryogenesis-abundant genes, including Em1 and Em6 and induce their expression [15,16]. On the other hand, ABA-induced miR159 accumulation requires ABI3 but is only partially dependent on ABI5. MYB33 and MYB101, two miR159 targets, are positive regulators of ABA responses during germination and are subject to ABA-dependent miR159 regulation [17].

The family of plant-specific WRKY transcription factors contains over 70 members in Arabidopsis thaliana [18-20]. WRKY proteins typically contain one or two domains composed of about 60 amino acids with the conserved amino acid sequence WRKYGQK, together with a novel zinc-finger motif. WRKY domain shows a high binding affinity to the TTGACC/T W-box sequence [21]. Based on the number of WRKY domains and the pattern of the zinc-finger motif, WRKY proteins can be divided into 3 different groups in Arabidopsis [20].

A growing body of studies has shown that WRKY genes are involved in regulating plant responses to biotic stresses. A majority of reported studies on WRKY genes address their involvement in disease responses and salicylic acid (SA)-mediated defense [20,22-25]. In addition, WRKY genes are involved in plant responses to wounding [26]. Although most WRKY proteins studied thus far have been implicated in regulating biotic stress responses, some WRKY genes regulate plant responses to freezing [27], oxidative stress [28], drought, salinity, cold, and heat [29-31].

There is also increasing evidence indicating that WRKY proteins are key regulators in certain developmental programs. Some WRKY genes regulate biosynthesis of anthocyanin [32], starch [33], and sesquiterpene [34]. Other WRKY genes may regulate embryogenesis [35], seed size [36], seed coat and trichome development [32,37], and senescence [38-40].

A number of studies have suggested that WRKY genes may mediate seed germination and postgermination growth. For example, wild oat WRKY proteins (ABF1 and ABF2) bind to the box2/W-box of the GA-regulated α-Amy2 promoter [41]. A barley WRKY gene, HvWRKY38, and its rice (Oryza sativa) ortholog, OsWRKY71 act as a transcriptional repressor of gibberellin-responsive genes in aleurone cells [42]. However, it is unclear whether WRKY genes mediate ABA-dependent seed germination and postgermination growth arrest. In this study, we report that wrky2-1 and wrky2-2 mutants are hypersensitive to ABA during germination and postgermination early growth. We have analyzed genetic interactions between the wrky2 mutant and the abi3-1 and abi5-1 mutants and found that ABA hypersensitivity of the wrky2 mutants is attributable to increased mRNA levels for ABI5, ABI3 and ABI5-induced Em1 and Em6. Furthermore, ABI5, ABI3, ABA2 and ABA3 are important regulators of ABA-induced expression of WRKY2.

Results

WRKY2 T-DNA Insertion Mutants are Hypersensitive to ABA During Seed Germination and Postgermination Early Growth

To analyze the molecular events in ABA-regulated germination and seedling growth, we sought to identify WRKY transcription factors associated with these growth and developmental stages. For this purpose, we screened T-DNA insertion mutants for a number of Arabidopsis WRKY genes for altered sensitivity to ABA based on their germination rates and seedling growth in MS media containing ABA. As shown in Figure 1, WRKY2 T-DNA insertion mutants wrky2-1 (Salk_020399) and wrky2-2 (Sail_739_F05) accumulated no WRKY2 transcript of expected size and exhibited increased sensitivity to ABA during seed germination and seedling growth. The two mutants display no other obvious phenotypes in morphology, growth, development or seed size (data not shown).

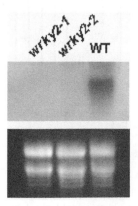

Figure 1. Identification of wrky2 mutants by northern blot analysis.
RNA was extracted from the seedlings that have grown on MS medium with 1.5 µM ABA 4 days after the end of stratification. Each lane contained 20 µg total RNA. Each experiment also was executed three times.

To determine the role of WRKY2 in seed germination and early seedling growth, wild-type and wrky2 mutant seeds were germinated on MS medium containing 0 µM, 0.5 µM, 1.0 µM, 1.5 µM, 2.0 µM ABA, and compared for differences in germination and postgerminative growth. In the absence of ABA, there was no significant difference in germination between wild-type and mutant seeds (Figure 2 and Figure 3A). In the presence of ABA, both mutants germinated later than wild type. On MS medium with 0.5 µM and 1.0 µM, 40% and 25% of wild-type seeds germinated after one day, respectively. At these two concentrations, the germination rates of the two mutants were only about half of wild type (Figure 2). On MS medium with 1.5 µM, 10% of wild-type seeds still germinated, but no wrky2-1 and wrky2-2 seeds germinated. Likewise, significantly more wild-type seeds germinated than the mutant seeds after two days on MS medium with 1.5 µM and 2.0 µM ABA (Figure 2). Early seedling growth of both mutants was also slower than that of wild type. After 7 d, 46% of wild-type but only 4% of wrky2-1 and none of wrky2-2 mutants had green cotyledons on MS medium with 1.5 µM ABA (Figure 4, 3B and 3C). These results show that wrky2 mutants are hypersensitive to ABA responses during germination and postgermination growth.

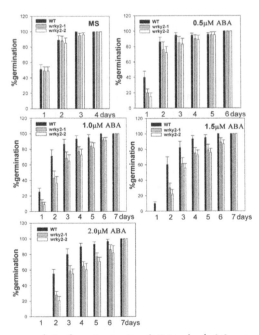

Figure 2. ABA dose-response analysis of germination in wrky2-1 and wrky2-2 mutants.
Seeds were germinated on MS plates containing 0 µM, 0.5 µM, 1.0 µM, 1.5 µM, and 2.0 µM ABA. Plates were routinely kept for 3 days in the dark at 4°C and transferred to a tissue culture room under constant light at 22°C. Germination efficiencies (radicle emergence) of wild type and wrky2 mutants seeds for 7 d after stratification. Three independent experiments are shown, and above 100 seeds were used in each experiment.

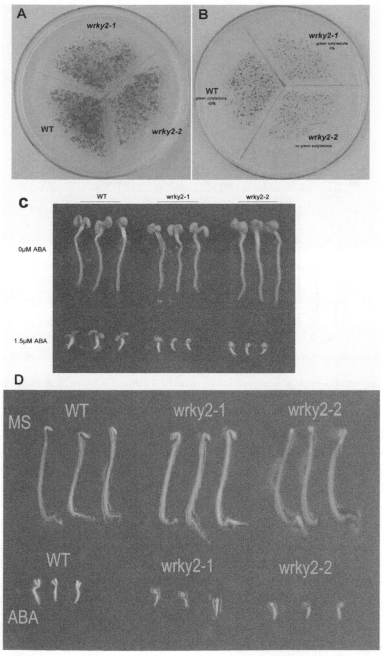

Figure 3. wrky2 mutants are hypersensitive to ABA responses during postgermination growth.

(A) Photographs of WT, wrky2-1 and wrky2-2 seedlings on MS medium at 7 d after the end of stratification. (B) Photographs of WT, wrky2-1 and wrky2-2 seedlings on MS medium with 1.5 μM ABA at 7 d after stratification. (C) Photographs of representative examples in A and B. (D) Photographs of representative examples in A and B in darkness.

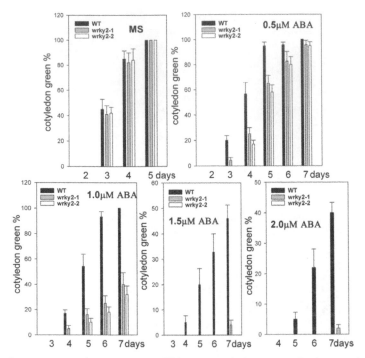

Figure 4. wrky2 mutants are hypersensitive to ABA responses during postgermination growth Seeds were germinated.

Postgermination growth efficiencies (green cotyledons) were scored for 7 d after stratification. Three independent experiments are shown, and above 100 seeds were used in each experiment.

To analyze the role of WRKY2 during the seedling growth stages, we first germinated the wild-type and wrky2 mutant seeds on MS medium for 4 d and then transferred the seedlings onto MS medium containing 0 µM, 0.5 µM, 1.0 µM, 1.5 µM, 2.0 µM, 5.0 µM, 10 µM, 20 µM, 40 µM and 80 µM ABA. No significant difference was observed between the wild type and wrky2 mutants 10 days after the transfer (data not shown).

When the germination experiments were carried out in dark, wrky2 mutants are again more sensitive to ABA responses than wild type (Figure 3). Taken together, these observations suggest that wrky2 mutants are hypersensitive to ABA responses only during seed germination and early seedling growth.

Response of WRKY2 Mutants to ABA Defines a Limited Developmental Window

The transcription factors ABI3 and ABI5 are known to be important regulators of ABA-dependent growth arrest during germination [14,15]. ABI3 and ABI5 are reactivated by ABA during a short development window. Given the ABA hypersensitivity of wrky2 mutants, we studied the effect of ABA on the early development of wrky2 mutants. Application of 5 μM ABA [5] within 48 h post-stratification maintained the germinated embryos of wrky2 mutants in an arrested state for several days, but did not prevent germination, whereas a significant percentage of wild-type germinated embryos escaped growth arrest and turned green (Figure 5A and 5B). ABA applied outside the 48-h time frame failed to arrest growth and prevent greening (Figure 5C). If the seeds were transferred to ABA-containing media immediately or one day after stratification, we observed that the wrky2 mutants were more sensitive to ABA responses than the wild type. However, if the seeds were transferred to ABA-containing media 2 days post-stratification, there was no significant difference between the wild type and wrky2 mutants. These results indicated that wrky2 mutants were hypersensitive to ABA responses for a short development window.

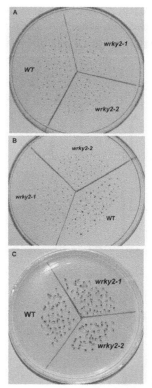

Figure 5. wrky2 mutants are hypersensitive to ABA responses in a short development window.

(A) Seeds were germinated on MS medium 1 d after stratification, all were transferred to MS medium with 5 μM ABA. Photographs were taken 6 d after transfer. (B) Seeds were germinated on MS medium and 2 d after stratification, all were transferred to MS medium with 5 μM ABA. Photographs were taken 5 d after transfer. (C) Seeds were germinated on MS medium and 3 d after stratification, all were transferred to MS medium with 5 μM ABA. Photographs were taken 4 d after transfer.

WRKY2 Mediates Signal Pathway of ABA-Dependent Germination and Postgermination Early Growth

During germination, ABA can activate de novo late embryogenesis programs to confer osmotic tolerance in arrested, germinated embryos [15]. During a short development window, ABI3, ABI5 and late embryogenesis genes are reactivated by ABA. ABI3 acts upstream of ABI5 and is essential for ABI5 gene expression. ABA induces ABI5 occupancy on the promoter of Em1 and Em6, and activates these late embryogenesis-abundant genes [15,16]. To analyze the expression of these genes in wrky2 mutants, we germinated the wild type and wrky2 mutants on MS medium with 0 and 1.5 μM ABA for 4 or 7 days. Total RNA was isolated and analyzed using quantitative RT-PCR with gene-specific primers. As shown in Figure 6, when these seedlings have grown on MS medium without ABA 4 days post-stratification, expression of ABI5, ABI3, Em1 and Em6 was reduced in the wrky2 mutants relative to that in the wild type. At 7 days post-stratification, wild type and wrky2 mutants accumulated similar levels of transcripts for ABI3, ABI5, Em1 and Em6 (Figure 6).

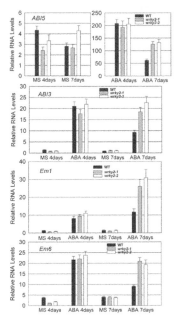

Figure 6. RNA levels of ABI5, ABI3, Em1 and Em6 in wrky2 mutants and wild type.

RNA was extracted from seedlings on MS medium without ABA or with 1.5 μM ABA 4 d or 7 d after the end of stratification. Relative RNA levels of 4 genes were analyzed using gene-specific primers by real-time PCR. Three independent experiments are shown by reextracting RNA from other samples. Each experiment also was executed three times.

On the other hand, on medium containing ABA, the expression levels of ABI5, ABI3, Em1 and Em6 were similar in the wild type and wrky2 mutants at 4 days post-stratification. However, at 7 days post-stratification, there were higher expression levels of these genes in wrky2 mutants than in the wild type, which was consistent with microarray analysis (Figure 6 and Table 1). By contrast at 4 days post-stratification, the expression of ABI3 and Em6 didn't decrease or decreased only slightly in wrky2 mutants, but decreased rapidly in the wild type within 7 days post-stratification. Within 7 days post-stratification, expression of AIB5 decreased faster in wild type than in wrky2 mutants. Expression of Em1 increased in wrky2 mutants 7 days post-stratification, but was unchanged in wild type (Figure 6).

Table 1. Microarray analysis of wrky2 mutants and wild type

	Microarray data of 6 genes	
Gene	Transcript ID	The ratio of wrky2 mutants vs WT
ABI5	At2G36270	2.30
ABI3	At3G24650	6.06
Em1	At3G51810	4.29
Em6	At2G40170	5.28
MYB33	At5G06100	1.15
MYB101	At2G32460	0.62

RNA was extracted from seedlings of wrky2 mutants and wild type on MS medium with 1.5 μM ABA 7 d post-stratification. Relative RNA levels of ABI5, ABI3, Em1, Em6, MYB33 and MYB101 were analyzed by Microarray data.

These results indicated that wrky2 mutants displayed delayed or decreased expression of ABI5 and ABI3, but increased or prolonged expression of Em1 and Em6.

The Expression of WRKY2 in ABA-Insensitive Mutants and ABA-Deficient Mutants

Because wrky2 mutants are more sensitive to ABA during seed germination and postgermination growth arrest than the wild type, and wrky2 mutants also affect important genes of the ABA signal pathway in the regulation of germination and postgermination growth, we analyze whether abi3-1, abi5-1, aba2-3 and aba3-1 mutants have an effect on the expression of WRKY2. We germinated all seeds on MS medium with 0 or 1.5 μM ABA for 4 days post-stratification. We performed quantitative RT-PCR with gene-specific primers. In the absence of ABA, the expression level of WRKY2 in the wild type (Ws) was 1.2 times of that in the abi5-1 mutants. In the presence of ABA, the level of WRKY2 was 2.5-fold. On the other hand, ABA treatment led to 13.6-fold increase in accumulation of WRKY2 in the wild type and 6.5-fold increase in abi5-1 mutants (Figure 7). These results suggest ABI5 is an important regulator of ABA-induced WRKY2 expression.

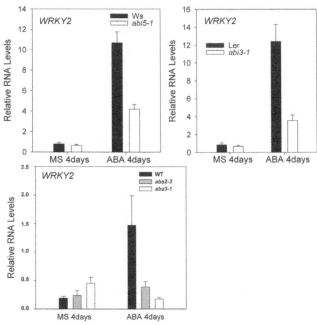

Figure 7. RNA levels of WRKY2 in abi5-1, abi3-1, aba2-3 and aba3-1 mutants.
RNA was extracted from seedlings on MS medium without ABA or with 1.5 µM ABA 4 d post-stratification. Relative RNA levels of WRKY2 were analyzed using gene-specific primers by real-time PCR. Three independent experiments are shown by reextracting RNA from other samples. Each experiment also was executed three times.

Without ABA, the level of WRKY2 in the wild type (Ler) was 1.2-fold higher than that in abi3-1 mutant. In the presence of ABA, the WRKY2 transcript level in wild type was 3.5-fold higher than that in abi3-1 mutant (Figure 7). These results indicate ABI3 maybe is a positive regulator of ABA-induced WRKY2. We also examined the expression level of WRKY2 in ABA-deficient aba2-3 and aba3-1 mutants [43,44]. As shown in Figure 7, with ABA, the expression level of WRKY2 in the wild type was 7.5 time higher than that without ABA, but the expression levels of WRKY2 in aba2-3 and aba3-1 mutants were only 1.6 and 0.38 times of those without ABA, respectively. On the other hands, without ABA, the expression levels of WRKY2 in aba2-3 and aba3-1 mutants were 1.2 and 2.3

times of that in the wild type. These observations show elevated expression of WRKY2 by ABA treatment requires ABA2 and ABA3. These results strongly suggest that ABI5, ABI3, ABA2 and ABA3 are important regulators of ABA-induced WRKY2 expression.

The Response of WRKY2 to ABA Is Independent of miR159, MYB33 and MYB101

It is known that ABA-induced miR159 accumulation requires ABI3 but is only partially dependent on ABI5. Furthermore, MYB33 and MYB101, which are miR159 targets, are positive regulators of ABA responses during germination and are subject to ABA-dependent miR159 regulation [17]. Figure 8A indicates that there was no significant difference in the level of mature miR159 between wrky2 mutants and the wild type. As shown in Figure 8B and Table 1, there was no difference in levels of transcripts of MYB33 and MYB101 either. These results indicate that WRKY2-mediated ABA signaling pathway is independent of miR159 and its target genes (MYB33 and MYB101) during seed germination and post-germination growth arrest.

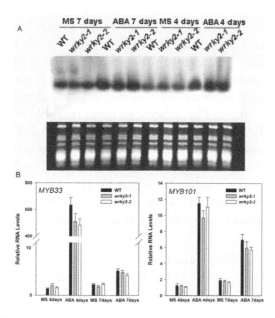

Figure 8. RNA levels of miR159, MYB33 and MYB101 in wrky2 mutants and wild type. RNA was extracted from seedlings on MS medium without ABA or with 1.5 μM ABA 4 d or 7 d post-stratification.

(A) Each lane contained 20 μg total RNA. Each experiment also was executed three times. (B) Relative RNA levels of WRKY2 were analyzed using gene-specific primers by real-time PCR. Three independent experiments are shown by reextracting RNA from other samples. Each experiment also was executed three times.

Discussion

The ability of exogenous ABA to arrest postgermination development has been used extensively to identify genes involved in ABA signaling [45]. In the present study, we discovered that wrky2-1 and wrky2-2 mutants were more sensitive than wild type to ABA responses during seed germination and postgermination early growth.

The wrky2 mutants are hypersensitive to ABA responses only within a short development window during seed germination and early seedling growth. During seed germination and early growth, the transcription factors ABI3 and ABI5 are known to be important regulators of ABA-dependent growth arrest, and their expression defines a narrow developmental checkpoint following germination [14,15]. ABI5 is a key player in ABA-triggered postgermination growth arrest, and the abi5 mutant seeds are insensitive to growth arrest by ABA, whereas seeds of ABI5-overexpressing lines are hypersensitive to ABA [5]. ABI3 acts upstream of ABI5 and is an important regulator of germination and postgermination growth by ABA [15]. ABA-induced ABI5 also occupies the promoter of Em1 and Em6 and activates these two late embryogenesis-abundant genes [15,16]. Lopez-Molina et al. (2001) have shown that application of 5 μM ABA within 60 h post-stratification in the wild type [5] (Ws) does not prevent germination, but arrest the germinated embryos for several days, during which both ABI5 transcripts and ABI5 proteins are detected at high levels. When applied outside the 60-h time frame, however, ABA fails to arrest growth and prevent greening and ABI5 transcripts and proteins are present at low levels [5]. Therefore, it is possible that the ABA-induced growth-arrest of wrky2 mutants might be associated with the expression level of ABI5. To test this hypothesis, we compared wild type and wrky2 mutants for analyzed ABA-regulated expression levels of ABI5 and ABI5-related important regulators during seed germination and postgermination development. We found that wrky2 mutants had higher mRNA levels of ABI5, ABI3 and ABI5-induced Em1 and Em6 than the wild type (Figure 6). The higher expression levels of these genes in the wrky2 mutants may be partly responsible for the enhanced growth arrest relative to that in the wild type in the presence of ABA (Figures 3 and 4).

With ABA treatment, miR159 accumulation requires ABI3 but is only partially dependent on ABI5 [17]. MYB33 and MYB101, which are miR159 targets, are positive regulators of ABA responses during germination [17]. Therefore we examined whether WRKY2 affected the mRNA levels of these genes. We found that the expression levels of these genes were not significantly different between wrky2 mutants and the wild type (Figure 8). These observations indicate that the response of WRKY2 to ABA during germination and early growth is independent of miR159 and its target genes (MYB33 and MYB101).

We also show that ABI5, ABI3, ABA2 and ABA3 are important positive regulators of ABA-induced WRKY2 accumulation (Figure 7). The expression levels of WRKY2 were elevated more drastically in the wild type than in abi5-1 and abi3-1 mutants after ABA treatment (Figure 7). This result indicates that these two genes are positive regulators of ABA-induced WRKY2 accumulation.

Genes encoding the enzymes for most of the steps of the ABA biosynthesis pathway have been cloned and their functions confirmed using ABA-deficient mutants for ABA1 [46], ABA2 [47,48], ABA3 [44,49,50] and ABA4 [51,52]. We analyzed the effect of ABA biosynthesis genes on WRKY2 transcripts using ABA-deficient aba2-3 and aba3-1 mutants [43,44]. We shows that elevated expression of WRKY2 after ABA treatment requires ABA2 and ABA3, indicating that these two genes are positive regulators of ABA-induced WRKY2 accumulation (Figure 7).

Several studies have shown that ABI5 binds to the ABA-responsive DNA elements (ABREs) with an ACGT core in the promoter of Em1 and Em6, and activates their expression [15,16]. On the other hand, other reports have shown that wild oat WRKY proteins (ABF1 and ABF2) bind to the box2/W-box of the GA-regulated α-Amy2 promoter [41], and GaWRKY1 highly activated the CAD1-A promoter by binding to W-box [34], and a barley WRKY gene, HvWRKY38, and its rice (Oryza sativa) ortholog, OsWRKY71 act as a transcriptional repressor of gibberellin-responsive genes in aleurone cells [42]. The promoter zone of WRKY2 has an ABA-responsive DNA element (CACGTGGC) containing an ACGT core, and the promoter zone of ABI5 has 6 W-box, whereas the promoter of ABI3 has 2 W-box. This raises the possibility that WRKY2, ABI3 and ABI5 are mutually regulated.

Conclusion

Transcription factors ABI5 is an important regulator of postgermination developmental arrest mediated by ABA. Postgermination proteolytic degradation of the essential ABI5 transcription factor is interrupted by perception of an increase in ABA concentration, leading to ABI5 accumulation and reactivation of embryonic genes. Here we report that wrky2-1 and wrky2-2 mutants are more sensitive to ABA responses than the wild type during seed germination and postgermination early seedling establishment. ABA hypersensitivity of the wrky2 mutants is attributable to elevated mRNA levels of ABI5, ABI3 and ABI5-induced Em1 and Em6 in the mutants. WRKY2-mediated ABA responses are independent of miR159 and its target genes MYB33 and MYB101. ABA-induced WRKY2 accumulation during germination and postgermination early growth requires ABI5, ABI3, ABA2 and ABA3, suggesting that they are important regulators of the transcripts of WRKY2 by ABA treatment. Our results suggest that WRKY2 transcription factor mediates seed germination and postgermination developmental arrest by ABA.

Methods

Plant Material and Growth Conditions

The Arabidopsis thaliana ecotypes Columbia, Wassilewskija and Landsberg erecta were used throughout this study. Seeds of the different genotypes of Arabidopsis thaliana were harvested from plants of the same age and stored for 5 weeks in the dark at 4°C. Seeds were surface-sterilized with 10% bleach and washed three times with sterile water. Sterile seeds were suspended in 0.1% agarose and plated on MS medium plus 1% sucrose. ABA (mixed isomers, Sigma) was added to the medium where indicated. Plates were routinely kept for 3 days in the dark at 4°C to break dormancy (stratification) and transferred thereafter to a tissue culture room under constant light at 22°C. Seeds of abi5-1, abi3-1, aba2-3 and aba3-1 were obtained from the Arabidopsis Biological Resource Center (ABRC) (Alonso et al., 2003).

Identification of the WRKY2 T-DNA Insertion Mutants

The wrky2-1 mutant (Salk_020399), obtained from the Arabidopsis Biological Resource Center (ABRC) (Alonso et al., 2003), contains a T-DNA insertion in the promoter of the WRKY2 gene, while wrky2-2 mutant (Sail_739_F05) is a gift of Dr. Zhixiang Chen (Department of Botany and Plant Pathology, Purdue University, West Lafayette, Indiana, USA). Homozygous plants of the wrky2-1 mutant were identified by two PCRs. In the first PCR, a pair of gene-specific primers designed to anneal outside of the T-DNA insertion were used, which in case of homozygosity does not produce a band of the predicted size (negative selection): forward primer 5'-ATCGTCATCATCTTCACCATTT-3' and reverse primer 5'-AACTGAAATCCTCAGTTCCGT-3'. In the subsequent PCR, the T-DNA border primer (5'-AAACGTCCGCAATGTGTTAT-3') in combination with forward primer in the first PCR. To confirm the nature and location of the T-DNA insertion, the PCR products were sequenced. To remove additional T-DNA loci or mutations from the mutants, backcrosses to wild-type plants were performed and plants homozygous for the T-DNA insertion were again identified.

Quantitative Real-Time PCR

We germinated seeds of the wild type and wrky2 mutants on MS medium with or without 1.5 μM ABA for 4 or 7 days, and germinated the seeds of abi5-1, abi3-1, aba2-3, aba3-1 mutants and the wild type (Col, Ws and Ler) on MS medium with or without 1.5 μM ABA for 4 days. Harvest samples were froze immediately in liquid nitrogen, and stored at 80°C. RNA was extracted from these samples using An RNeasy Plant Mini kit (QIAgen, Valencia, CA), and DNA was removed via an on-column DNase treatment. For real-time PCR, the First Strand cDNA Synthesis kit (Roche, Diagnostics, Mannheim, Germany) was used to make cDNA from 1 μg of RNA in a 20 μL reaction volume. Each cDNA sample was diluted 1:20 in water, and 2 μL of this dilution was used as template for qPCR. Half-reactions (10 μL each) were performed with the Lightcycler FastStart DNA Master SYBR Green I kit (Roche, Mannheim, Germany) on a Roche LightCycler real-time PCR machine, according to the manufacturer's instructions. ACT2 (AT3G18780) was used as a control in qPCR. Gene-specific primers for detecting transcripts of ACT2, WRKY2, ABI5, ABI3, Em1 and Em6 are listed in Table 2. Gene-specific primers of MYB33 and MYB101 are as described by Allen et al. (2007) [53]. The qPCR reactions (10 μL each) for these genes contained the following: 1 μL SYBR Green I reaction mix, 3 mM $MgCl_2$, 0.5 μM forward and reverse primers and 2 μL cDNA. The annealing temperature was 52°C in all cases. A no-template control was routinely included to confirm the absence of DNA or RNA contamination. The mean value of four replicates was normalized using the ACT2 gene as the control. Standard curves were generated using linearized plasmid DNA for each gene of interest. A second set of experiments was conducted on an independent set of tissue as a control.

Table 2. List of quantitative RT-PCR primer sequences

Gene	Primer sequences (5'->3')	
		Quantitative RT-PCR primers
ABI5	Primer forward	AGATGACACTTGAGGATTTCTTGGT
AT2G36270	Primer reverse	TGGTTCGGGTTTGGATTAGG
ABI3	Primer forward	CTGATTCTTGAATGGGTC
AT3G24650	Primer reverse	TTGTTATTAGGGTTAGGGT
Em1	Primer forward	CGGAGGAAGAAGGGATTGAGA
AT3G51810	Primer reverse	TGCCAAACACGGAACCTACA
Em6	Primer forward	GCAAAGAAGGGCGAGACC
AT2G40170	Primer reverse	TCCTCCTCAGCGTGTTCC
WRKY2	Primer forward	TTTCTTTGGGTTACGATG
AT5G56270	Primer reverse	CACAACAACTCTTGGCTC
ACT2	Primer forward	TGTGCCAATCTACGAGGGTTT
AT3G18780	Primer reverse	TTTCCCGCTCTGCTGTTGT

Quantitative RT-PCR primer sequences of ABI5, ABI3, Em1, Em6, WRKY2 and ACT2.

Northern Blotting

Total RNA was isolated using the TRIZOL reagent (BRL Life Technologies, Rockville, MD). 20 μg RNA was separated by electrophoresis on denaturing 17% polyacrylamide gels, and electroblotted onto a Hybond-N+ membrane. The membrane was UV cross-linked and hybridized with PerfectHyb plus hybridization buffer (Sigma). DNA oligonucleotides complementary to miR159 were end-labeled using T4 polynucleotide kinase (Roche Applied Science, Penzberg, Germany). For RNA gel blot analysis of WRKY2, 20 μg total RNA was separated on 1.5% agarose-formaldehyde gels and blotted to nylon membranes. Blots were hybridized with [α-32P]dATP labeled gene-specific probes. Hybridization was performed in PerfectHyb plus hybridization buffer (Sigma) overnight at 68°C. The membrane was then washed for 10 minutes twice with 2× SSC (1× SSC is 0.15 M NaCl and 0.015 M sodium citrate) and 1% SDS and for 10 minutes with 0.1× SSC and 1% SDS at 68°C. Transcripts for WRKY2 were detected with about 1 kb before stop codon of WRKY2 cDNA as probe.

Authors' Contributions

WJ carried out all experiments of WRKY2 gene, participated in the design of the study, drafted and edit the manuscript. DY conceived of the study, participated in the design and helped to draft and edit the manuscript. All authors read and approved the final manuscript.

Acknowledgements

We thank the ABRC at the Ohio State University (Columbus, OH) for the Arabidopsis mutants. We are grateful to Dr. Zhixiang Chen (Department of Botany and Plant Pathology, Purdue University, West Lafayette, Indiana, USA) for Arabidopsis mutants and his critical reading of the manuscript. This work was supported by the Science Foundation of the Chinese Academy of Sciences (grant no. KSCX2-YW-N-007), and the Hundred Talents Program of the Chinese Academy of Sciences, the Natural Science Foundation of Yunnan Province (grant no. 2003C0342M), and the Ministry of Science and Technology of China (grant no. 2006AA02Z129).

References

1. Hetherington AM: Guard cell signaling. Cell 2001, 107(6):711–714.

2. Assmann SM, Wang X-Q: From milliseconds to millions of years: guard cells and environmental responses. Current Opinion in Plant Biology 2001, 4(5):421–428.

3. MacRobbie EAC: Signal transduction and ion channels in guard cells. Philos Trans R Soc Lond B Biol Sci 1998, 353:1475–1488.

4. Himmelbach A, Iten M, Grill E: Signalling of abscisic acid to regulate plant growth. Philos Trans R Soc Lond B Biol Sci 1998, 353:1439–1444.

5. Lopez-Molina L, Mongrand S, Chua NH: A postgermination developmental arrest checkpoint is mediated by abscisic acid and requires the ABI5 transcription factor in Arabidopsis. Proc Natl Acad Sci USA 2001, 98(8):4782–4787.

6. Ingram J, Bartels D: The molecular basis of dehydration tolerance in plants. Annual Review of Plant Physiology and Plant Molecular Biology 1996, 47(1):377–403.

7. Lopez-Molina L, Chua N-H: A null mutation in a bZIP factor confers ABA-insensitivity in Arabidopsis thaliana. Plant Cell Physiol 2000, 41:541–547.

8. Finkelstein RR, Lynch TJ: The Arabidopsis abscisic acid response gene ABI5 encodes a basic leucine zipper transcription factor. Plant Cell 2000, 12(4):599–610.

9. Gosti F, Beaudoin N, Serizet C, Webb AA, Vartanian N, Giraudat J: ABI1 protein phosphatase 2C is a negative regulator of abscisic acid signaling. Plant Cell 1999, 11(10):1897–1910.

10. Finkelstein RR, Wang ML, Lynch TJ, Rao S, Goodman HM: The Arabidopsis abscisic acid response locus ABI4 encodes an APETALA 2 domain protein. Plant Cell 1998, 10(6):1043–1054.

11. Meyer K, Leube MP, Grill E: A protein phosphatase 2C involved in ABA signal transduction in Arabidopsis thaliana. Science 1994, 264(5164):1452–1455.

12. Leung J, Bouvier-Durand M, Morris PC, Guerrier D, Chefdor F, J G: Arabidopsis ABA response gene ABI1: features of a calcium-modulated protein phosphatase. Science 1994, 264:1448–1452.

13. Giraudat J, Hauge BM, Valon C, Smalle J, Parcy F, Goodman HM: Isolation of the Arabidopsis ABI3 gene by positional cloning. Plant Cell 1992, 4(10):1251–1261.

14. Zhang XR, Garreton V, Chua NH: The AIP2 E3 ligase acts as a novel negative regulator of ABA signaling by promoting ABI3 degradation. Genes & Development 2005, 19(13):1532–1543.

15. Lopez-Molina L, Mongrand S, McLachlin DT, Chait BT, Chua N-H: ABI5 acts downstream of ABI3 to execute an ABA-dependent growth arrest during germination. The Plant Journal 2002, 32(3):317–328.

16. Bensmihen S, Rippa S, Lambert G, Jublot D, Pautot V, Granier F, Giraudat J, Parcy F: The homologous ABI5 and EEL transcription factors function antagonistically to fine-tune gene expression during late embryogenesis. Plant Cell 2002, 14(6):1391–1403.

17. Reyes JL, Chua N-H: ABA induction of miR159 controls transcript levels of two MYB factors during Arabidopsis seed germination. The Plant Journal 2007, 49(4):592–606.

18. Eulgem T, Somssich IE: Networks of WRKY transcription factors in defense signaling. Current Opinion in Plant Biology 2007, 10(4):366–371.

19. Dong J, Chen C, Chen Z: Expression profiles of the Arabidopsis WRKY gene superfamily during plant defense response. Plant Molecular Biology 2003, 51(1):21–37.

20. Eulgem T, Rushton PJ, Robatzek S, Somssich IE: The WRKY superfamily of plant transcription factors. Trends in Plant Science 2000, 5(5):199–206.

21. Ulker B, Somssich IE: WRKY transcription factors: from DNA binding towards biological function. Current Opinion in Plant Biology 2004, 7(5):491–498.

22. Asai T, Tena G, Plotnikova J, Willmann MR, Chiu W-L, Gomez-Gomez L, Boller T, Ausubel FM, Sheen J: MAP kinase signalling cascade in Arabidopsis innate immunity. Nature 2002, 415(6875):977–983.

23. Zheng Z, Mosher SL, Fan B, Klessiq DF, Chen Z: Functional analysis of Arabidopsis WRKY25 transcription factor in plant defense against Pseudomonas syringae. BMC Plant Biology 2007, 7:2.

24. Dellagi A, Heilbronn J, Avrova AO, Montesano M, Palva ET, Stewart HE, Toth IK, Cooke DE, Lyon GD, Birch PR: A potato gene encoding a WRKY-like transcription factor is induced in interactions with Erwinia carotovora subsp. atroseptica and phytophthora infestans and is coregulated with class I endochitinase expression. Molecular Plant-Microbe Interactions 2000, 13(10):1092–1101.

25. Lai Z, Vinod K, Zheng Z, Fan B, Chen Z: Roles of Arabidopsis WRKY3 and WRKY4 transcription factors in plant responses to pathogens. BMC Plant Biology 2008, 8:68.

26. Hara K, Yagi M, Kusano T, Sano H: Rapid systemic accumulation of transcripts encoding a tobacco WRKY transcription factor upon wounding. Molecular and General Genetics MGG 2000, 263(1):30–37.

27. Huang T, Duman JG: Cloning and characterization of a thermal hysteresis (antifreeze) protein with DNA-binding activity from winter bittersweet nightshade, Solanum dulcamara. Plant Molecular Biology 2002, 48(4):339–350.

28. Rizhsky L, Davletova S, Liang H, Mittler R: The zinc finger protein Zat12 is required for cytosolic ascorbate peroxidase 1 expression during oxidative stress in Arabidopsis. J Biol Chem 2004, 279(12):11736–11743.

29. Seki M, Narusaka M, Ishida J, Nanjo T, Fujita M, Oono Y, Kamiya A, Nakajima M, Enju A, Sakurai T, et al.: Monitoring the expression profiles of 7000 Arabidopsis genes under drought, cold and high-salinity stresses using a full-length cDNA microarray. Plant J 2002, 31(3):279–292.

30. Li S, Fu Q, Huang W, Yu D: Functional analysis of an Arabidopsis transcription factor WRKY25 in heat stress. Plant Cell Rep 2009, 28(4):683–693.

31. Qiu Y, Yu D: Over-expression of the stress-induced OsWRKY45 enhances disease resistance and drought tolerance in Arabidopsis. Environmental and Experimental Botany 2009, 65:35–47.

32. Johnson CS, Kolevski B, Smyth DR: TRANSPARENT TESTA GLABRA2, a trichome and seed coat development gene of Arabidopsis, encodes a WRKY transcription factor. Plant Cell 2002, 14(6):1359–1375.

33. Sun C, Palmqvist S, Olsson H, Boren M, Ahlandsberg S, Jansson C: A novel WRKY transcription factor, SUSIBA2, participates in sugar signaling in Barley by binding to the sugar-responsive elements of the iso1 promoter. Plant Cell 2003, 15(9):2076–2092.

34. Xu YH, Wang JW, Wang S, Wang JY, Chen XY: Characterization of GaWRKY1, a cotton transcription factor that regulates the sesquiterpene synthase gene (+)-{delta}-cadinene synthase-A. Plant Physiol 2004, 135(1):507–515.

35. Lagace M, Matton DP: Characterization of a WRKY transcription factor expressed in late torpedo-stage embryos of Solanum chacoense. Planta 2004, 219(1):185–189.

36. Luo M, Dennis ES, Berger F, Peacock WJ, Chaudhury A: MINISEED3 (MINI3), a WRKY family gene, and HAIKU2 (IKU2), a leucine-rich repeat (LRR) KINASE gene, are regulators of seed size in Arabidopsis. Proc Natl Acad Sci USA 2005, 102(48):17531.

37. Ishida T, Hattori S, Sano R, Inoue K, Shirano Y, Hayashi H, Shibata D, Sato S, Kato T, Tabata S, et al.: Arabidopsis TRANSPARENT TESTA GLABRA2 is

directly regulated by R2R3 MYB transcription factors and is involved in regulation of GLABRA2 transcription in epidermal differentiation. Plant Cell 2007, 19(8):2531–2543.

38. Miao Y, Zentgraf U: The antagonist function of Arabidopsis WRKY53 and ESR/ESP in leaf senescence is modulated by the jasmonic and salicylic acid equilibrium. Plant Cell 2007, 19(3):819–830.

39. Robatzek S, Somssich IE: A new member of the Arabidopsis WRKY transcription factor family, AtWRKY6, is associated with both senescence- and defence-related processes. Plant J 2001, 28(2):123–133.

40. Jing S, Zhou X, Song Y, Yu D: Heterologous expression of OsWRKY23 gene enhances pathogen defense and dark-induced leaf senescence in Arabidopsis. Plant Growth Regul 2009, 58:181–190.

41. Rushton PJ, Macdonald H, Huttly AK, Lazarus CM, Hooley R: Members of a new family of DNA-binding proteins bind to a conserved cis-element in the promoters of alpha-Amy2 genes. Plant Molecular Biology 1995, 29(4):691–702.

42. Zou X, Neuman D, Shen QJ: Interactions of two transcriptional repressors and two transcriptional activators in modulating gibberellin signaling in aleurone cells. Plant Physiol 2008, 148(1):176–186.

43. Lin PC, Hwang SG, Endo A, Okamoto M, Koshiba T, Cheng WH: Ectopic expression of ABSCISIC ACID 2/GLUCOSE INSENSITIVE 1 in Arabidopsis promotes seed dormancy and stress tolerance. Plant Physiol 2007, 143(2):745–758.

44. Xiong L, Ishitani M, Lee H, Zhu J-K: The Arabidopsis LOS5/ABA3 locus encodes a molybdenum cofactor sulfurase and modulates cold stress- and osmotic stress-responsive gene expression. Plant Cell 2001, 13(9):2063–2083.

45. Lu C, Han MH, Guevara-Garcia A, Fedoroff NV: Mitogen-activated protein kinase signaling in postgermination arrest of development by abscisic acid. Proc Natl Acad Sci USA 2002, 99(24):15812–15817.

46. Marin E, Nussaume L, Quesada A, Gonneau M, Sotta B, Hugueney P, Frey A, Marion-Poll A: Molecular identification of zeaxanthin epoxidase of Nicotiana plumbaginifolia, a gene involved in abscisic acid biosynthesis and corresponding to the ABA locus of Arabidopsis thaliana. the EMBO Journal 1996, 15:2331–2342.

47. Cheng WH, Endo A, Zhou L, Penney J, Chen HC, Arroyo A, Leon P, Nambara E, Asami T, Seo M, et al.: A unique short-chain dehydrogenase/reductase in Arabidopsis glucose signaling and abscisic acid biosynthesis and functions. Plant Cell 2002, 14(11):2723–2743.

48. Rook F, Corke F, Card R, Munz G, Smith C, Bevan MW: Impaired sucrose-induction mutants reveal the modulation of sugar-induced starch biosynthetic gene expression by abscisic acid signalling. The Plant Journal 2001, 26(4):421–433.

49. Bittner F, Oreb M, Mendel RR: ABA3 is a molybdenum cofactor sulfurase required for activation of aldehyde oxidase and xanthine dehydrogenase in Arabidopsis thaliana. J Biol Chem 2001, 276(44):40381–40384.

50. Schwartz SH, Leon-Kloosterziel KM, Koornneef M, Zeevaart JA: Biochemical characterization of the aba2 and aba3 mutants in Arabidopsis thaliana. Plant Physiol 1997, 114(1):161–166.

51. North HM, De Almeida A, Boutin JP, Frey A, To A, Botran L, Sotta B, Marion-Poll A: The Arabidopsis ABA-deficient mutant aba4 demonstrates that the major route for stress-induced ABA accumulation is via neoxanthin isomers. The Plant Journal 2007, 50(5):810–824.

52. Dall'Osto L, Cazzaniga S, North H, Marion-Poll A, Bassi R: The Arabidopsis aba4-1 mutant reveals a specific function for neoxanthin in protection against photooxidative stress. Plant Cell 2007, 19(3):1048–1064.

53. Allen RS, Li J, Stahle MI, Dubroué A, Gubler F, Millar AA: Genetic analysis reveals functional redundancy and the major target genes of the Arabidopsis miR159 family. Proc Natl Acad Sci USA 2007, 104(41):16371–16376.

CITATION

Originally published under the Creative Commons Attribution License. Jiang W, Yu D Arabidopsis WRKY2 Transcription Factor Mediates Seed Germination and Postgermination Arrest of Development by Abscisic Acid. BMC Plant Biology 2009, 9:96 doi:10.1186/1471-2229-9-96.

DNA Methylation Causes Predominant Maternal Controls of Plant Embryo Growth

Jonathan FitzGerald, Ming Luo, Abed Chaudhury and
Frédéric Berger

ABSTRACT

The parental conflict hypothesis predicts that the mother inhibits embryo growth counteracting growth enhancement by the father. In plants the DNA methyltransferase MET1 is a central regulator of parentally imprinted genes that affect seed growth. However the relation between the role of MET1 in imprinting and its control of seed size has remained unclear. Here we combine cytological, genetic and statistical analyses to study the effect of MET1 on seed growth. We show that the loss of MET1 during male gametogenesis causes a reduction of seed size, presumably linked to silencing of the paternal allele of growth enhancers in the endosperm, which nurtures the embryo. However, we find no evidence for a similar role of MET1 during female gametogenesis. Rather, the reduction of MET1 dosage in the maternal somatic tissues causes seed size increase. MET1 inhibits seed growth by restricting cell division and elongation in the maternal integuments that surround the seed. Our data demonstrate new controls of seed growth linked to the mode of reproduction typical of flowering plants. We conclude that the regulation of embryo growth by MET1 results from a combination of predominant maternal controls, and that DNA methylation maintained by MET1 does not orchestrate a parental conflict.

Introduction

In flowering plants, meiosis is followed by the production of haploid structures, the male pollen and the female embryo sac, each containing two gametes. After double-fertilization, the female gametes, the egg cell and central cell, respectively give rise to the embryo and its nurturing annex, the endosperm. The embryo and the endosperm develop within the maternally derived seed integuments. Seed size is controlled primarily by interactions between the endosperm and integuments [1], [2] although the embryo also contributes [3].

The parental contributions to seed size were identified in crosses involving diploid and tetraploid plants. Tetraploid mothers produced smaller seeds when crossed to diploid fathers, however tetraploid fathers crossed to diploid mothers produced larger seeds [4], [5]. Hence seed size is enhanced by an excess of paternal genomes and restricted by an excess of maternal genomes. These phenomena were linked to the DNA methyltransferase MET1, using a dominant antisense construct, MET1a/s 6–9. Maternal inheritance of MET1a/s causes an increase of seed size whereas paternal inheritance has an opposite effect. MET1 is a key player in the control of parental genomic imprinting, which restricts gene expression from one of the two parental alleles [10]. In Arabidopsis, it was proposed that MET1 controls the expression of two pools of imprinted genes: maternally expressed inhibitors and paternally expressed enhancers of endosperm growth [11]. In Arabidopsis two imprinted genes dependent on MET1 have been identified [12]. MET1 silences the genes FWA and FERTILIZATION INDEPENDENT SEED 2 (FIS2) in the male gametes [12]. FIS2 and FWA are expressed in the female central cell [9], [13]. After fertilization FIS2 and FWA are expressed in the endosperm from their maternal allele, while MET1 maintains silencing on the paternal allele [12], [13]. The parental imbalance of expression thus defines FIS2 and FWA as imprinted genes.

It was expected that the contrasting effects of MET1a/s were mediated by removal of silencing of the paternal allele of endosperm growth inhibitors, thus causing seed size increase and vice versa [11]. However, MET1a/s has a dominant effect, which does not allow distinguishing whether seed size variations in wild type (wt)×MET1a/s crosses originated from the loss of MET1 in the previous parental generation (sporophyte) or in the haploid generation producing the gametes (gametophyte). In addition, MET1a/s lines accumulate epimutations [6] and abnormal methylation profiles [14], which could be partially responsible of the phenotypes observed. A study based on a recessive loss-of-function allele, met1-6 [15] showed clearly that the loss of met1 during male gametogenesis reduces seed size. This result was also in agreement with the demonstration of a gametophytic effect of met1-3 on the silencing of the paternal alleles of the imprinted genes FIS2 and FWA [12]. However the existence of a gametophytic maternal effect of met1-6 on seed size remained unclear [15] and a potential effect on met1-6 loss of function on the diploid parental sporophytic generation was not tested explicitly. To address these concerns, we restricted our analysis to homozygous and heterozygous mutants derived from a self-fertilized heterozygous met1-3/+ mother and compared the effects on seed development of met1-3 loss of function during male gametogenesis, female gametogenesis and the parental diploid generation.

Results and Discussion

A Distinctive Paternal Effect Is Associated to MET1 Loss-of-Function During Male Gametogenesis

The null recessive allele met1-3 causes a loss of DNA methylation in first generation homozygous plants [16]. The loss of met1 function is caused by a T-DNA insert linked to a gene conferring resistance to the herbicide BASTA. To confirm specific parental contributions of met1-3 to seed size, we analyzed digital images of seeds from crosses that varied MET1 genotype and parent of transmission (Figure 1, Table 1). Seeds produced by crosses between wild-type ovules and pollen from met1-3/met1-3 plants were smaller than seeds produced between wild type ovules and wild type pollen (Figure 1A). Quantitative analysis resolved these two genotypes into two distinct populations based on seed width and length (n = 108; P<.0001 for ANOVA, t-test and Mann Whitney) (Figure 1B, Table 1). This verified that met1-3 has a paternal effect on seed size as observed in previous studies [7], [9], [15]. We then conducted the same experiment with heterozygous met1-3/+ plants. Half of the pollen from met1-3/+ plants carries the met1-3 allele causing re-activation of imprinted genes [12] and other silenced loci [17]. It is thus possible to predict a gametophytic paternal effect of met1 with size reduction in only 50% of the seeds produced by wild type ovules crossed to met1-3/+ pollen. Accordingly, we observed both large and small seeds by visual inspection (Figure 1A; 45.4% small seeds; n = 900) and quantitative analysis (wt×met1/+, n = 374; wt×wt, n = 257; P<.0001 for ANOVA, t-test and Mann Whitney) (Figure 1C, Table 1). In the small seeds from crosses between wild type ovules and pollen from met1-3/+ plants, embryo growth was relatively normal as compared to the endosperm, which exhibited reduced growth.

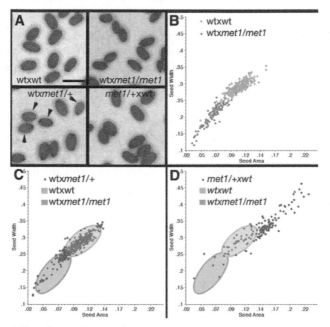

Figure 1. Parental effect of met1-3/+ on seed size.

(A) Seed populations produced by crosses between wild-type (wt) ovules and pollen from met1-3/+ or met1-3/met1-3 plants. The scale bar represents 0.5 mm. (B) Morphometric parameters of seeds from crosses between wt ovules and pollen from wt or from met1/met1 plants. (C) Morphometric parameters of seeds from crosses between ovules from wt plants and met1-3/+ pollen. The green and red ovals represent the extent of the populations of seeds shown in B. (D) Morphometric parameters of seeds from crosses between ovules from met1/+ plants and wild-type pollen. The green and red ovals represent the extent of the populations of seeds shown in B.

Table 1. Morphometric measurements of seeds from various crosses reported in Figure 1.

Cross genotype (mat×pat)	n	Seed area mean	s.d.	s.e.m	Seed width mean	s.d.	s.e.m
wt×wt	257	0.105	0.018	0.001	0.283	0.033	0.002
wt×met1/met1	108	0.65	0.014	0.001	0.207	0.031	0.003
wt×met1/+	374	0.98	0.022	0.001	0.273	0.036	0.002
Met1/+ ×wt	138	0.142	0.037	0.003	0.321	0.056	0.005

To confirm the link between the small seeds and paternal inheritance of met1-3, seeds from wt×met1-3/+ crosses were visually sorted according to their size relative to a wild type control, and BASTA resistance associated to met1-3 was tested. Two populations of seeds were distinguished. All smallest seeds were resistant to BASTA (n = 323) while all largest seeds were sensitive to BASTA (n = 336). The 1:1 proportion supported the predicted association of the paternal effect of met1-3 to gametogenesis (p = 0.6126 χ2). As we did not analyze the entire population we may have missed a complex genetic component regulating seed size. To ensure that abnormally small seeds or seed lethality were not missing from our bulked seed population, we analyzed all seeds from single crosses between wild-type mothers and pollen from met1-3/+ plants (Figure 2A, Table 2). In this analysis we also ensured that crosses with pollen from wt and met1-3/+ plants were performed on the same mother plant to allow an absolute size comparison. BASTA resistance correlated with the smallest seeds of the population (p<0.0001 ANOVA and Mann-Whitney) demonstrating that paternal inheritance of met1-3 causes seed size reduction as a result of the loss of MET1 activity during male gametogenesis. The loss of MET1 during male gametogenesis may allow paternal expression of imprinted growth inhibitors and cause a decrease of endosperm and seed size. Loss-of-function paternal effects are uncommon and until now have only been linked to defects in fertilization in Drosophila [18], [19], C.elegans [20] and Arabidopsis [3]. We thus conclude that met1-3 causes a paternal effect associated with defects after fertilization and thus representing a distinct class of paternal effect mutations.

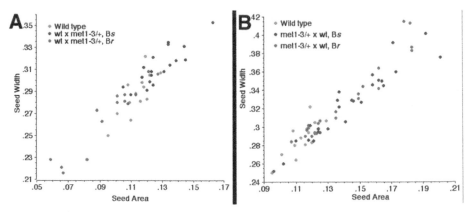

Figure 2. Correlation between seed size and the inheritance of met1-3 associated to BASTA resistance.

(A) BASTA resistance (Br) and sensitivity (Bs) are correlated with seed size in seeds from crosses between wild-type ovules and met1-3/+ pollen. Segregation of the BASTA marker remains 1:1 (p = 0.4795 χ²), so although some seed lethality was observed (n = 11) it is not linked to met1-3. (B) Br and Bs are not correlated with seed size in seeds from crosses between met1-3/+ ovules and wild-type pollen.

Table 2. Morphometric measurements of seeds correlated to BASTA R as reported in

Cross Genotype	Seed Genotype								
	BASTA R			BASTA S					
	n	mean	s.d.	n	mean	s.d.	ANOVA	M-W
mat wt×pat *met1-3/+*								
area	14	0.095	0.021	18	0.128	0.014	<0.0001	<0.0001
width	14	0.267	0.031	18	0.313	0.017	<0.0001	<0.0001
mat *met1-3/+*×pat wt								
Silique 1 area	13	0.122	0.007	12	0.121	0.013	0.7953	0.8066
width	13	0.298	0.009	12	0.290	0.019	0.1787	0.3270
Silique 2 area	9	0.167	0.015	14	0.161	0.019	0.4847	0.4120
width	9	0.368	0.033	14	0.352	0.024	0.1815	0.2567

Loss of MET1 During Female Gametogenesis does not Impact on Seed Size

While crosses between wild type ovules and the MET1a/s pollen caused a decrease of seed size, a symmetrical increase of seed size was observed in seeds from the reciprocal crosses MET1a/s×wt. [7]–[9]. We tested whether maternal inheritance of met1-3 from met1-3/+ mothers would increase size in 50% of the seeds. Crosses between ovules from met1-3/+ plants and wild-type pollen did exhibit increased seed size relative to wild type controls (Figure 1A; n = 900) correlated with an increased in endosperm size. However, this increase in size affected the whole population of seeds (Figure 1D, Table 1, n = 138). Largest seeds selected by visual inspection from a population of 900 seeds from wt×met1-3/+ crosses did not show a preferential resistance to BASTA (55.1% BASTA Resistant in a population of n = 84 largest seeds). This is contrary to the expected consequence of a maternal gametophytic effect of met1-3/+, which should produce a greater proportion BASTA resistance among the largest seeds in a population derived from met1-3/+×wt crosses. To confirm this finding we compared BASTA resistance and seed size in an entire population of seeds from met1-3/+×wt crosses from a single plant. We observed that larger seeds did not always inherit the met1-3 allele and the means of size measurements did not differ between seed genotypes (Figure 2B, Table 2). These results were in clear contrast to the results obtained from crosses involving pollen from met1-3/+ plants. The inheritance of met1-3 from met1-3/+ plants through the female gametes did not cause the increase of size in 50% of the seed population as expected for a gametophytic maternal effect. However we observed an overall increase of seed size in the entire population of seeds (Fig. 1D, Table 1). Thus, it was possible that either the gametophytic effect was not fully penetrant and could not be detected clearly. Alternatively it was possible that the maternal effect of met1 was mediated from the maternal tissues surrounding the seed.

Maternal Effects Linked to Loss-of-Function of MET1 in Vegetative Tissues

In seeds derived from met1-3/+ fathers we expected that genetically wild-type seeds would have a wild-type seed size. In contrast, both the seed area and width of the BASTA sensitive wild type seeds derived from met1-3/+ fathers are significantly larger than the wild-type controls pollinated after emasculation and grown in the same conditions, even though these seeds are genetically identical (Table 2). This effect on seed size likely originates from the reduced dosage of active MET1 in the heterozygous met1-3/+ vegetative tissues. Similarly the average seed size of wild type seed produced from crosses between ovules from met1/+ plants and wild type pollen were also larger than wild type ovules from controls emasculated wild type plants crossed with wild type pollen (Table 2). Since we failed to detect a gametophytic component in the genetic maternal control of seed size by met1-3/+ plants, we concluded that the size increase observed in met1/+×wt crosses originated from the effect of met1 in vegetative tissues. Thus, plants heterozygous for met1-3 enhanced seed growth both maternally and paternally with no evidence for antagonism between the two parents. In addition our results suggest that an overall reduction of MET1 levels in met1-3/+ plants could lead to a reduced level of DNA methylation activity prior to meiosis and promote seed size increase.

MET1 Controls Embryo Size Through Its Action on the Maternal Tissues

The maternal inheritance of the dominant MET1a/s construct caused a dramatic increase of seed size [7]. Similarly, seeds from crosses between ovules from met1-6 [15] or met1-3 homozygous crossed to wild type pollen are much larger than seeds produced from met1/+ heterozygous mothers crossed to wild type pollen. The range of phenotypes suggested that seed size and development were influenced by MET1 dosage in the maternal sporophyte. All seeds were affected, indicating that defects could originate from the maternal tissues responsible for supplying maternal nutrients to the seed or the maternal seed integuments. Deregulation of cell proliferation and cell elongation of integuments influences seed size [1], [21], [22]. We thus investigated whether MET1 controls integuments development. We observed that met1-3/met1-3 integuments contain 50% more cells than in the wild type (Figures 3A and 3B and Table 3). We thus conclude that MET1 represses cell proliferation in the integuments. In addition, we observed that in the absence of fertilization, the fruits of met1-3/met1-3 plants elongated (Figure 3C and Table 3), resulting in production of seed-like structures devoid of embryo and endosperm (Figure 3, D and E and Table 3). Similar observations were made with MET1a/s plants (Table 3). The autonomous seed-like structures are devoid of endosperm or embryo and develop only from ovules that are deficient of MET1 in the sporophytic integuments but not from ovules from met1/+ plants, 50% of which are deficient of MET1 in the female gametophyte. We conclude that autonomous growth of seed-like structures did not originate from the loss of MET1 activity in the central cell or the egg cell. Rather, MET1 thus controls seed size maternally through its action on cell proliferation and elongation in the seed integuments. Double fertilization causes enhanced cell division followed by elongation in the wild type [1]. Our results thus suggest that double-fertilization releases MET1-inhibited controls. Hence we show that mechanisms acting in the integuments in addition to the endosperm [23] and the embryo [3], [24] prevent seed development in absence of fertilization.

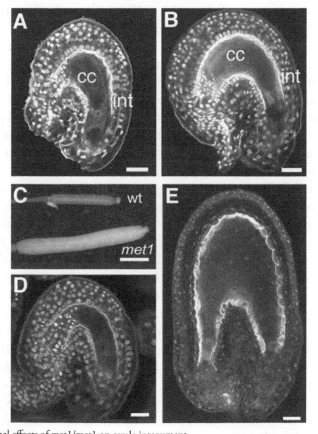

Figure 3. Maternal effects of met1/met1 on ovule integument.

(A) Wild-type ovule at the mature stage shows four or five cell layers of integuments (int) surrounding the central cell (cc). (B) A similar confocal section of a met1/met1 ovule. (C) Fruits from met1-3/met1-3 plants elongate in absence of fertilization (10 Days After Emasculation, (DAE)) in comparison to wild-type fruits. (D) Wild-type ovule with collapsed central cell at 8 DAE. (E) Seed-like structure in elongated fruits from met1-3/met1-3 plants at 8 DAE. Scale bars represent 20 μm (A, B, D and E) and 1.5 mm (C).

Table 3. Morphometric measurements of autonomous fruits and seeds produced by plants deficient for MET1.

Genotype	Fruit elongation at 10 DAP/ 11DAE			Autonomous seed-like structures at 8DAE			Integument cell number in the endothelium at 5 DAE			Integument length at 5 DAE		
	Length (mm)	s.d.	n	%	s.d.	n	#	s.d.	n	Length (μm)	s.d.	n
+/+	3.8	0.2	5	0	0	124	27.6	2.1	8	190.2	23.1	4
met1-3/met1-3	8.5	0.7	8	17.8	3.2	134	43.4	2.7	7	370.5	51.2	4
MET1a/s	8.9	0.3	20	13.1	5.5	355	41.6	2.5	8	366.2	49.5	4

Conclusions

MET1 independently controls both endosperm growth and cell division and elongation of the integuments. Presumably MET1 silences maternal genes in the integuments and restricts seed growth through this maternal sporophytic control. In addition MET1 restricts the expression of imprinted genes in endosperm to the maternal alleles, resulting eventually in a different type of maternal control of endosperm growth. Our results also suggest that a memory of the maternal epigenetic status prior to meiosis is recorded during gametogenesis and influences seed size. Overall the epigenetic control of seed size by MET1 appears to result primarily from maternal controls. These derive directly from the action of MET1 on the sporophytic vegetative tissues and indirectly from the restriction of expression of imprinted inhibitors of seed size to their maternal allele by MET1 acting during male gametogenesis. This conclusion does not support MET1-mediated antagonism between imprinted loci expressed from the paternal or maternal genomes as originally predicted by the parental conflict hypothesis. It is unlikely that CMT3 and DRM2 involved in global de novo DNA methylation control seed size since they do not appear to impact the expression of imprinted genes [25], [26]. However we do not exclude that other epigenetic controls such as histone methylation by Polycomb group complexes [26]–[28] are responsible for an opposite action of the expression between paternally and maternally expressed imprinted genes. In mammals, the function and regulation of some imprinted genes support the parental conflict theory [11], [12], [17], [29]. However some results also suggest a predominant maternal control of placental and embryo growth [30]–[32]. In conclusion, in plants and mammals a complex series of maternal controls balance the unequal parental contributions to the offspring and may mimic a parental conflict without involving symmetrical antagonistic molecular controls.

Materials and Methods

Plant Lines and Growth Conditions

The wild-type control lines C24 and Col were supplied by the ABRC stock center. The line met1a/s (C24) was supplied by J. Finnegan [6]. The line met1-3 (Col) was supplied by J. Paszkowsky and contains a TDNA insert conferring resistance to BASTA [17]. The met1-3 line was maintained as heterozygous by repeated backcrosses to wild-type plants in order to avoid accumulation of epigenetic defects. Once allowed to self, the resulting segregating homozygous plants were used for emasculation for crosses to wild-type plants and for observation of autonomous development.

Plants were grown at 22 C and 60% hygrometry in short days (16 h night) for three weeks followed by long days (8 h night) in Conviron Growth chambers.

Microscopy and Measurements

Developing seeds were cleared with Hoyer's medium and observed with DIC optics with a Leica microscope (DM600). Images were recorded with a Snapshot camera and processed with Metamorph for morphometric measurements. For confocal microscopy, material was prepared and observed as described previously [1].

Experimental Strategy

In order to evaluate the relationship between seed size and parental inheritance of met1 we performed a series of four experiments. We produced crosses between wild type and met1-3/+ plants grown in the same conditions and obtained two populations of 900 seeds with inheritance of met1 from the mother or from the father. We visually separated seeds according to size categories in each population and tested BASTA resistance in a subset representing the largest or smallest seeds. In a second series of crosses we produced crosses between wild type emasculated plants and wild-type or met1-3/+ plants or met1-3/met1-3 plants grown together. The seeds obtained were imaged and seed size was measured as detailed below and the data are reported in Figure 1 and Table 1. We obtained a third series of crosses from single plants in order to have an ideal wild type control to compare seed size with and to establish correlation with BASTA resistance. The dataset is reported in Figure 2.

Statistical Analysis

To determine seed area and height, digital images of seeds on a white background were thresholded in Adobe Photoshop CS2. These black and white images were analyzed by ImageJ. We set a threshold on the grayscale such that the seed appears uniformly black against a white background. The black areas are detected automatically and converted as ellipsoids with the measurement of area and minor axes. To test the differences between the means of two seed populations, both analysis of variation (ANOVA) and the non-parametric Mann-Whitney test (M-W) were employed as certain portions of very small seeds in some experiments may have violated the normality assumption in ANOVA. 1:1 ratios of small and large seeds were tested by the Pearson's χ^2 test (χ^2). Finally, we used the Kalmagorov-Smirnoff Normality test (K-S) to determine whether seed size phenotypes fit a normal distribution based on comparison to a generated ideal normal distribution of similar mean and standard deviation. Calculations were performed using StatView 5.0.1 (SAS Institute, Cary NC), except for χ^2, which was calculated on Excel×(Microsoft). p-values provided in the text are followed by the abbreviation of the test used.

Authors' Contributions

Conceived and designed the experiments: FB AC JF. Performed the experiments: FB JF ML. Analyzed the data: FB AC JF ML. Wrote the paper: FB JF.

References

1. Garcia D, FitzGerald JN, Berger F (2005) Maternal Control of Integument Cell Elongation and Zygotic Control of Endosperm Growth Are Coordinated to Determine Seed Size in Arabidopsis. Plant Cell 17: 52–60.

2. Sundaresan V (2005) Control of seed size in plants. Proc Natl Acad Sci USA 102: 17887–17888.

3. Nowack MK, Grini PE, Jakoby MJ, Lafos M, Koncz C, et al. (2006) A positive signal from the fertilization of the egg cell sets off endosperm proliferation in angiosperm embryogenesis. Nat Genet 38: 63–67.

4. Kermicle JL, Allemand M (1990) Gametic imprinting in maize in relation to the angiosperm life cycle. Development (Suppl) 1: 9–14.

5. Scott RJ, Spielman M, Bailey J, Dickinson HG (1998) Parent-of-origin effects on seed development in Arabidopsis thaliana. Development 125: 3329–3341.

6. Finnegan EJ, Peacock WJ, Dennis ES (1996) Reduced DNA methylation in Arabidopsis thaliana results in abnormal plant development. Proc Natl Acad Sci USA 93: 8449–8454.

7. Adams S, Vinkenoog R, Spielman M, Dickinson HG, Scott RJ (2000) Parent-of-origin effects on seed development in Arabidopsis thaliana require DNA methylation. Development 127: 2493–2502.

8. Garcia D, Saingery V, Chambrier P, Mayer U, Jurgens G, et al. (2003) Arabidopsis haiku mutants reveal new controls of seed size by endosperm. Plant physiology 131: 1661–1670.

9. Luo M, Bilodeau P, Dennis ES, Peacock WJ, Chaudhury A (2000) Expression and parent-of-origin effects for FIS2, MEA, and FIE in the endosperm and embryo of developing Arabidopsis seeds. Proc Natl Acad Sci USA 97: 10637–10642.

10. Feil R, Berger F (2007) Convergent evolution of genomic imprinting in plants and mammals. Trends Genet 23: 192–199.

11. Spielman M, Vinkenoog R, Dickinson HG, Scott RJ (2001) The epigenetic basis of gender in flowering plants and mammals. Trends Genet 17: 705–711.

12. Jullien PE, Kinoshita T, Ohad N, Berger F (2006) Maintenance of DNA Methylation during the Arabidopsis Life Cycle Is Essential for Parental Imprinting. Plant Cell 18: 1360–1372.

13. Kinoshita T, Miura A, Choi Y, Kinoshita Y, Cao X, Jacobsen SE, Fischer RL, Kakutani T (2004) One-way control of FWA imprinting in Arabidopsis endosperm by DNA methylation. Science 303: 521–3.

14. Mathieu O, Reinders J, Caikovski M, Smathajitt C, Paszkowski J (2007) Trans-generational stability of the Arabidopsis epigenome is coordinated by CG methylation. Cell 130: 851–862.

15. Xiao W, Brown RC, Lemmon BE, Harada JJ, Goldberg RB, et al. (2006) Regulation of seed size by hypomethylation of maternal and paternal genomes. Plant Physiol 142: 1160–1168.

16. Tariq M, Saze H, Probst AV, Lichota J, Habu Y, et al. (2003) Erasure of CpG methylation in Arabidopsis alters patterns of histone H3 methylation in heterochromatin. Proc Natl Acad Sci U S A 100: 8823–8827.

17. Saze H, Scheid OM, Paszkowski J (2003) Maintenance of CpG methylation is essential for epigenetic inheritance during plant gametogenesis. Nat Genet 34: 65–69.

18. Loppin B, Lepetit D, Dorus S, Couble P, Karr TL (2005) Origin and neofunctionalization of a Drosophila paternal effect gene essential for zygote viability. Curr Biol 15: 87–93.

19. Fitch KR, Yasuda GK, Owens KN, Wakimoto BT (1998) Paternal effects in Drosophila: implications for mechanisms of early development. Curr Top Dev Biol 38: 1–34.

20. Browning H, Strome S (1996) A sperm-supplied factor required for embryogenesis in C. elegans. Development 122: 391–404.

21. Canales C, Bhatt AM, Scott R, Dickinson H (2002) EXS, a putative LRR receptor kinase, regulates male germline cell number and tapetal identity and promotes seed development in Arabidopsis. Curr Biol 12: 1718–1727.

22. Schruff MC, Spielman M, Tiwari S, Adams S, Fenby N, et al. (2006) The AUXIN RESPONSE FACTOR 2 gene of Arabidopsis links auxin signalling, cell division, and the size of seeds and other organs. Development 133: 251–261.

23. Chaudhury AM, Ming L, Miller C, Craig S, Dennis ES, et al. (1997) Fertilization-independent seed development in Arabidopsis thaliana. Proc Natl Acad Sci USA 94: 4223–4228.

24. Guitton AE, Berger F (2005) Loss of function of MULTICOPY SUPPRESSOR OF IRA 1 produces nonviable parthenogenetic embryos in Arabidopsis. Curr Biol 15: 750–754.

25. Kinoshita T, Miura A, Choi Y, Kinoshita Y, Cao X, et al. (2004) One-way control of FWA imprinting in Arabidopsis endosperm by DNA methylation. Science 303: 521–523.

26. Gehring M, Huh JH, Hsieh TF, Penterman J, Choi Y, et al. (2006) DEMETER DNA Glycosylase Establishes MEDEA Polycomb Gene Self-Imprinting by Allele-Specific Demethylation. Cell 124: 495–506.

27. Jullien PE, Katz A, Oliva M, Ohad N, Berger F (2006) Polycomb Group Complexes Self-Regulate Imprinting of the Polycomb Group Gene MEDEA in Arabidopsis. Curr Biol 16: 486–492.

28. Makarevich G, Leroy O, Akinci U, Schubert D, Clarenz O, et al. (2006) Different Polycomb group complexes regulate common target genes in Arabidopsis. EMBO Rep 7: 947–952.

29. Haig D (2004) Genomic imprinting and kinship: how good is the evidence? Annu Rev Genet 38: 553–585.

30. Ferguson-Smith AC, Moore T, Detmar J, Lewis A, Hemberger M, et al. (2006) Epigenetics and imprinting of the trophoblast - a workshop report. Placenta 27 Suppl 122–126.

31. Constancia M, Angiolini E, Sandovici I, Smith P, Smith R, et al. (2005) Adaptation of nutrient supply to fetal demand in the mouse involves interaction between the Igf2 gene and placental transporter systems. Proc Natl Acad Sci USA 102: 19219–19224.

32. Lin SP, Youngson N, Takada S, Seitz H, Reik W, et al. (2003) Asymmetric regulation of imprinting on the maternal and paternal chromosomes at the Dlk1-Gtl2 imprinted cluster on mouse chromosome 12. Nat Genet 35: 97–102.

CITATION

Originally published under the Creative Commons Attribution License. FitzGerald J, Luo M, Chaudhury A, Berger F (2008) DNA Methylation Causes Predominant Maternal Controls of Plant Embryo Growth. PLoS ONE 3(5): e2298. doi:10.1371/journal.pone.0002298.

Gibberellin Acts Through Jasmonate to Control the Expression of MYB21, MYB24, and MYB57 to Promote Stamen Filament Growth in Arabidopsis

Hui Cheng, Susheng Song, Langtao Xiao, Hui Meng Soo,
Zhiwei Cheng, Daoxin Xie and Jinrong Peng

ABSTRACT

Precise coordination between stamen and pistil development is essential to make a fertile flower. Mutations impairing stamen filament elongation, pollen maturation, or anther dehiscence will cause male sterility. Deficiency in plant hormone gibberellin (GA) causes male sterility due to accumulation of

DELLA proteins, and GA triggers DELLA degradation to promote stamen development. Deficiency in plant hormone jasmonate (JA) also causes male sterility. However, little is known about the relationship between GA and JA in controlling stamen development. Here, we show that MYB21, MYB24, and MYB57 are GA-dependent stamen-enriched genes. Loss-of-function of two DELLAs RGA and RGL2 restores the expression of these three MYB genes together with restoration of stamen filament growth in GA-deficient plants. Genetic analysis showed that the myb21-t1 myb24-t1 myb57-t1 triple mutant confers a short stamen phenotype leading to male sterility. Further genetic and molecular studies demonstrate that GA suppresses DELLAs to mobilize the expression of the key JA biosynthesis gene DAD1, and this is consistent with the observation that the JA content in the young flower buds of the GA-deficient quadruple mutant ga1-3 gai-t6 rga-t2 rgl1-1 is much lower than that in the WT. We conclude that GA promotes JA biosynthesis to control the expression of MYB21, MYB24, and MYB57. Therefore, we have established a hierarchical relationship between GA and JA in that modulation of JA pathway by GA is one of the prerequisites for GA to regulate the normal stamen development in Arabidopsis.

Introduction

Arabidopsis flowers are organized into four concentric whorls of distinct organs (sepals, petals, stamens and pistils). Stamens, the male reproductive organs of flowering plants, form the third whorl. Processes of stamen filament elongation and anthesis are precisely controlled so that they coincide with the pistil development to determine the fertility [1]. Mutations that impair stamen development such as filament elongation, pollen maturation or anther dehiscence will result in male sterility [2],[3]. Many genes have been found to control stamen development [4],[5]. Stamen development is also subjected to hormonal control. For example, mutations affecting biosynthesis of two plant hormones gibberellin (GA) (e.g ga1-3 mutation) and jasmonate (JA) (e.g opr3 mutation) both confer male sterile phenotype due to failure of stamen filament elongation and of completion of anthesis and anther dehiscence [6],[7].

A severe Arabidopsis GA-deficient mutant, ga1-3 exhibits retarded growth at both vegetative and reproductive stages [7]. The development of floral organs, especially petals and stamens, is impaired in the ga1-3 mutant. Detailed anatomical analysis showed that the male sterile phenotype of ga1-3 is due to the arrestment of stamen filament cell elongation and failure of completion of anthesis [8]. Application of exogenous GA can restore all the floral defects of ga1-3 [7]. Further studies revealed that the arrested floral development in ga1-3 is mediated by DELLA proteins [8],[9]. DELLAs are a subfamily of the plant GRAS family of putative transcription regulators [10],[11] and have been revealed to function as negative regulators of GA response in diverse plant species including Arabidopsis, barley, rice and wheat etc [12]–[17]. There are five DELLAs in Arabidopsis, namely GAI, RGA, RGL1, RGL2 and RGL3 [18],[19]. Genetic studies have shown that RGA, RGL2 and RGL1 act synergistically in repressing petal and stamen development and GA triggers the degradation of these DELLAs to promote floral development [8], [9], [20]–[22]. Severe JA deficient mutant opr3 and JA-signaling mutant coi1 also displayed retarded filament elongation, delayed anther dehiscence, and reduced pollen viability. As a consequence, the opr3 and coi1 mutants are male sterile [6],[23]. Application of exogenous JA can fully restore the stamen development to opr3 [6].

It is intriguing to know whether GA-mediated and JA-mediated stamen development are via two parallel pathways or in a hierarchical way to control stamen development. In Arabidopsis, the known GA-response genes encoding transcription factors involved in stamen development are GAMYBs (MYB33 and MYB65), a subset of MYB genes [24]. GAMYB is the best characterized GA-regulated transcription factor and was first identified in barley. GAMYB was found to bind to the GA-response elements (GARE) in the promoter of the ?–amylase gene in cereals [25],[26]. Genetic studies showed that Arabidopsis GAMYBs (MYB33 and MYB65) are essential to anther maturation but not for the elongation of stamen filament in Arabidopsis [24]. Previous studies have shown that GA regulates GAMYB through DELLA protein SLN1 and SLR1 in barley and rice, respectively [27],[28]. However, several reports failed to identify MYB33 and MYB65 as GA-inducible genes in Arabidopsis and these two MYB genes are in fact regulated at the post-transcriptional level by miRNA159 [24], [29]–[31]. Two recent reports showed that three MYB genes (MYB21, MYB24 and MYB108) are responsive to JA treatment in opr3 mutant and loss-of-function of MYB21 and MYB24 resulted in a short stamen phenotype [32] whereas MYB108 is involved in stamen and pollen maturation but not stamen filament elongation [33]. Interestingly, in an expression profiling study, we identified several MYBs including MYB21, MYB24, and MYB57 as DELLA-downregulated genes in ga1-3 flower buds [30]. This fact prompted us to investigate if there might be a cross-talk between GA signaling and JA signaling during stamen development.

MYB21 and MYB24 have been shown to be expressed in all four whorls of the flower [32],[34],[35]. In this report, we showed that MYB21, MYB24, and MYB57 are down-regulated in the ga1-3 single mutant and the sterile quadruple mutant ga1-3 gai-t6 rga-t2 rgl1-1 (loss-of-function of GAI, RGA, RGL1 three DELLA genes but RGL2 is normal) but restored to wild type levels in the fertile penta mutant ga1-3 gai-t6 rga-t2 rgl1-1 rgl2-1 (loss-of-function of GAI, RGA,

RGL1 and RGL2 four DELLA genes). We also showed that absence of the four DELLAs (GAI, RGA, RGL1 and RGL2) cannot suppress the short stamen phenotype conferred by the loss-of-function of MYB21 and MYB24. In addition, we observed that application of exogenous JA onto the ga1-3 gai-t6 rgl1-1 rgl2-1 quadruple mutant flower buds could restore the expression of MYB21, MYB24 and MYB57 whereas application of exogenous GA onto opr3 mutant flower buds failed to increase the expression of these three MYBs. Most importantly, we showed that GA upregulates JA-biosynthetic genes DAD1 and LOX1 and the JA content in the young flower buds of the GA-deficient quadruple mutant ga1-3 gai-t6 rga-t2 rgl1-1 is much lower than that in the WT and penta mutant. Therefore, we conclude that GA upregulates the DAD1 and LOX1 expression to promote JA production to promote the expression of the three MYBs necessary for stamen filament development.

Results

Identification of DELLA-Repressed Stamen-Enriched Genes

The ga1-3 mutant is retarded in floral development, suggesting that the transcriptome for floral development in the ga1-3 mutant must be kept at a repressive state. Conversely, the fact that the ga1-3 gai-t6 rga-t2 rgl1-1 rgl2-1 mutant (penta mutant) confers GA independent flowering suggests that the transcriptome responsible for floral development must have been constitutively activated in the penta mutant. We compared the expression profiles between ga1-3 and ga1-3 gai-t6 rga-t2 rgl1-1 rgl2-1 and identified 360 DELLA-repressed and 273 DELLA-activated genes essential for floral development [30]. To identify DELLA-repressed stamen-enriched genes, we examined expression of 43 DELLA-repressed genes in the sepal, petal, stamen and pistil via semi-quantitative RT-PCR. These 43 genes were chosen based on two criteria: 1) they are homologous to transcription factors known to regulate GA-response (e.g MYB gene family) and 2) genes whose expression showed drastic changes between the ga1-3 and penta mutant [30]. Only genes whose expression are either enriched in the stamen or highly expressed in the stamen and also in some other floral organs but not ubiquitously highly expressed in all four floral organs were classified as the stamen-enriched genes. A total of 34 genes, including two APG-like genes (At1g75880, At1g75900) and three genes (IRX1, IRX3, IRX5) encoding the cellulose synthase subunits which are known to be enriched in the stamen, were identified as DELLA-repressed stamen-enriched genes (Figure 1; Table 1).

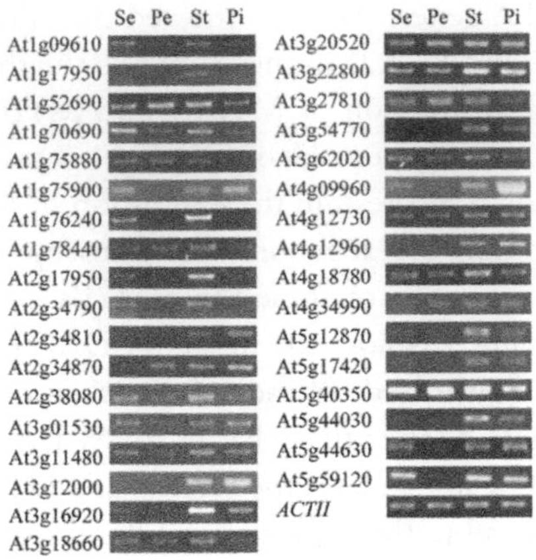

Figure 1. Identification of DELLA-Repressed Stamen-Enriched Genes.

At least three independent samples were used for RT-PCR analysis for each individual gene and a representative gel picture for each gene was shown here. Total 34 genes were identified as DELLA-repressed stamen-enriched genes (summarized in Table 1) based on their relative more abundant expression in the stamen than in one or more of the rest of the floral organs.

Table 1. RT-PCR examination of DELLA-down genes in different floral organs.

Gene ID	Gene description	DELLA-Down
At1g09610	Hypothetical protein	confirmed
At1g17950	MYB52	confirmed
At1g52690	Late embryogenesis abundant protein	confirmed
At1g70690	Unknown	confirmed
At1g75880	APG-like	confirmed
At1g75900	APG-like	confirmed
At1g76240	Hypothetical protein	confirmed
At1g78440	Gibberellin 2-oxidase	confirmed
At2g17950	Homeodomain transcription factor	confirmed
At2g34790	Berberine bridge enzyme	confirmed
At2g34810	Berberine bridge enzyme	confirmed
At2g34870	Unknown	confirmed
At2g38080	Putative diphenol oxidase	confirmed
At3g01530	MYB57	confirmed
At3g11480	Hypothetical	confirmed
At3g12000	S-locus related	confirmed
At3g15270	Squamose promoter binding 5	confirmed
At3g16920	Chitinase(GHF19)	confirmed
At3g18660	Hypothetical protein	confirmed
At3g20520	Hypothetical protein	confirmed
At3g22800	Extensin-like	confirmed
At3g27810	MYB21	confirmed
At3g54770	RNA binding protein	confirmed
At3g62020	Germin-like protein	confirmed
At4g12730	Putative pollen surface protien	confirmed
At4g12960	Unknown	confirmed
At4g18780	Cellulose synthase (IRX1)	confirmed
At4g34990	MYB32	confirmed
At5g12870	MYB46	confirmed
At5g17420	Cellulose synthase (IRX3)	confirmed
At5g40350	MYB24	confirmed
At5g44030	Cellulose synthase (IRX5)	confirmed
At5g44630	Terpene synthase	confirmed
At5g59120	Subtilisin-like serine protease	confirmed

DELLAs Repress the Expression of MYB21, MYB24, and MYB57

Three MYB genes, namely MYB21, MYB24 and MYB57, were among the identified DELLA-repressed stamen-enriched genes (Figure 1; Table 1). Based on the phylogenetic tree, MYB24 and MYB21 are classified into the subgroup 19 of R2R3-MYB family [36]. MYB57 shares high similarity with this subfamily and is a close member to this subfamily [37]. Overall, MYB21 shares 61.6% and 51.0% identity with MYB24 and MYB57 at the amino acid level, respectively. The expression of these three MYBs in the young flower buds were reduced to a very low level in ga1-3 but restored to the wild type (WT) level in the ga1-3 gai-t6 rga-t1 rgl1-1 rgl2-1 penta mutant (Figure 2A). In order to find out which DELLA (RGL1, RGL2, RGA and GAI) is more effective in repressing the expression of

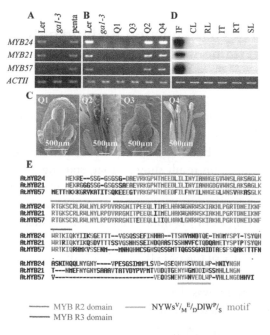

Figure 2. MYB21, MYB24I, and MYB57 Are RGA- and RGL2-Repressible Floral Specific Genes.

(A) RT-PCR analysis shows that the expression of the three MYB genes in the young flower buds are greatly reduced in ga1-3 but restored to the WT level in ga1-3 gai-t6 rgat2 rgl1-1 rgl2-1 (penta). (B–C) RT-PCR analysis shows that the repressed expression of the three MYB genes in ga1-3 was restored in ga1-3 gai-t6 rga-t2 rgl2-1 (Q2) and ga1-3 rga-t2 rgl1-1 rgl2-1 (Q4) two quadruple mutants but not in ga1-3 gai-t6 rgl1-1 rgl2-1 (Q1) and ga1-3 gai-t6 rga-t2 rgl1-1 (Q3) two quadruple mutants (B). This restoration of MYB expression nicely correlates with the recovery of fertility in Q2 and Q4 (C). Total RNA used in RT-PCR analysis was extracted from the young flower buds. (D) RT-PCR analysis shows that the three MYB genes are floral specific genes. IF, inflorescence; CL, cauline leaves; RL, rosette leaves; IT, internodes; RT, roots; SL, siliques. (E) Amino acid alignment of MYB21, MYB24 and MYB57 proteins. The conserved R2 and R3 domains and the NYWSV/ME/DDIWP/S motif are highlighted in red, blue and green, respectively.

MYB21, MYB24 and MYB57, transcript levels of each individual MYB gene were studied in four quadruple mutants in which only one of the four DELLA genes remains intact. All three MYB genes were almost undetectable in the Q1 (ga1-3 gai-t6 rgl1-1 rgl2-1, wild type for RGA) and barely detectable in the Q3 (ga1-3 gai-t6 rgl1-1 rga-t2, wild type for RGL2) mutants but were detected at high levels in the Q2 (ga1-3 rga-t2 rgl1-1 rgl2-1, wild type for GAI) and Q4 (ga1-3 gai-t6 rga-t2 rgl2-1, wild type for RGL1) mutants (Figure 2B), suggesting that RGA and RGL2, but not GAI nor RGL1, were the more effective DELLAs in repressing the expression of these three MYB genes. Interestingly, we showed previously that while Q1 and Q3 mutants, as the ga1-3 mutant, were retarded in floral development both Q2 and Q4 mutants produced normal fertile flowers (Figure 2C) [8]. Therefore, it seems there is a nice correlation between normal floral development and the expression of MYB21, MYB24 and MYB57, suggesting that these three MYBs are probably necessary for normal floral development.

MYB21, MYB24, and MYB57 Function Redundantly in Controlling Stamen Filament Elongation

Expression analysis showed that MYB21 and MYB24 [32],[34] as well as MYB57 are flower-specific genes (Figure 2D). To determine if the spatial and temporal expression patterns of MYB21 and MYB24 correlate with their proposed role during stamen filament elongation, we examined MYB21 expression via in situ hybridization and generated pMYB24:GUS transgenic for examining MYB24 expression. Our in situ hybridization result showed that, starting from floral stage 12 [1],[38], MYB21 is expressed in the anther vascular tissue and in cells at the junction between anther and stamen filament where rapid filament elongation is hypothesized to occur starting from the floral stage 13 after a successful pollination [1]. MYB21 expression is also detected in the nectaries and ovules. Similarly, staining the young inflorescence of the pMYB24::GUS plants revealed that strong GUS activity was detected in the vascular tissue of stamen filament and sepals whereas only weak GUS activity was detectable in the petals starting from floral stage 12. GUS activity was also detected in the upper part of the pistils.

To investigate their roles in GA-mediated floral organ development, we identified T-DNA insertional mutant lines corresponding to these three MYB genes from the Salk Institute Genomic Arabidopsis Laboratory (SIGnAL) database. Mutant alleles were confirmed (data not shown) and designated as myb21-t1 (SALK_042711) for MYB21, myb24-t1 (SALK_017221) for MYB24, and myb57-t1 (SALK_065776) for MYB57 (Figure 3A). myb24-t1 and myb57-t1 are both likely null alleles since MYB24 and MYB57 transcripts were undetectable in myb24-t1 and myb57-t1 mutant flower buds, respectively (Figure 3B). On the other hand, MYB21 transcripts were still detectable in myb21-t1 although its level was greatly reduced in the mutant, suggesting that myb21-t1 is likely a leaky allele (Figure 3B). After two rounds of backcross, we found that myb24-t1 and myb57-t1 mutant plants were phenotypically indistinguishable from the WT control plant (Figure 3C; Table 2). However, in myb21-t1 the early developed flowers (~the first 10 flowers) bore short stamens with greatly reduced fertility and only the late developed flowers yielded proper seed settings (Figure 3C; Table 2). A close look at the matured early flowers in myb21-t1 showed that the stamens did produce pollens (Figure 3D, panel d). Cross-pollinating the pollens onto the myb21-t1 stigma yielded seeds that were homozygous for myb21-t1 and onto the WT stigma yielded myb21-t1 heterozygous seeds (data not shown), demonstrating that the short stamen is responsible for the partial sterile phenotype. Although myb21-t1 is likely a leaky allele, the short stamen phenotype conferred by the myb21-t1 mutation is identical to a MYB21 null allele we obtained later from Gabi-Kat stock (stock number N311167, data not shown).

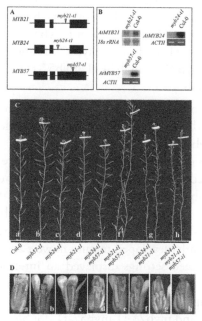

Figure 3. MYB21, MYB24, and MYB57 Function Redundantly in Regulating the Stamen Filament Development.

(A) Schematic diagram shows the respective T-DNA insertions in the three MYB genes. Black box: exon; black line: intron; triangle: T-DNA insertion site. (B) RT-PCR analysis of MYB24 transcripts in myb24-t1 and MYB57 transcripts in myb57-t1 and northern analysis of MYB21 transcripts in myb21-t1. Total RNA for RT-PCR and northern analysis was extracted from the young flower buds. (C) Comparison of main shoots bearing siliques among different mutant lines as indicated. (D) Comparison of the stamen phenotype among different mutant lines as indicated. Genotypes for flowers a-h in (D) corresponds to that showed in (C).

Table 2. Seed settings in different mutants grown under LD condition.

Col-*0*/mutant	Number of Siliques	Number of Siliques With Seeds	Percentage of Siliques With Seeds
Col-0	21.1±5.1	20.7±5.0	98.1±2.7
myb57	22.3±5.7	20.3±6.6	89.4±1.27
myb24	28.1±3.8	27.3±3.8	97.3±2.9
myb21	26.3±4.8	17.1±5.4	64.2±10.7
myb24myb57	26.3±6.9	22.6±6.8	85.1±6.3
myb21myb57	30.6±6.4	9.4±6.5	29.5±17.3
myb21myb24	30.8±9.5	5.6±4.3	16.9±11.8
myb21myb24myb57	33.4±7.5	1.6±1.5	4.1±3.6

The WT-like phenotype displayed by myb24-t1 and myb57-t1 and mild floral phenotype displayed by myb21-t1 suggest that these MYB genes might function redundantly during stamen development. To prove this hypothesis, crosses were made among homozygous myb24-t1, myb21-t1 and myb57-t1 plants. Three double mutants (myb21-t1 myb24-t1, myb21-t1 myb57-t1, and myb24-t1 myb57-t1) and one triple mutant (myb21-t1 myb24-t1 myb57-t1) were generated and used in our phenotypic analysis.

The flower development of myb24-t1 myb57-t1 double mutant at all stages was indistinguishable from the WT control (Figure 3D; Table 2) [1]. Stamens in mature flowers of the myb21-t1 myb24-t1 double mutant were shorter than that of the myb21-t1 single mutant and shorter stamens were also observed in majority of the late developed mature flowers in the double mutant. As a result, the myb21-t1 myb24-t1 double mutant is more severely sterile than myb21-t1 by having fewer siliques with seed settings (Figure 3C and 3D; Table 2), an observation also reported by Mandaokar et al [32]. The myb21-t1 myb57-t1 double mutant had shorter stamens in early developed mature flowers and some of the later flowers (Figure 3D) and its seed settings displayed an intermediate phenotype between myb21-t1 single and myb21-t1 myb24-t1 double mutants (Figure 3C and 3D; Table 2). Interestingly, cross-pollination showed that the short stamens in both myb24-t1 myb21-t1 and myb21-t1 myb57-t1 double mutant plants produced viable pollens (data not shown), suggesting that the short stamen is responsible for the reduced fertility in these mutants.

The myb21-t1 myb24-t1 myb57-t1 triple mutant, as the myb21-t1 myb24-t1 double mutant, had short stamens but was even more severely sterile than myb21-t1 myb24-t1 (Figure 3C and 3D; Table 2). For myb21-t1 myb24-t1 double and myb21-t1 myb24-t1 myb57-t1 triple mutant plants, we occasionally observed that while, in the same inflorescence, most of the flowers did not set or set very few seeds, some were able to develop normal siliques filled up with seeds (Figure 3C and 3D). Cross pollination showed the pollens produced by myb21-t1 myb24-t1 myb57-t1 triple mutant plants were partial viable (data not shown), suggesting the short stamens in the triple mutants are the main cause of the sterility. It is possible that environmental factors (e.g. temperature) may influence male fertility in these mutants, an observation also reported for MYB33 and MYB65 [24]. Therefore, MYB21, MYB24 and MYB57 function redundantly to control the stamen filament development in the late developed flowers.

Since MYB21 and MYB24 are also expressed in sepals and petals, we examined the sepal and petal development in the single, double and triple myb mutants. As shown in Figure 3D, sepal development appeared normal in all mutants whereas petal development varied in different mutants. Petals in the myb24-t1 and myb57-t1 two single mutants grew to a final length longer than the pistils, as that did the WT petals. Petals in the myb21-t1 single, myb24-t1 myb57-t1 and myb21-t1 myb57-t1 two double mutants grew to a final height parallel to the pistil (Figure 3D). Petals in the myb21-t1 myb24-t1 double mutant grew just out of the sepals but ended at a lower level than the stigma (Figure 3D). The growth of petals in the myb21-t1 myb24-t1 myb57-t1 triple mutant was arrested and the petals never grew out of the sepals (Figure 3D).

myb21-t1 myb24-t1 Is Epistatic to gai-t6 rga-t2 rgl1-1 rgl2-1 in Controlling Stamen Filament Elongation

MYB21 and MYB24 were repressed in ga1-3 but their expressions were restored to the WT level in the ga1-3 gai-t6 rga-t2 rgl1-1 rgl2-1 penta mutant, suggesting that GA regulates MYB21 and MYB24 through inactivating DELLA proteins. Application of exogenous GA could not rescue the stamen development in myb21 myb24 mutant (data not shown), suggesting that MYB21 and MYB24 are needed in GA-mediated stamen development. To further confirm this hypothesis, we crossed myb21 myb24 with ga1-3 gai-t6 rga-t2 rgl1-1 rgl2-1 to generate two hexa mutants (hexa1: ga1-3 gai-t6 rga-t2 rgl1-1 rgl2-1 myb21-t1; hexa2: ga1-3 gai-t6 rga-t2 rgl1-1 rgl2-1 myb24-t1) and one hepta mutant (ga1-3 gai-t6 rga-t2 rgl1-1 rgl2-1 myb21-t1 myb24-t1). The two hexa mutants overall appeared similar to each other and had wildtype-like stamens and were largely fertile (Figure 4A). The hepta mutant plant displayed no difference from the penta mutant plant in its vegetative growth. However, its mature flowers showed a short filament phenotype identical to that in the myb21-t1 myb24-t1 double mutant (Figure 4A). This observation demonstrated that myb21-t1 myb24-t1 double mutations are epistatic to DELLA mutations. SEM analysis showed that the short stamen phenotype in the hepta mutant was due to reduced cell length (Figure 4B and 4C), rather than to a reduction in cell number (Figure 4D). Therefore, MYB21 and MYB24 act downstream of DELLAs in GA signaling pathway to control the stamen filament development.

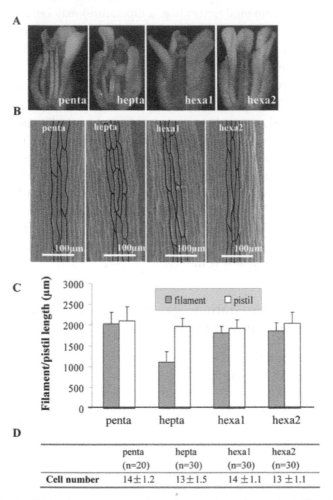

Figure 4. myb21-t1 myb24-t1 Is Epistatic to ga1-3 gai-t6 rga-t2 rgl1-1 rgl2-1 in Controlling Stamen Filament Elongation.

(A) Comparison of the stamen phenotype among ga1-3 gai-t6 rga-t2 rgl1-1 rgl2-1(penta), ga1-3 gai-t6 rga-t2 rgl1-1 rgl2-1 myb21-t1 myb24-t1 (hepta), ga1-3 gai-t6 rga-t2 rgl1-1 rgl2-1 myb21-t1 (hexa1) and ga1-3 gai-t6 rga-t2 rgl1-1 rgl2-1 myb24-t1 (hexa2). (B) SEM of stamen filament epidermal cells in the penta, hepta, hexa1 and hexa2 mutants. Segments shown were all from the middle part of the filament. Some individual cells were outlined with black lines for easy visualization. (C) Comparison of stamen and pistil lengths among different genotypes. Filament and pistil lengths were measured from SEM pictures (n = 30). (D) Average number of epidermal cells per stamen filament in penta, hepta, hexa1 and hexa2. n: number of stamens used in counting.

GA Application Fails to Induce the Expression of MYB21, MYB24 and MYB57 in JA-Deficient Mutant

We showed in the above that the expression of MYB21, MYB24 and MYB57 was repressed in the ga1-3 gai-t6 rga-t2 rgl1-1 quadruple mutant (wild type for RGL2) but restored to normal in the ga1-3 gai-t6 rga-t2 rgl1-1 rgl2-1 penta mutant (Figure 2A and 2B). Mandaokar et al reported that the expression of MYB21 and MYB24 was downregulated in opr3 mutant and application of exogenous JA could restore their expression [32]. These results suggest that there might be a crosstalk between the GA and JA pathways in regulating the expression of MYB21, MYB24 and MYB57 during stamen development. Genetically, there are three possible ways of interaction between GA and JA. Firstly, GA might act through the JA pathway to regulate the expression of these MYB genes. In this case it is expected that JA application onto ga1-3 gai-t6 rga-t2 rgl1-1 would induce the expression of MYB21, MYB24 and MYB57 whilst GA application onto opr3 would have no effect on their expression. Conversely, JA may act upstream of the GA pathway to regulate the expression of these three MYB genes. In this case, GA application onto opr3 would induce whilst JA application onto ga1-3 gai-t6 rga-t2 rgl1-1 would have no effect on the expression of MYB21, MYB24 and MYB57. The third possibility is that GA and JA may not act in a hierarchical manner but rather via parallel pathways to regulate the expression of the three MYB genes. If this is the case, GA application onto opr3 and JA application onto ga1-3 gai-t6 rga-t2 rgl1-1 would probably both induce the expression of the three MYB genes. To find out which is the likely case, we first examined the effect of GA application on JA-deficient mutant opr3 and found that GA application failed to rescue the opr3 mutant phenotype and failed to induce the expression of MYB21, MYB24 and MYB57 in opr3 even at 96 hrs after GA treatment (Figure 5A). Failure in induction of expression of MYB21, MYB24 and MYB57 in GA-treated opr3 mutants could be due to inactivation of GA signaling in JA-deficient background. GA3ox1 and GA20ox2 are two key genes that contribute to the biosynthesis of bioactive GA and these two genes are under negative feedback regulation by GA signaling pathway (GA-down) [30]. On the other hand, GA2ox1 is a GA-up gene responsible for GA catabolism [30]. Examination of the GA3ox1 and GA20ox2 and GA2ox1 expression in GA-treated opr3 mutants showed expected GA-response (Figure 5B). Meanwhile, expression of GA3ox1 and GA2ox1 appeared normal in opr3 (Figure 6A). These results suggest that JA-deficiency specifically blocks the GA-signaling leading to the induction of MYB21, MYB24 and MYB57 expression but not the negative feedback pathway for GA-biosynthesis.

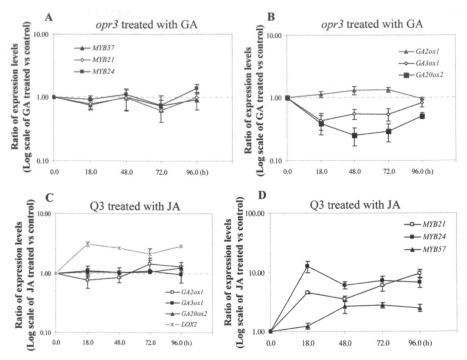

Figure 5. JA-Deficiency Specifically Blocks GA-Signaling Leading to the Induction of Expression of MYB21, MYB24, and MYB57.

(A–B) Semi-quantitative analysis of MYB21, MYB24, MYB57 (A), GA2ox1, GA3ox1 and GA20ox2 (B) expression in the opr3 mutant flowers at 18, 48, 72 and 96 hrs after GA treatment. Data were averaged from 2–4 batches of independently treated samples and ACTII was used as the normalization control. The graph was drawn based on Log10 scale of the ratio of the expression levels of GA treated versus untreated samples. (C–D) Semi-quantitative analysis of LOX2 (in red line), GA2ox1, GA3ox1 and GA20ox2 (C), MYB21, MYB24 and MYB57 (D) expression in the ga1-3 gai-t6 rga-t2 rgl1-1 (Q3) mutant flowers at 18, 48, 72 and 96 hrs after JA treatment. Data were averaged from 2–4 batches of independently treated samples and ACTII was used as the normalization control. The graph was drawn based on Log10 scale of the ratio of the expression levels of JA treated versus untreated samples.

JA Application Restores the Expression of MYB21, MYB24 and MYB57 in GA-Deficient Mutant

We then studied the effect of JA application on ga1-3 gai-t6 rga-t2 rgl1-1 (GA-deficient) by examining the expression of MYB21, MYB24 and MYB57 in the young flower buds at 18, 48, 72 and 96 hrs post-treatment. As expected, LOX2, a JA-response gene, was strongly upregulated by JA application at 18 hrs post treatment (Figure 5C) [39]. Interestingly, we observed that JA-treatment induced high expression of MYB21 and MYB24 and weak expression of MYB57 in the ga1-3 gai-t6 rga-t2 rgl1-1 quadruple mutant at 18 hrs post treatment (Figure 5D). However, examination of GA3ox1 and GA20ox2 (two GA-down genes) and GA2ox1 (GA-up gene) showed that JA treatment did not obviously change the expression patterns of these three GA-response genes in the ga1-3 gai-t6 rga-t2 rgl1-1 quadruple mutant (Figure 5C). These data suggested that JA signaling might mediate a specific branch of GA signaling to regulate the expression of the three MYB genes.

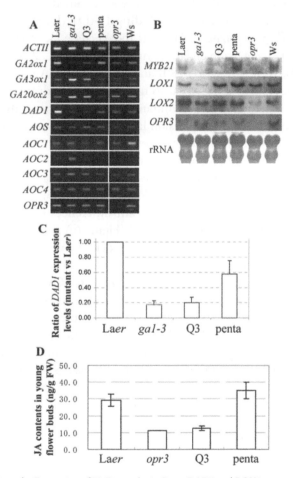

Figure 6. GA Regulates the Expression of JA Biosynthesis Genes DAD1 and LOX1.

(A) RT-PCR analysis of GA- and JA-biosynthesis genes in the young flower buds of La-er WT control, ga1-3, ga1-3 gai-t6 rga-t2 rgl1-1 (Q3), ga1-3 gai-t6 rga-t2 rgl1-1rgl2-1 (penta), opr3 (in WS background) and Ws WT control. (B) Northern analysis of MYB21, GA20ox2, LOX1, LOX2 and OPR3 in the young flower buds of the La-er WT control, ga1-3, Q3, penta, opr3 and Ws WT control. (C) Semi-quantitative analysis of DAD1 expression in ga1-3, Q3 and penta relative to that in WT (Laer), respectively. Data were averaged from three independent batches of samples and ACTII was used as the normalization control. The expression level of WT is set as 1. (D) Comparison of JA contents among WT (La-er), opr3, ga1-3 gai-t6 rga-t2 rgl1-1 (Q3) and ga1-3 gai-t6 rga-t2 rgl2-1 rgl1-1 (penta). For WT and the penta mutant JA contents were averaged from four repeats. For the Q3 mutant, JA was detected in three out of the four repeats. For opr3, JA was detected only in one out of the four repeats. FW, fresh weight.

GA Suppresses DELLA to Upregulate the JA-Biosynthesis Gene LOX1 and DAD1

Considering the fact that JA application was able to induce the expression of MYBs in the ga1-3 gai-t6 rga-t2 rgl1-1 mutant it is reasonable to argue that JA biosynthesis, instead of JA signaling pathway, is likely affected in the ga1-3 gai-t6 rga-t2 rgl1-1 mutant. To test this hypothesis, we examined the expression of known or putative JA biosynthesis genes including DAD1 (Defective in anther dehiscence 1), LOX1 (Lipoxygenase 1), LOX2 (Lipoxygenase 2), AOS (Allene oxide synthase), AOC1 (Allene oxide cyclase 1, At3g25760), AOC 2 (At3g25770), AOC 3 (At3g25780), AOC 4 (At1g13280) and OPR3 (OPDA reductase 3) in La-er WT, Ws WT, ga1-3 single, ga1-3 gai-t6 rga-t2 rgl1-1 quadruple, ga1-3 gai-t6 rga-t2 rgl1-1 rgl2-1 penta, and opr3 mutants. We found that only MYB21, LOX2 and AOC1 showed reduced expression in the opr3 mutant whereas all the other genes, including GA-biosynthesis genes, expressed similarly in the opr3 mutant and Ws WT control (Figure 6A and 6B), suggesting that JA-deficiency does not affect GA biosynthesis. Expression of these genes in GA-related mutants was more complicated. We found that the expression levels of AOS, AOC1, AOC3, AOC4, LOX2 and OPR3 did not show significant differences in all GA-related mutants when compared to the La-er WT control (Figure 6A and 6B), suggesting these genes are probably regulated in a GA-independent fashion. The expression of AOC2 was obviously induced in ga1-3 and then reduced to the WT level in the quadruple and penta mutants (Figure 6A), suggesting AOC2 is a GA-down gene. In contrast, the expression of LOX1 was significantly reduced in ga1-3 but was restored both in the ga1-3 gai-t6 rga-t2 rgl1-1 quadruple and penta mutants (Figure 6B), suggesting that although LOX1 is a GA-up gene and its expression is not repressed by RGL2. Interestingly, DAD1 expression was found to be down-regulated to approximately 20% of the WT level in both ga1-3 and the ga1-3 gai-t6 rga-t2 rgl1-1 quadruple mutant whereas was restored to approximately 60% of the WT level in the penta mutant (approximately three folds increase in penta versus Q3) (Figure 6A and 6C), indicating that GA may regulate DAD1 expression via suppression of RGL2.

JA Levels Are Greatly Reduced in the Young Flower Buds of the ga1-3 gai-t6 rga-t2 rgl1-1 Quadruple Mutant

One expected consequence of downregulation of DAD1 expression by RGL2 is the reduction of JA levels in the ga1-3 gai-t6 rga-t2 rgl1-1 quadruple mutant (Q3). To text this hypothesis, we measured the JA contents in the young flower buds in WT, opr3, the Q3 quadruple mutant and the ga1-3 gai-t6 rga-t2 rgl1-1 rgl2-1 penta mutant (penta). The data obtained clearly showed that the JA content was greatly reduced in the young flower buds of the quadruple Q3 mutant whereas was restored in the penta mutant when compared to that in the WT and opr3 mutant (Figure 6D).

GA Application Induces DAD1 Expression Prior to the Induction of MYB21, MYB24, and MYB57

DAD1 is a stamen specific gene encoding chloroplastic phospholipase A1 protein that catalyzes the first step of JA biosynthesis. Mutation in DAD1 resulted in a typical JA-deficient phenotype in stamen development [40], a phenotype similar to that of myb21-t1 myb24-t1 double mutant. As mentioned earlier, JA likely acts downstream of GA to regulate the expression of MYB21, MYB24 and MYB57. To study whether there is a correlation between GA-regulated DAD1 expression and MYB21, MYB24 and MYB57 expression, we treated the ga1-3 gai-t6 rga-t2 rgl1-1 quadruple mutant with GA. We first confirmed the GA-responsiveness in the quadruple mutant plants by examining the expression of known GA-response genes GA3ox1, GA20ox2 and GA2ox1 (Figure 7A). Then we examined the expression of DAD1 and the three MYB genes MYB21, MYB24 and MYB57.

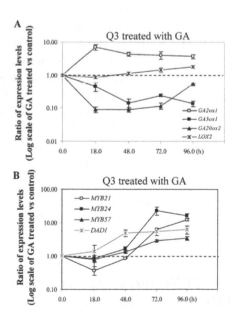

Figure 7. GA Induces DAD1 Expression Prior To Induction of Expression of MYB21, MYB24, and MYB 57. (A–B) Semi-quantitative analysis of LOX2, GA2ox1, GA3ox1 and GA20ox2 (A), DAD1 (in red line), MYB21, MYB24 and MYB57 (B) expression in the ga1-3 gai-t6 rga-t2 rgl1-1 (Q3) mutant flowers at 18, 48, 72 and 96 hrs after GA treatment. Data were averaged from 2–4 batches of independently treated samples and ACTII was used as the normalization control. The graph was drawn based on Log10 scale of the ratio of the expression levels of GA treated versus untreated samples.

Surprisingly, compared to the induction of MYB21 and MYB24 expression by JA treatment which is detectable at 18 hrs post treatment (Figure 5D), GA induction of the expression of these two MYB genes in ga1-3 gai-t6 rga-t2 rgl1-1 happens much later and only became detectable at 72 hrs (Figure 7B). More interestingly, GA induction of the expression of DAD1 is obviously detectable at 48 hrs which is prior to GA-induced expression of MYB21 and MYB24 in the ga1-3 gai-t6 rga-t2 rgl1-1 quadruple mutant (Figure 7B). Our data suggest that GA might first induce the expression of DAD1 to promoter JA production then via JA signaling to regulate the expression of MYB21 and MYB24.

Expression of MYB21, MYB24, and MYB57 Is Necessary But Insufficient for Normal Stamen Filament Elongation in ga1-3 gai-t6 rga-t2 rgl1-1

As shown in the above, MYB21, MYB24 and MYB57 act downstream of DEL-LAs in controlling stamen filament elongation. Expression of MYB21, MYB24 and MYB57 was repressed and floral development was arrested in the ga1-3 gai-t6 rga-t2 rgl1-1 Q3 quadruple mutant (Figure 2B and 2C). Regarding the fact that JA content is reduced in the young flower buds of Q3 we questioned whether restoration of expression of these MYBs by exogenous application of JA could rescue the stamen development to the ga1-3 gai-t6 rga-t2 rgl1-1 Q3 plants. We analyzed the flowers of JA-treated ga1-3 gai-t6 rga-t2 rgl1-1 plants and found that repeated JA application was unable to rescue the stamen development (Figure 8) though JA could restore the expression of the three MYB genes (Figure 5D), indicating that expression of MYB21, MYB24 and MYB57 alone was insufficient for normal stamen development in the ga1-3 gai-t6 rga-t2 rgl1-1 mutant. Furthermore, we found that exogenous application of GA to the ga1-3 gai-t6 rga-t2 rgl1-1 plants was able to induce the expression of MYB21, MYB24 and MYB57 (Figure 7B) and recover normal floral development (Figure 8). Taken together, our results demonstrate that besides these JA-inducible MYBs, other important GA-regulated JA-independent factors are needed for normal stamen filament development in ga1-3 gai-t6 rga-t2 rgl1-1.

Overexpression of MYB21 Restored Stamen Filament Elongation and Fertility to opr3 Flowers

To test our hypothesis that GA acts through JA to control expression of the MYB genes to promote filament elongation, we fused MYB21 gene with the CaMV35S promoter (pCAMBIA1301 vector) and this construct was used to generate transgenic plants in the opr3 mutant background. Semi-quantitative RT-PCR showed that MYB21 was overexpressed in the transgenic plants in the opr3 background (Figure 9A). We found that overexpression of MYB21 could restore the stamen filament growth (Figure 9B) and restore the fertility (Figure 9C and 9D) to the opr3 mutant partially. Together with the fact that loss of function of four DELLA (GAI, RGA, RGL1 and RGL2) could not restore the fertility and filament elongation to the coil1 mutant, we have now provided strong evidence to show that GAs act through JA to control expression of the MYBs and promote stamen filament elongation.

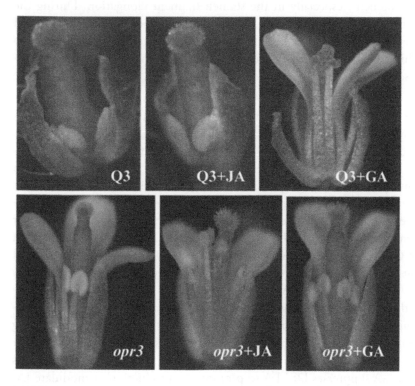

Figure 8. MYB21, MYB24, and MYB57 Are Necessary but Insufficient to Complete the Normal Stamen Filament Development.

Pictures are shown to compare the stamen phenotype in JA or GA repeatedly treated ga1-3 gai-t6 rga-t2 rgl1-1 (Q3) and opr3 plants with respective untreated controls.

Figure 9. Overexpression of MYB21 Rescues the Stamen Filament Growth and Fertility to the opr3 Mutant.

(A) RT-PCR analysis of MYB21 and OPR3 gene expression in WT, opr3 and opr3 MYB21OE-1. Total RNA was extracted from the young flower buds. ACTIN was used as the normalization control. (B) Comparison of the flowers at stage 14 in different genotypes. The flower in opr3 MYB21OE-1 shows elongated filament than that in opr3. (C and D) Comparison of seed set in different genotypes as shown (C) and of plant growth of WT (Col-0) (50 days old), opr3 (50 days old) and opr3 MYB21OE-1. The third plant from left was an opr3 MYB21OE-1 plant with primary shoot (50 days old) whereas the last plant was a 60-day-old opr3 MYB21OE-1 with axillary shoots after its primary influence has been removed earlier. White arrows highlight siliques with seed set, red arrows highlight sterile siliques.

Discussions

In this report, we first identified 34 DELLA-repressed stamen-enriched genes by RT-PCR analysis of candidate genes selected from microarray data [30]. We then selected MYB21, MYB24 and MYB57 for detailed genetic analysis because GAMYBs are the best characterized transcription factors involved in GA-response. We showed that Arabidopsis MYB21, MYB24 and MYB57 are highly expressed in the stamen. The stamen-enriched expression pattern is consistent with the observation that the myb21-t1 myb24-t1 myb57-t1 triple mutant is impaired in the stamen development, especially in the stamen filament elongation. During the course of studying these three MYBs, Mandaokar et al reported that MYB21 and MYB24 are also JA-inducible [32] which immediately attracted our attention to study the hierarchical relationship between GA and JA in regulating the expression of these three MYBs.

We first tested the responses of GA- and JA-deficient mutants (i.e. ga1-3 gai-t6 rga-t2 rgl1-1 Q3 quadruple mutant and opr3 mutant, respectively) to GA and JA treatments and found that JA- treatment induced the expression of MYB21, MYB24 and MYB57 in the GA-deficient plants whereas GA-treatment failed to do so in the JA-deficient plant. This result suggests that JA likely acts downstream of GA pathway to control the expression of these three MYBs. It is possible that JA acts downstream by modulating the stability or activity of DELLA proteins to induce the expression of the three MYBs. If this is the case, we would expect that JA-treatment would lead to RGL2 degradation or would change the expression patterns of GA-response genes in ga1-3 gai-t6 rga-t2 rgl1-1. However, we found that neither the RGL2 protein level (data not shown) nor the expression patterns of three GA-response genes GA2ox1, GA3ox1 and GA20ox1 were obviously altered in the JA-treated ga1-3 gai-t6 rga-t2 rgl1-1 plants at 18 hrs post treatment although the three MYBs are highly expressed at this time point, suggesting that destabilization or inactivation of DELLA proteins is unlikely the cause for JA-induced expression of MYBs in ga1-3 gai-t6 rga-t2 rgl1-1. Alternatively, it is possible that GA suppresses DELLA to promote JA production or modulate JA-signaling to induce the expression of the three MYBs. The fact that JA application can restore the expression of the three MYBs in the GA-deficient background strongly suggests that, at least in part, JA biosynthesis is impaired in the ga1-3 gai-t6 rga-t2 rgl1-1 mutant.

JA biosynthesis is accomplished by a sequential biochemical reactions mediated by JA-biosynthesis genes including DAD1, LOX1, 2, AOS, AOC1, 2, 3, 4 and OPR3 and is regulated by OPDA compartmentalization and a JA-mediated positive feedback loop [41]. Biotic and abiotic stresses also induce JA formation [42]–[44]. In our experiment, we found that JA biosynthesis gene DAD1 was greatly down-regulated in both ga1-3 single and ga1-3 gai-t6 rga-t2 rgl1-1 quadruple mutants but partially restored to a relatively high level in the penta mutant, suggesting that GA is required for the expression of DAD1 to control the production of JA via repression of DELLA proteins. In flowers of dad1 null mutant, the JA levels were only 22% of that of WT [40], demonstrating that limited initial substrate generation by DAD1 reaction acts as a control point for JA biosynthesis in flowers. Therefore, it is highly possible that reduced expression of DAD1 in ga1-3 gai-t6 rga-t2 rgl1-1 or ga1-3 mutant may result in relative low JA production. This hypothesis is strongly supported by the observation that JA content was greatly reduced in the young flower buds of the GA-deficient quadruple mutant ga1-3 gai-t6 rga-t2 rgl1-1 (Q3 mutant). Furthermore, the fact that the induction of DAD1 expression happens prior to the expression of MYBs by GA in the ga1-3 gai-t6 rga-t2 rgl1-1 mutant strongly support our hypothesis that GA may regulate the MYBs' expression via mobilization of the biosynthesis of JA. A recent report showed that DAD1 expression is directly controlled by AGAMOUS (AG) [45]. Interestingly, Yu et al reported that AG expression was downregulated in the GA-deficient mutant ga1-3 and exogenous GA application promoted the AG expression [20]. It will be interesting to study if there is a relationship among DELLAs, AG and DAD1 in the future. High level of JA would induce the expression of the three MYB genes essential for stamen development. In addition to DAD1, we also observed that expression of LOX1 was down-regulated in ga1-3 mutant and restored to the WT level in the penta mutant. On the other hand, another JA biosynthesis gene AOC2 was up-regulated in ga1-3 mutant. These observations suggested that GA may be one of the endogenous signal involved in the regulation of JA biosynthesis genes.

Genetic studies have shown that MYB21, MYB24 and MYB57 are indispensable for stamen development. The stamen phenotype of myb21-t1 myb24-t1 myb57-t1 triple mutant is similar to that of JA-deficient mutants including opr3 and dad1 mutants. Overexpression of MYB21 restored the stamen filament elongation and fertility to the opr3 flowers, strongly suggesting that JA-mediated stamen filament growth is mainly through the MYB pathway. Both ga1-3 single and ga1-3 gai-t6 rga-t2 rgl1-1 quadruple mutants showed a more severe flower phenotype than myb21-t1 myb24-t1 myb57-t1 triple mutant. The fact that expression of these MYBs in ga1-3 gai-t6 rga-t2 rgl1-1 plants was not enough to rescue the mutant flower phenotype indicates that these MYBs are necessary but not sufficient for GA-mediated floral development. These data also indicate that modulation of JA pathway may be only one of the branches of GA function in regulating stamen development.

Active cross-talk between different hormone signaling pathways have been revealed in many developmental processes [46]. For example, it was reported that auxin was necessary for GA-mediated Arabidopsis root growth by promoting GA-dependent degradation of DELLA proteins [47]. In contrast, ethylene inhibits

Arabidopsis root growth by delaying the GA-induced destabilization of DELLA [48]. Recently, the complexity of interactions between ethylene and GA signal transduction pathways were analyzed by using combinations of different ethylene and GA related mutants [49]. Hormone-hormone interaction also plays an important role in controlling flowering. For example, it was found that stress induced hormone ethylene control floral transition via DELLA-dependent regulation of floral meristem-identity genes LEAFY and SUPPRESSOR OF OVEREXPRESSION OF CONSTANS 1 (SOC1) [50]. We have here established a linear relationship between GA and JA in that GA modulates the expression of DAD1 that in a likely scenario to promote JA biosynthesis and in return JA induces the expression of MYB21, MYB24 and MYB57 to control the normal stamen development in Arabidopsis.

Materials and Methods

Plant Materials

Plants were grown as described previously [19]. Mutant lines (La-er background) ga1-3, Q3 (ga1-3 gai-t6 rga-t2 rgl1-1) and penta (ga1-3 gai-t6 rga-t2 rgl1-1 rgl2-1) were described previously [8]. Mutant lines (Col-0 background) myb21-t1 (SALK_042711), myb24-t1 (SALK_017221) and myb57-t1 (SALK_065776) were obtained from Arabidopsis Biological Resource Centre at the Ohio State University [51] and verified using primer pairs listed in Table 3. These lines were backcrossed twice to purify the genetic background and were then used for all experiments described in this paper. Double mutants were generated from crosses between the relevant single mutants. Triple mutant myb21-t1 myb24-t1 myb57-t1 was obtained from cross between myb21-t1 myb24-t1 and myb24-t1 myb57-t1. Hexa1 (ga1-3 gai-t6 rga-t2 rgl1-1 rgl2-1 myb21-t1), hexa2 (ga1-3 gai-t6 rga-t2 rgl1-1 rgl2-1 myb24-t1) and hepta (ga1-3 gai-t6 rga-t2 rgl1-1 rgl2-1 myb21-t1 myb24-t1) mutants were in Laer background via cross-pollination of myb21-t1 myb24-t1 to the ga1-3 gai-t6 rga-t2 rgl1-1 rgl2-1 penta mutant four times. SEM of the penta, hexa1, hexa2 and hepta mutants was performed as described previously [8]. The opr3 mutant is in the Ws background [6].

Table 3. Primer pairs used for genotyping MYB mutants.

Mutant lines	Primer pairs used for T-DNA insertion verification	Primer pairs used to amplify fragment spanning T-DNA insertion
myb21-t1	LBa1: TGGTTCACGTAGTGGGCCATCG	4334F: ATCGTGCCTATTTCTCCTCCAT
	5355R: TTGATATGATGTCGGTGTAGGAGA	5577R: CGCGGCCGAATAGTTACCATAGT
myb24-t1	LBa1: TGGTTCACGTAGTGGGCCATCG	4566F: TGCCGATTCTACCACAAC
	4975R: CTACATCTACGTCGAGCAATAA	4975R: CTACATCTACGTCGAGCAATAA
myb57-t1	LBa1: TGGTTCACGTAGTGGGCCATCG	3411F: CATGGTGAAGGTCTTTGGAACT
	3411F: CATGGTGAAGGTCTTTGGAACT	4511R: TAAACAATAACAACGTCCCTTCCT

Hormone Treatment

Both the ga1-3 gai-t6 rga-t2 rgl1-1 (Q3) and opr3 mutant plants (~27 days old) were sprayed with mock (0.1% ethanol v/v), GA3 (10?4 M) (Sigma) or MeJA (0.015% v/v) (Sigma). After treatment, young inflorescences were collected at different time point (18 hrs, 48 hrs, 72 hrs and 96 hrs) for total RNA extraction. For observing rescue of stamen development, mutant plants were repeatedly treated (once a week) with GA or JA.

RT-PCR and Northern Analysis

Different organs (sepal, petal, stamen, pistil and peduncle) of stage 11–12 flowers were dissected under microscope and pooled for RNA extraction. Flowers younger than stage 11 were pooled as young flower buds for RNA extraction. Total RNA was extracted from the young flower buds of respective genotypes treated with or without GA and JA using Tri Reagent (Molecular Research Center, Cincinnati, OH). The residue genomic DNA in the total RNA was removed via treatment with RNase-free DNase I (Roche, Germany) and the total RNA further purified with the RNeasy Mini kit (QIAgen, Valencia, CA, USA). First strand cDNA was synthesized using SuperScript™II RNase H? Reverse Transcriptase (Invitrogen, USA). First strand cDNA was used as the template in PCR using gene specific primers. Primer pairs used in identification of DELLA-repressed stamen-enriched genes were listed in Table S1. Primer pairs for RT-PCR analysis of GA2ox1, GA3ox1, GA20ox2, DAD1, AOS, OPR3, LOX1, 2 and AOC1, 2, 3, 4 were listed in Table 4. For quantifying the gene expression levels, PCR products were stained with ethidium bromide and the intensity was quantified using software Molecular Analyst (Bio-Rad). The gene expression level was normalized to the expression level of ACTII and then displayed as a ratio of expression levels of GA (or JA) treated samples versus untreated control.

Table 4. Gene specific primer pairs used in RT-PCR analysis.

Genes	Primers	Genes	Primers
AtMYB21	5′ AAAATCGCCAAACATCTTCC 3′	LOX2	5′ CCCGGCCGTTTATGGTG 3′
	5′ AATTATAACCCCAAACCTCTACAA 3′		5′ GTCTATTTGCCGCTATTATGTATG 3′
AtMYB24	5′ ATGCAAAATGGGGAAATAGGTG 3′	AOS	5′ GGCGGGCGGGTCATCAAGT 3′
	5′ AAGATCATCGACGCTCCAATAGTT 3′		5′ TCGCCGGAAAATCTCAATCACAAA 3′
AtMYB57	5′ GTGCGGCGAGGGAACATAA 3′	AOC1	5′ CACGCCCAAGAAGAAACTCACTC 3′
	5′ TCAGCAATAGAAAAACCAAATAAC 3′		5′ GCTGGCTCCACGTCCTTAGA 3′
GA2ox1	5′ CGGTTCGGGTCCACTATTTC 3′	AOC2	5′ CTCGGAGATCTCGTACCATTCAC 3′
	5′ ACCTCCCATTTGTCATCACCTG 3′		5′ ACTTATAACTCCGCTAGGCTCCAG 3′
GA3ox1	5′ GGCCCCAACATCACCTCAACTACT 3′	AOC3	5′ CAATGGCTTCTTCTTCTGCTGCTA 3′
	5′ GGACCCCAAAGGAATGCTACAGA 3′		5′ CTTCGAATCTGTCACCGCTCTTTT 3′
GA20ox2	5′ CCGGCAGAGAAAGAACACGAA 3′	AOC4	5′ TCCCCTTCACAAACAAACTCTACA 3′
	5′ TACGCCTAAACTTAAGCCCAGAA 3′		5′ GGACGGGACACATTACGCTTACG 3′
DAD1	5′ GGGCCTACTGGAGCAAATCTAAAC 3′	OPR3	5′ ACGGCGGCACAAGGGAACTCTAAC 3′
	5′ GTCTCCTCCACGCGTCTCTGTAT 3′		5′ GGGAACCATCGGGCAACAAAACTC 3′
LOX1	5′ GGGCTTGAGGTTTGGTATGCTATT 3′		
	5′ AACGCCTCCAACGCTTCTTTCT 3′		

Northern blot hybridization was performed as described [19]. Fragments of MYB21 (+294 to +801 nt, the A of the start codon ATG = 1), GA20ox2 (+28 to +627 nt), LOX1 (+1903 to +2408 nt), LOX2 (+1278 to +1714 nt), and OPR3 (+4 to +439 nt) were labeled using PCR DIG probe synthesis kit (Roche, Germany) and used as probes in Northern blot hybridization.

Quantification of JA

500 mg young flower buds harvested from different genotypes were frozen in liquid N2 and ground to a fine powder with a mortar and pestle. Following addition of 600 µL methanol, homogenates were mixed and kept at 4°C overnight, then centrifuged at 4,800 g for 10 min. The supernatant was transferred to a new 5 mL glass tube and the residue was re-extracted with 200 µL of methanol. 3000 µL ddH2O was added to the combined extracts and this solution was applied onto the Sep-pak C18 cartridge. The cartridge was washed with 200 µL 20% methanol and 250 µL 30% methanol 300 µL, respectively. Finally, the cartridge was eluted with 300 µL 100% methanol and the eluted solution was collected and used as the samples. Pre-prepared JA solutions (three concentrations were used: 10 ng/mL; 100 ng/mL; 1000 ng/mL) were used as the internal normalization standard. Samples were analyzed by a Thermo TSQ Quantum Ultra LC-MS-MS system. 10 µL of sample was injected onto a Hypersil Gold column (150?2.1, 3 µm). The mobile phase comprised solvent A (0.1% formic acid) and solvent B (methanol) used in a gradient mode [time/concentration of A/concentration of B(min/%/%) for 0/90/10; 1/90/10; 10/10/90; 15/10/90; 16/90/10; 28/90/10]. The machine was run with a spray voltage 4800 v, atomization flow 30 mL/min, auxiliary flow: 2 mL/min, capillary transfer temperature 380°C, lens compensation voltage 77 v, molecular ions m/z 133 (JA), collision energy 15 eV and signal collection interval 15–19 min.

CaMV35S::MYB21 Transgenic Plants

For MYB21 overexpression construct, the Arabidopsis MYB21 was cloned into an overexpression vector using a primerF: (5¢-agctctagaAtggagaaaag aggaggaggaag-3¢) and a primerR: (5¢-atcgagctctcaattaccattcaataaatgca-3¢) through XbaI and SacI sites. The overexpression vector, which was derived from pCAMBIA1301, contains the CaMV35S promotor to drive the expression of MYB21. The plasmids was confirmed by sequencing and introduced into Agrobacterium tumefaciens by electroporation and then introduced into heterzygous OPR3/opr3 plant by flower dip method [19]. More than 20 transformed lines were obtained based on PCR analysis. Homozygous opr3 mutants were identified using Opr3-RP (5¢-ctcaaatattggcgagacctg-3¢) and Opr3-LP (5¢-GGCAGAGTATTATGCTCAACG-3¢).

pMYB24::GUS Transgenic Plants

To make the pMYB24::GUS construct, a 3098 bp (68 bp upstream of MYB24 start codon ATG) genomic DNA fragment was PCR amplified from Col-0 genomic DNA using primers 18F (PstI, 5¢ TTCTAGGCTGCAGCTAAAC-GACTTC 3¢) and 2934R (5¢ GTAATAGAAAGGGAGAGTTGTGAAAG 3¢). PCR amplifications of promoter regions were performed using PfuTurbo DNA polymerase (Stratagene). The amplified DNA fragment was digested with PstI and then cloned into PstI/NcoI-cleaved pCambia 1301 vector and their sequences were confirmed by sequencing. The pMYB24::GUS fusion construct was then introduced into Arabidopsis thaliana ecotype Col-0 plants using flower dip method [19]. More than three independent lines were examined at various stages of floral development in this study.

In Situ Hybridization

Whole inflorescences which included unopened flower buds were fixed and in situ hybridization was carried out as described before [8]. Antisense and sense probes of MYB21 (+294 to +801 nt, nt stands for nucleotides, the A of the start codon ATG = 1) for in situ hybridization were DIG-labeled by in vitro transcription.

Authors' Contributions

Conceived and designed the experiments: DX JP. Performed the experiments: HC SS LX HMS ZC. Analyzed the data: HC SS LX ZC DX JP. Contributed reagents/materials/analysis tools: LX. Wrote the paper: DX JP.

References

1. Smyth DR, Bowman JL, Meyerowitz EM (1990) Early flower development in Arabidopsis. Plant Cell 2: 755–767.

2. Chaudhury AM (1993) Nuclear genes controlling male fertility. Plant Cell 5: 1277–1283.

3. Taylor PE, Glover JA, Lavithis M, Craig S, Singh MB, et al. (1998) Genetic control of male fertility in Arabidopsis thaliana: structural analyses of postmeiotic developmental mutants. Planta 205: 492–505.

4. Nakayama N, Arroyo JM, Simorowski J, May B, Martienssen R, et al. (2005) Gene trap lines define domains of gene regulation in Arabidopsis petals and stamens. Plant Cell 17: 2486–2506.

5. McCormick S (2004) Control of male gametophyte development. Plant Cell 16: SupplS142–S153.

6. Stintzi A, Browse J (2000) The Arabidopsis male-sterile mutant, opr3, lacks the 12-oxophytodienoic acid reductase required for jasmonate synthesis. Proc Natl Acad Sci USA 97: 10625–10630.

7. Koornneef M, van der Veen JH (1980) Induction and analysis of gibberellin sensitive mutants in Arabidopsis thaliana (L.) Heynh. Theor Appl Genet 58: 257–263.

8. Cheng H, Qin L, Lee S, Fu X, Richards DE, et al. (2004) Gibberellin regulates Arabidopsis floral development via suppression of DELLA protein function. Development 131: 1055–1064.

9. Tyler L, Thomas SG, Hu J, Dill A, Alonso JM, et al. (2004) Della proteins and gibberellin-regulated seed germination and floral development in Arabidopsis. Plant Physiol 135: 1008–1019.

10. Pysh LD, Wysocka-Diller JW, Camilleri C, Bouchez D, Benfey PN (1999) The GRAS gene family in Arabidopsis: sequence characterization and basic expression analysis of the SCARECROW-LIKE genes. Plant J 18: 111–119.

11. Richards DE, King KE, Ait-Ali T, Harberd NP (2001) How gibberellin regulates plant growth and development: A molecular genetic analysis of gibberellin signaling. Annu Rev Plant Physiol Plant Mol Biol 52: 67–88.

12. Boss PK, Thomas MR (2002) Association of dwarfism and floral induction with a grape 'green revolution' mutation. Nature 416: 847–850.

13. Chandler PM, Marion-Poll A, Ellis M, Gubler F (2002) Mutants at the Slender1 locus of barley cv Himalaya. Molecular and physiological characterization. Plant Physiol 129: 181–190.

14. Ikeda A, Ueguchi-Tanaka M, Sonoda Y, Kitano H, Koshioka M, et al. (2001) slender rice, a constitutive gibberellin response mutant, is caused by a null mutation of the SLR1 gene, an ortholog of the heightregulating gene GAI/RGA/RHT/D8. Plant Cell 13: 999–1010.

15. Peng J, Carol P, Richards DE, King KE, Cowling RJ, et al. (1997) The Arabidopsis GAI gene defines a signaling pathway that negatively regulates gibberellin responses. Genes Dev 11: 3194–3205.

16. Peng J, Richards DE, Hartley NM, Murphy GP, Devos KM, et al. (1999) 'Green revolution' genes encode mutant gibberellin response modulators. Nature 400: 256–261.

17. Silverstone AL, Ciampaglio CN, Sun T (1998) The Arabidopsis RGA gene encodes a transcriptional regulator repressing the gibberellin signal transduction pathway. Plant Cell 10: 155–169.

18. Dill A, Sun T (2001) Synergistic derepression of gibberellin signaling by removing RGA and GAI function in Arabidopsis thaliana. Genetics 159: 777–785.

19. Lee S, Cheng H, King KE, Wang W, He Y, et al. (2002) Gibberellin regulates Arabidopsis seed germination via RGL2, a GAI/RGA-like gene whose expression is up-regulated following imbibition. Genes Dev 16: 646–658.

20. Yu H, Ito T, Zhao Y, Peng J, Kumar P, et al. (2004) Floral homeotic genes are targets of gibberellin signaling in flower development. Proc Natl Acad Sci USA 101: 7827–7832.

21. Griffiths J, Murase K, Rieu I, Zentella R, Zhang ZL, et al. (2006) Genetic characterization and functional analysis of the GID1 gibberellin receptors in Arabidopsis. Plant Cell 18: 3399–3414.

22. Willige BC, Ghosh S, Nill C, Zourelidou M, Dohmann EM, et al. (2007) The DELLA domain of GA INSENSITIVE mediates the interaction with the GA INSENSITIVE DWARF1A gibberellin receptor of Arabidopsis. Plant Cell 19: 1209–1220.

23. Xie DX, Feys BF, James S, Nieto-Rostro M, Turner JG (1998) COI1: an Arabidopsis gene required for jasmonate-regulated defense and fertility. Science 280: 1091–1094.

24. Millar AA, Gubler F (2005) The Arabidopsis GAMYB-like genes, MYB33 and MYB65, are microRNA-regulated genes that redundantly facilitate anther development. Plant Cell 17: 705–721.

25. Gubler F, Kalla R, Roberts JK, Jacobsen JV (1995) Gibberellin-regulated expression of a myb gene in barley aleurone cells: evidence for Myb transactivation of a high-pI alpha-amylase gene promoter. Plant Cell 7: 1879–1891.

26. Gubler F, Raventos D, Keys M, Watts R, Mundy J, Jacobsen JV (1999) Target genes and regulatory domains of the GAMYB transcriptional activator in cereal aleurone. Plant J 17: 1–9.

27. Gubler F, Chandler PM, White RG, Llewellyn DJ, Jacobsen JV (2002) Gibberellin signaling in barley aleurone cells. Control of SLN1 and GAMYB expression. Plant Physiol 129: 191–200.

28. Kaneko M, Itoh H, Inukai Y, Sakamoto T, Ueguchi-Tanaka M, et al. (2003) Where do gibberellin biosynthesis and gibberellin signaling occur in rice plants? Plant J 35: 104–115.

29. Achard P, Herr A, Baulcombe DC, Harberd NP (2004) Modulation of floral development by a gibberellin-regulated microRNA. Development 131: 3357–3365.

30. Cao D, Cheng H, Wu W, Soo HM, Peng J (2006) Gibberellin mobilizesdistinct DELLA-dependent transcriptomes to regulate seed germination and floral development in Arabidopsis. Plant Physiol 142: 509–525.

31. Tsuji H, Aya K, Ueguchi-Tanaka M, Shimada Y, Nakazono M, et al. (2006) GAMYB controls different sets of genes and is differentially regulated by microRNA in aleurone cells and anthers. Plant J 47: 427–444.

32. Mandaokar A, Thines B, Shin B, Markus LB, Choi G, et al. (2006) Transcriptional regulators of stamen development in Arabidopsis identified by transcriptional profiling. Plant J 46: 984–1008.

33. Mandaokar A, Browse J (2009) MYB108 Acts Together with MYB24 to Regulate Jasmonate-Mediated Stamen Maturation in Arabidopsis. Plant Physiol 149: 851–862.

34. Shin B, Choi G, Yi H, Yang S, Cho I, et al. (2002) AtMYB21, a gene encoding a flower-specific transcription factor, is regulated by COP1. Plant J 30: 23–32.

35. Yang XY, Li JG, Pei M, Gu H, Chen ZL, et al. (2006) Over-expression of a flower-specific transcription factor gene AtMYB24 causes aberrant anther development. Plant Cell Rep 26: 219–228.

36. Stracke R, Werber M, Weisshaar B (2001) The R2R3-MYB gene family in Arabidopsis thaliana. Curr Opin Plant Biol 4: 447–456.

37. Kranz HD, Denekamp M, Greco R, Jin H, Leyva A, et al. (1998) Towards functional characterisation of the members of the R2R3-MYB gene family from Arabidopsis thaliana. Plant J 16: 263–276.

38. Sanders PM, Bui AQ, Weterings K, McIntire KN, Hsu Y-C, et al. (1999) Anther developmental defects in Arabidopsis thaliana male-sterile mutants. Sex Plant Reprod 11: 297–322.

39. Bell E, Mullet JE (1993) Characterization of an Arabidopsis Lipoxygenase Gene Responsive to Methyl Jasmonate and Wounding. Plant Physiol 103: 1133–1137.

40. Ishiguro S, Kawai-Oda A, Ueda J, Nishida I, Okada K (2001) The DEFECTIVE IN ANTHER DEHISCIENCE gene encodes a novel phospholipase A1 catalyzing the initial step of jasmonic acid biosynthesis, which synchronizes pollen maturation, anther dehiscence, and flower opening in Arabidopsis. Plant Cell 13: 2191–2209.

41. Sasaki Y, Asamizu E, Shibata D, Nakamura Y, Kaneko T, et al. (2001) Monitoring of methyl jasmonate-responsive genes in Arabidopsis by cDNA macroarray: self-activation of jasmonic acid biosynthesis and crosstalk with other phytohormone signaling pathways. DNA Res 8: 153–161.

42. Howe GA, Lee GI, Itoh A, Li L, DeRocher AE (2000) Cytochrome P450- dependent metabolism of oxylipins in tomato. Cloning and expression of allene oxide synthase and fatty acid hydroperoxide lyase. Plant Physiol 123: 711–724.

43. Maucher H, Hause B, Feussner I, Ziegler J, Wasternack C (2000) Allene oxide synthases of barley (Hordeum vulgare cv. Salome): tissue specific regulation in seedling development. Plant J 21: 199–213.

44. Ziegler J, Stenzel I, Hause B, Maucher H, Hamberg M, et al. (2000) Molecular cloning of allene oxide cyclase. The enzyme establishing the stereochemistry of octadecanoids and jasmonates. J Biol Chem 275: 19132–19138.

45. Ito T, Ng KH, Lim TS, Yu H, Meyerowitz EM (2007) The homeotic protein AGAMOUS controls late stamen development by regulating a jasmonate biosynthetic gene in Arabidopsis. Plant Cell 19: 3516–3529.

46. Nemhauser JL, Hong F, Chory J (2005) Different plant hormones regulate similar processes through largely nonoverlapping transcriptional responses. Cell 126: 467–75.

47. Fu X, Harberd NP (2003) Auxin promotes Arabidopsis root growth by modulating gibberellin response. Nature 421: 740–743.

48. Achard P, Vriezen WH, Van Der SD, Harberd NP (2003) Ethylene regulates arabidopsis development via the modulation of DELLA protein growth repressor function. Plant Cell 15: 2816–2825.

49. De Grauwe L, Vriezen WH, Bertrand S, Phillips A, Vidal AM, et al. (2007) Reciprocal influence of ethylene and gibberellins on responsegene expression in Arabidopsis thaliana. Planta 226: 485–498.

50. Achard P, Baghour M, Chapple A, Hedden P, Van Der SD, et al. (2007) The plant stress hormone ethylene controls floral transition via DELLA-dependent regulation of floral meristem-identity genes. Proc Natl Acad Sci U S A 104: 6484–6489.

51. Alonso JM, Stepanova AN, Leisse TJ, Kim CJ, Chen H, et al. (2003) Genome-wide insertional mutagenesis of Arabidopsis thaliana. Science 301: 653–657.

CITATION

Originally published under the Creative Commons Attribution License. Cheng H, Song S, Xiao L, Soo HM, Cheng Z, Xie D, Peng J. Gibberellin Acts through Jasmonate to Control the Expression of MYB21, MYB24, and MYB57 to Promote Stamen Filament Growth in Arabidopsis. PLoS Genet. 2009 Mar;5(3):e1000440. doi:10.1371/journal.pgen.1000440.

Expressions of ECE-CYC$_2$ Clade Genes Relating to Abortion of Both Dorsal and Ventral Stamens in Opithandra (Gesneriaceae)

Chun-Feng Song, Qi-Bing Lin, Rong-Hua Liang and Yin-Zheng Wang

ABSTRACT

Background

ECE-CYC2 clade genes known in patterning floral dorsoventral asymmetry (zygomorphy) in Antirrhinum majus are conserved in the dorsal identity function including arresting the dorsal stamen. However, it remains uncertain whether the same mechanism underlies abortion of the ventral stamens, an important morphological trait related to evolution and diversification of zygomorphy in Lamiales sensu lato, a major clade of predominantly zygomor-phically flowered angiosperms. Opithandra (Gesneriaceae) is of particular in-terests in addressing this question as it is in the base of Lamiales s.l., an early representative of this type zygomorphy.

Results

We investigated the expression patterns of four ECE-CYC2 clade genes and two putative target cyclinD3 genes in Opithandra using RNA in situ hybrid-ization and RT-PCR. OpdCYC gene expressions were correlated with abor-tion of both dorsal and ventral stamens in Opithandra, strengthened by the negatively correlated expression of their putative target OpdcyclinD3 genes. The complement of OpdcyclinD3 to OpdCYC expressions further indicated that OpdCYC expressions were related to the dorsal and ventral stamen abor-tion through negative effects on OpdcyclinD3 genes.

Conclusion

These results suggest that ECE-CYC2 clade TCP genes are not only function-ally conserved in the dorsal stamen repression, but also involved in arresting ventral stamens, a genetic mechanism underlying the establishment of zygo-morphy with abortion of both the dorsal and ventral stamens evolved in an-giosperms, especially within Lamiales s.l.

Background

One important event during the evolution of angiosperms is the emergence of flower bilateral symmetry, i.e. zygomorphy, a key innovation associated with im-portant adaptive radiations [1]. Several zygomorphic clades have independently evolved successfully from actinomorphic ancestors in angiosperms, including La-miales sensu lato that includes a major genetic model organism snapdragon (An-tirrhinum majus) [2,3].

In A.majus, CYCLOIDEA (CYC) and DICHOTOMA (DICH) are essential for the development of dorsoventral asymmetry in flowers due to their dorsal identity function, i.e. controlling the fate of the dorsal floral organs in the second and third whorls [4,5]. CYC promotes cell expansion in the dorsal petals, while it arrests the growth of the dorsal stamen to become a staminode [4,5]. Mean-while, DICH activity affects the internal asymmetry of the dorsal petals [4,5]. The ability of CYC to arrest the dorsal stamen depends on its negative effect on expression of cell-cycle genes, such as cyclin D3b [3,6]. CYC and DICH encode proteins within the ECE-CYC2 clade (ECE lineage, CYC/TB1 subfamily) in the TCP family of transcription factors with TCP domain related to cell proliferation

[3,7-12]. In legumes, distantly related to A. majus, several CYC homologues, such as LjCYC2 in Lotus and PsCYC2 in pea, also have the function in establishing dorsal identity in legume flowers [13,14]. In Arabidopsis thaliana, a model eudicot species with ancestrally actinomorphic flowers, and its close relative Iberis amara, CYC homologues, TCP1 and IaTCP1 genes are characteristic of dorsal identity function, in which IaTCP1 dorsal-specific expression represses the two dorsal petal development in Iberis amara [15,16]. Recent studies in the sunflower family (Asteraceae) show that CYC-like genes, i.e. RAY1, RAY2 in Senecio and GhCYC2 in Gerbera, have played a key role in the establishment and evolution of the capitulate inflorescence [17,18]. Therefore, it is suggested that CYC-like TCP genes have been recruited multiple times for a role of dorsal identity and its modifications in establishing zygomorphy in core eudicots [3,19]. Even though the genetic control for the floral dorsoventral asymmetry has been intensively studied in model systems, it is still a great challenge to explain how modifications of development led to the transformation among different types of zygomorphy and the morphological diversification of zygomorphy in angiosperms, especially in Lamiales s.l., a major clade of predominantly zygomorphically flowered angiosperms.

Zygomorphy is believed to be ancestral in Lamiales s.l. [2,19,20]. Most zygomorphic groups in Lamiales s.l. have a pentamerous perianth with four stamens plus a dorsal staminode and two carpels as in A.majus. However, there is a great variation in morphology and number of corolla lobes and stamens [1]. The dorsal staminode can be completely lost as in Rehmannia and Veronica (Scrophulariaceae sensu lato) [2,21,22] and the two lateral stamens may become aborted instead of one dorsal staminode as in Mohavea (Scrophulariaceae s.l.) and Chirita (Gesneriaceae) [23-25]. In some cases, the two ventral stamens may become staminodes rather than the lateral stamens and the dorsal one, such as in Opithandra and Epithema (Gesneriaceae) [24]. In extreme cases, each flower may have only a single stamen as in Hippuris (Scrophulariaceae s.l.) [20]. In Mohavea, a close relative of A. majus, there is a derived floral morphology with abortion of both the dorsal and lateral stamens unlike the flowers of A.majus with abortion of only the dorsal stamen [23]. The derived floral morphology of Mohavea is correlated with the expression changes of McCYC/McDICH via CYC/DICH, i.e. expansion from the dorsal to both the dorsal and lateral stamens [23]. A similar correlation of expanded expression of CYC-like genes with abortion of both the dorsal and lateral stamens is also observed in Chirita (Gesneriaceae) [25]. However, we are still not clear about the abortion of the ventral stamens that has been involved in the evolutionary shifts of stamen number during the morphological diversification of zygomorphy in Lamiales s.l. [1].

In addition, in the backpetals mutant in A.majus, the ectopic expression of CYC in the lateral and ventral positions results in a dorsalized corolla. However, it seems likely that the androecial development is not affected by the ectopic expression of CYC because the two lateral and two ventral stamens are still fertile [5,26]. In the actinomorphic flower of legume Cadia, no LegCYC1B mRNA is detected in stamens [27]. It is hard to determine whether LegCYC1B or other CYC homologues in legumes have a role in controlling androecial development from data to date [13,14,21]. RAY2 in Senecio and GhCYC2 in Gerbera (Asteraceae) mainly promote the growth of the ligule (the ventral petals) in ray florets and are excluded from the dorsal rudimentary petals [17,18]. In Veronica and Gratiola (Scrophulariaceae s.l.), some of CYC-like genes have dorsal-specific expressions while some have lost this feature, but their expressions do not correlate with ventral stamen arrest [21]. Therefore, expression data correlated with the ventral stamen abortion have not been reported yet for members of the ECE-CYC2 clade. It is uncertain whether abortion of the ventral stamens is related to CYC-like gene activities or to the effect of an unknown analogous counterpart of CYC-like genes, such as members of ECE-CYC3 clade or other factors [3,12,20].

The family Gesneriaceae is sister to the remainder of Lamiales s.l. [28] and has diverse forms of zygomorphy relating to the floral organ differentiation early in the order [1,2,24]. In Gesneriaceae, Opithandra exhibits a peculiar floral morphology, where only the two lateral stamens are fertile and both the dorsal and ventral stamens are aborted in the third whorl (Figure 1). Phylogenetic analyses suggest that the floral morphology of Opithandra is likely derived from a weakly

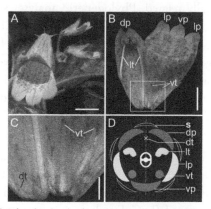

Figure 1. Flower morphology of Opithandra dinghushanensis.

A) Inflorescence with flowers near or at anthesis, showing strongly zygomorphic corolla; B) An opened corolla at anthesis showing two dorsal petals smaller than two lateral and one ventral petals, and androecium with two fertile lateral stamens and two ventral and one dorsal staminodes; C) Magnification of the framed part in (B), showing two infertile lateral stamens with short filaments and small sterile anthers, and a tiny dorsal staminode that is barely visible; D) Floral diagram; Scale bars, 10 mm (A), 7 mm (B) and 3 mm (C). dp, dorsal petal (in blue); dt, dorsal staminode; lp, lateral petal (in yellow); lt, lateral stamen (in yellow); s, sepal; vp, ventral petal (in pink); vt, ventral staminode (in pink).

zygomorphic flower with four fertile stamens and a dorsal staminode [24]. Therefore, Opithandra represents an ideal candidate for exploring a potentially novel genetic mechanism underlying the establishment of zygomorphy with ventral stamen arrest in angiosperms, especially in Lamiales s.l.

Here we report that there is a correlation between OpdCYC gene expressions and abortion of both the dorsal and ventral stamens in Opithandra, strengthened by the negatively correlated expression of their putative direct target OpdcyclinD3 genes. The novel patterns of CYC-like gene expressions in Opithandra indicate that ECE-CYC2 clade TCP genes are involved in the ventral stamen repression evolved within Lamiales s. l.

Results

Sequence and Phylogenetic Analyses of OpdCYC and OpdcyclinD3

We isolated four CYC-like genes from Opithandra dinghushanensis, named OpdCYC1C, OpdCYC1D, OpdCYC2A and OpdCYC2B. The full length open reading frames (ORFs) of OpdCYC1C, OpdCYC1D, OpdCYC2A and OpdCYC2B are 1017 base pair (bp), 1038 bp, 1044 bp and 993 bp, respectively. Sequence analyses show that they share 43-48% and 45-51% identity with AmCYC at nucleotide and amino acid levels, respectively. When comparing the TCP and R domains, they share 90-95% identity with AmCYC at the amino acid level, suggesting they are functionally related. Phylogenetic analyses show that OpdCYC genes have a close relationship with AmCYC and AmDICH, and, along with AmCYC and TCP1, belong to the ECE-CYC2 clade in the ECE lineage (CYC/TB1 subfamily) of TCP gene family [12] (Figure 2A) (We have not found a member of the ECE-CYC3 clade yet in Gesneriaceae, probably failed to amplify them because of difficulty in designing specific primers for this clade). OpdCYC genes are closely related to GCYC from Oreocharis among GCYC genes in Gesneriaceae (Figure 2B).

Two D3-type cyclin genes, designated as OpdcyclinD3a and OpdcyclinD3b, were isolated from O. dinghushanensis with full length ORFs of 1563 bp and 1200 bp, respectively. The two D3-type cyclin genes contain a cyclin box [29] and the putative (Rb)-binding motif (LxCxE, where x is any amino acid) which are found both in animals [30,31] and plants [6,32,33]. Phylogenetic analyses show that OpdcyclinD3 genes belong to cyclinD3a and cyclinD3b clade, respectively, in the cyclinD3 lineages, in which they have close relations with AmcyclinD3a and AmcyclinD3b from A.majus, respectively (Figure 2C).

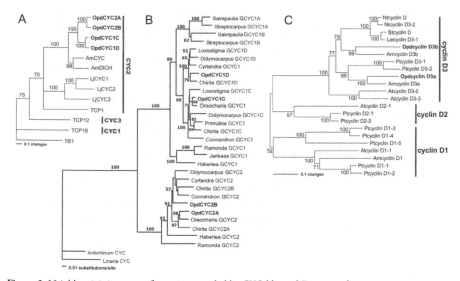

Figure 2. Neighbor-joining trees of proteins encoded by CYC-like and D-type cyclin genes.
A) Neighbor-joining tree of proteins encoded by the ECE lineage genes in CYC/TB1 subfamily, showing that OpdCYC1C/1D and OpdCYC2A/2B form a branch that is sister to AmCYC/AmDICH from Antirrhinum majus, which belong to the CYC2 clade in the ECE lineage. B) Phylogram of GCYC, showing the phylogenetic relations of OpdCYC genes with other GCYC in Gesneriaceae. C) Neighbor-joining tree of proteins encoded by D-type cyclin genes, showing that OpdcyclinD3a and OpdcyclinD3b are clustered with cyclinD3a and cyclinD3b clades, respectively, in the cyclinD3 lineage. For sequence information see Methods. Phylogenetic analyses were conducted using PAUP*4.0b4a, and bootsrap values over 50% (1,000 replicates) are indicated for each branch.

Gene mRNA Expression Patterns

To assess the potential role of CYC-like genes in floral development, we conducted in situ hybridization complemented by gene-specific RT-PCR on O. dinghushanensis. As petal and stamen primordia began to emerge, OpdCYC1C mRNA was detected in all five petal and stamen primordia (Figure 3A). Weak mRNA signals were also detectable in the lateral edges and vascular tissue of sepals (Figure 3A). After primordial initiation of petals and stamens, OpdCYC1C expression signals were gradually weakened in the two lateral stamens (Figure 3B-D) with weak mRNA detected in the ring meristem of the corolla-tube outside the stamen primordia (Figure 3B-C). Figures 3C and 3D were the successive sections from the same individual flower across the base of stamen primordia (3C) and over their upper parts (3D), respectively, which showed a size reduction from the base to the upper part of the dorsal and ventral stamens. The mRNA signal of OpdCYC1C was weak in lateral stamens while strong in both dorsal and ventral staminodes in which its mRNA signal was stronger in the upper part than at the base (Figure 3C-D). OpdCYC1C mRNA also accumulated less in lateral petals than in dorsal and ventral petals (Figure 3D). As the lateral stamens enlarged laterally, weak

OpdCYC1C mRNA shifted to peripheries and gradually became undetectable in the two lateral stamens, while OpdCYC1C transcripts continued to accumulate to a high level in the dorsal and ventral staminodes (Figure 3E-F). Meanwhile, OpdCYC1C mRNA became weak and not easily detectable in petals (Figure 3E-F). Weak mRNA signals of OpdCYC1C were also detected in the gynoecial primordium (Figure 3E). In the middle developmental stages with flower buds about 8 mm long, as stamens began filament elongation and anther differentiation, OpdCYC1C was strongly expressed in the dorsal region (the dorsal petals and staminode) and ventral staminode shown in RT-PCR while its weak mRNA signal was detected in sepals, lateral and ventral petals, and lateral stamens, in which the signal was much weaker in sepals and lateral stamens (Figure 4A, C). In the late stages, OpdCYC1C transcripts declined in the dorsal region and ventral staminodes, and were undetectable in other regions (Figure 4B-C). The Opd-CYC1C mRNA was undetectable in stamens as pollen sacs began development in the two lateral anthers while the two ventral anthers became sterile (Figure 3G).

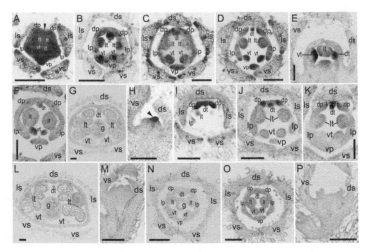

Figure 3. Tissue-specific expression of OpdCYC1 and OpdCYC2 during floral development in Opithandra dinghushanensis.

A-G) RNA in situ hybridizations with antisense probe of OpdCYC1C. A) Its mRNA is first detected in all five petal and stamen primordia with weak signals in lateral edges and vascular tissue of sepals. B-D) Its mRNA then weakened in two lateral stamens but strong in both dorsal and ventral staminodes with weak mRNA in the ring meristem of corolla-tube. C-D) Successive sections from the same individual flower across base (C) and upper parts (D) of stamens. E-F) Its expression shifts to peripheries and becomes undetectable in two enlarged lateral stamens while remains strong in dorsal and ventral staminodes. G) Its mRNA is undetectable in stamens as pollen sacs begin development. H-L) RNA in situ hybridizations with antisense probe of OpdCYC2A. H) Its dense transcript accumulation first restricted to the dorsal side of the floral apex (arrow). I-K) Its strong expression then restricted to two dorsal petals and the dorsal staminode as they are initiated. Note its mRNA later becomes restricted to the dorsal-most parts in two dorsal petals (K). L) Its mRNA is undetectable in stamens as pollen sacs begin development. OpdCYC1D (M) and OpdCYC2B (N) mRNA is not detected in floral tissues. As a negative control, RNA in situ hybridizations with sense probes of OpdCYC1C (O) and OpdCYC2A (P) detect no signal in floral tissues. dp, dorsal petal; ds, dorsal sepal; dt, dorsal staminode; g, gynoecium; lp, lateral petal; ls, lateral sepal; lt, lateral stamen; vp, ventral petal; vs, ventral sepal; vt, ventral staminode. Scale bars, 150 μm.

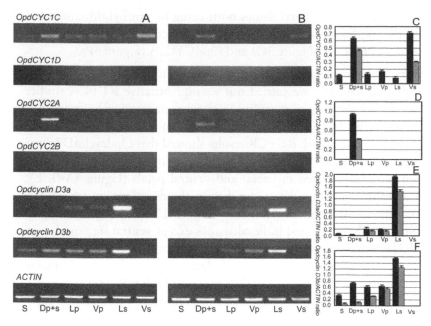

Figure 4. Gene-specific semiquantitative RT-PCR on RNA prepared from dissected Opithandra dinghushanensis flowers.

A) sepal (S), dorsal petal+staminode (Dp+s), lateral/ventral petals (Lp/Vp), lateral stamens (Ls) and ventral staminode (Vs) were dissected from flower buds of middle-stage (less than 1 cm long). B) sepal (S), dorsal petal+staminode (Dp+s), lateral/ventral petals (Lp/Vp), lateral stamens (Ls) and ventral staminode (Vs) were dissected from flowers of late-stage (3-4 cm long). C) Relative level of OpdCYC1C mRNA expression in middle-stage (black) vs. late-stage (grey) compared with ACTIN. D) Relative level of OpdCYC2A mRNA expression in middle-stage (black) vs. late-stage (grey) compared with ACTIN. E) Relative level of Opdcyclin D3a mRNA expression in middle-stage (black) vs. late-stage (grey) compared with ACTIN. F) Relative level of Opdcyclin D3b mRNA expression in middle-stage (black) vs. late-stage (grey) compared with ACTIN. ACTIN was used for RT template control. The values (means ± SD) shown are determined from five independent experiments.

In contrast to OpdCYC1C, OpdCYC2A mRNA densely accumulated in the dorsal region of the floral apex as petals and stamens became visible (Figure 3H). Then, the OpdCYC2A expression signal was specifically concentrated in the dorsal petals and the dorsal staminode (Figure 3I-J). As floral organs developed, OpdCYC2A transcripts continued to be highly concentrated at the dorsal staminode and the dorsal-most parts of the two dorsal petals (Figure 3J-K). In the middle stages, the strong mRNA signal of OpdCYC2A was detected in the dorsal region (the dorsal petals and staminode) shown in RT-PCR that declined in late stages with no signal in other regions (Figure 4A-B, D). Its mRNA was undetectable in stamens as pollen sacs began development in the two lateral anthers while the two ventral anthers became sterile (Figure 3L). Even though OpdCYC1D expression was undetectable in floral tissues using in situ hybridization (Figure 3M), its very weak mRNA signals were detected in the dorsal region (the dorsal petals and staminode) from middle to late stages using RT-PCR (Figure 4A-B). OpdCYC2B mRNA was not detectable in floral tissues both using in situ hybridization and RT-PCR (Figure 3N, Figure 4A-B). No signal was detected in floral tissues with sense probes of OpdCYC1C (Figure 3O) and OpdCYC2A (Figure 3P).

To further elucidate the role of OpdCYC genes in floral development of Opithandra dinghushanensis, we carried out semiquantitative RT-PCR studies of two OpdcyclinD3 genes because cyclinD3 genes were previously revealed to be negatively controlled by CYC as shown in the mid-to-late stage flowers in the model organism snapdragon [6]. RT-PCR results showed that OpdcyclinD3a was strongly expressed in lateral stamens from middle to late stages while its weak mRNA signal was also detected in lateral and ventral petals (Figure 4A-B, E). OpdcyclinD3a mRNA was not detected either in the dorsal region (the dorsal petals and staminode) and ventral staminodes (Figure 4A-B, E). OpdcyclinD3b transcripts were widely distributed in floral tissues except ventral staminodes, in which its mRNA signal was strong in lateral stamens from middle to late stages (Figure 4A-B, F). Transcripts of OpdcyclinD3b detected in the dorsal region were likely mainly distributed in the dorsal petals (OpdcyclinD3b mRNA was uneasily detectable in the dorsal staminode using in situ hybridization (data not shown)).

Discussion

The androecium of Opithandra only has two fertile stamens at the lateral positions with three sterile stamens (staminodes) at the dorsal and ventral sides (Figure 1) (also see [24]). The dorsal aborted stamen is tiny and barely detectable at anthesis while the two infertile ventral stamens have short filaments with very small and sterile anthers (Figure 1B-C). Correlative with the differentiation along the dorsoventral axis of the morphologically peculiar androecium, the OpdCYC2A strong expression is restricted to the dorsal staminode while OpdCYC1C transcripts are initially distributed in all five stamen primordia but later are concentrated in the dorsal and ventral staminodes to late stages. In the ECE lineage of CYC/TB1 subfamily, the TCP proteins in ECE-CYC2 clade studied to date function as negative regulators in stamen development, whereas they appear to vary in petal development according to the trait concerned [3-5,9,10,12,17,23,25,34,35]. The abortion of the dorsal stamen in Antirrhinum comes from CYC and DICH activities [4,5]. The CYC-like gene expression expansion from the dorsal to both the dorsal and lateral stamens is correlated with abortion of both the dorsal and lateral stamens in Mohavea and Chirita [23,25]. TB1 gene exhibits a mix feature of ECE-CYC1 and ECE-CYC2 clades and functions to suppress axillary meristem (CYC1) while retard stamen growth (CYC2) in maize [3,12,34]. In Asteraceae, a CYC homologue GhCYC2 from Gerbera functions by disrupting stamen development [17]. Given that CYC-like gene (ECE-CYC2 clade) function is conserved in repressing stamen development, OpdCYC2A expression restricted to the dorsal stamen and OpdCYC1C expression later concentrated in the dorsal and ventral stamens in the third whorl might be related to abortion of the dorsal and ventral stamens in Opithandra. In fact, the successive sections from the same individual flower indicate that correlative with OpdCYC gene strong expression, the early primordial growth have already been retarded in the dorsal and ventral staminodes in comparison with that in the lateral stamens.

Evidence shows that CYC functions to repress stamen development in the third whorl through its negative effect on expression of D3-type cyclin genes, including cyclinD3b, which usually play an important role in locally regulating cell proliferation in floral development [3,6,36]. The negative effects on cell cycle progression have been reported from other TCP genes, such as IaTCP1 from Iberis (Brassicaceae), TCP2 and TCP4 from Arabidopsis, and CIN from Antirrhinum [16,37,38]. We, therefore, investigated the expression pattern of OpdcyclinD3a and D3b to test for further correlation between CYC expression and stamen abortion through cell-cycle regulation, especially the ventral stamens. In strengthening the above suggestion, OpdcyclinD3 genes have expression patterns in floral tissues negatively correlated with those of OpdCYC genes and stamen abortion in Opithandra. Both Opdcyclin D3a and D3b transcripts are not detected, or weakly detected (i.e. dorsal), in the dorsal and ventral staminodes where both OpdCYC2A and OpdCYC1C or OpdCYC1C are strongly and continuously expressed throughout floral development, while their transcripts are much more concentrated in the lateral stamens where there is only a weak expression of OpdCYC1C in early stages. These factors indicate that OpdCYC gene activities may suppress the development of the dorsal and ventral stamens through negatively regulating OpdcyclinD3 genes (Figure 5). Our recent findings of consensus-binding sites of the TCP transcription factor in the 5' upstream regions of OpdcyclinD3 homologues, i.e. ChcyclinD3a and D3b in Chirita heterotricha, further support the direct regulatory relation between CYC-like and cyclinD3 genes in Gesneriaceae (Yang, Xia and Wang, Yin-Zheng, unpublished).

It seemingly remains a question whether OpdCYC activity is related to the lateral stamen development or not because OpdCYC1C has a weak expression in the early developing lateral stamens but has no obvious effect on their growth. The evidence that CYC-like genes regulate lateral stamen development comes from both functional analyses in Antirrhinum and expression data in Mohavea and Chirita [3,4,21,23,25]. A gradient of CYC effect along the dorsoventral axis results in abortion of the dorsal stamen and reduced size of lateral stamens in comparison with ventral ones in A. majus, while CYC-like gene strong and continuous expressions in both the dorsal and lateral stamens is correlated with abortion of both of them in Mohavea and Chirita [4,23,25,26]. The TCP1 gene is expressed early in floral buds of Arabidopsis, but the mature flowers of Arabidopsis are acti-nomorphic because they lack later effects of TCP1 [12,15]. In Iberis amara closely related to Arabidopsis, the IaTCP1 strong dorsal-specific expression represses the two dorsal petal growths to become much smaller than the ventral ones in size, while IaTCP1 has only a weak expressional signal in the natural actinomorphic variants [16]. In the tetrandrous flowers of Oreocharis that is closely related to Opithandra in both morphology and GCYC phylogeny [24] (Figure 2B), there is also a weak expression of CYC homologue ObCYC in the early developing lateral stamens that are reduced in size compared with ventral ones at anthesis as in A. majus [39]. In Bournea, another Opithandra's close relative with actinomorphic flowers in Gesneriaceae, the CYC homologue BlCYC1 is strongly expressed in the dorsal petal and stamen in early development and is downregulated later, which is correlated with the floral development undergoing a morphological transition from initial zygomorphy to actinomorphy at anthesis with five fertile stamens in Bournea [35]. According to Cubas [1], the maintenance of CYC expression after early floral development should be important for generating the morphological asymmetries in the flowers. Preston and Hileman [3] also suggest that early ex-pression of CYC-like genes may be unimportant for establishing mature flower symmetry. The high concentration of OpdcyclinD3a and D3b transcripts in the lateral stamens also indicates that OpdCYC1C has lost negative effects on their expression after early floral development; therefore, the two lateral stamens are fertile at anthesis in Opithandra (Figure 5).

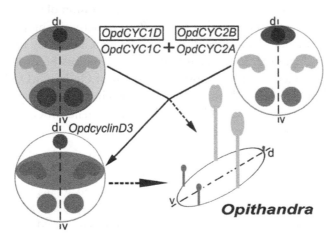

Figure 5. Diagram showing OpdCYC gene expressions correlated with the floral morphology, especially the infertility of the two ventral stamens in Opithandra, complemented by the expression of their putative negative target OpdcyclinD3 genes.

Notes: dorsal staminode is in blue, lateral stamens in green and ventral staminodes in pink; the gene that is not boxed indicates this gene is expressed in a pattern as shown by the shaded parts within the circle (shaded degrees indicate relative levels of gene expression); the gene boxed indicates this gene is not expressed or has very little expression signals in the floral tissues.

Zygomorphic flowers with three staminodes at the dorsal and lateral positions or at the dorsal and ventral sides have been considered to be derived in the family Gesneriaceae [24,40-42]. In the primitive zygomorphic taxa, such as Haberlea and Oreocharis characteristic of tetrandrous flowers with four didynamous stamens plus a dorsal staminode, and the derived actinomorphic groups (definition see note in [35]), such as Ramonda and Bournea, there is only one single copy of GCYC1 and GCYC2, respectively, found to date (Figure 2B) [35,39,43,44]. However, two copies of GCYC1 are frequently found in the advanced zygomorphic taxa, especially in the zygomorphic genera characterized by diandrous flowers with three staminodes, such as two African genera Streptocarpus and Saintpaulia with GCYC1A and GCYC1B [44,45] and Asian genera Didymocarpus, Chirita and Loxostigma with GCYC1C and GCYC1D [25,46] as well as Opithandra herein (Figure 2B). Recent studies show that two copies of GCYC2 (GCYC2A/2B) are also found in the Asian genera with three staminodes, such as Chirita [25] and Opithandra in this study (Figure 2B). The derived morphology of diandrous flowers might have resulted from subsequent expression differentiation after gene duplication events. In the diandrous flowers of Chirita (also Gesneriaceae) that differs from Opithandra in abortion of both the dorsal and lateral stamens rather than ventral stamens, ChCYC1C is strongly expressed both in the dorsal and lateral stamens while ChCYC1D maintains strong expressions in the dorsal floral regions, and ChCYC2A/2B have no expression signals in floral tissues [25]. No expression of GCYC2 detected in floral tissue is frequently found in Gesneriaceae while GCYC1 is usually conserved in dorsal-specific expression in this family, such as Orocharis and Bournea [35,39]. Therefore, the peculiar diandrous flowers established in Opithandra might involved not only gained or enhanced expression of OpdCYC1C in ventral staminodes but also the reactivated expression of OpdCYC2A specific to the dorsal staminode accompanied with the downregulation of OpdCYC1D in the dorsal region in the third whorl, a more complicated mechanism than that in another diandrous flowers of Chirita.

Phylogenetic analyses show that the CYC-like genes isolated from Opithandra and other Gesneriaceae belong to ECE-CYC2 clade as CYC and TCP1 from Antirrhinum and Arabidopsis (Figure 2B) [25,35,39,45]. As outlined above, ECE-CYC2 clade genes are characteristic of dorsal identity function which sometimes expands to lateral stamens [4,5,21,23,25,35,39]. It would be especially interesting to know whether or how CYC-like gene activities are related to abortion of the ventral stamens [3,12,20]. Even though not tested functionally, this positive correlation between CYC-like gene expression and ventral stamen abortion and the complement of cyclinD3 to CYC-like gene expressions suggests a genetic mechanism underlying the establishment of zygomorphy with abortion of both the dorsal and ventral stamen evolved within Lamiales s.l. However, it has been shown for Veronica and Gratiola (also Lamiales s.l.) that the CYC-like gene expression does not positively correlate with the ventral stamen abortion (Preston et al., 2009). These facts inconsonant in the expression data of ECE-CYC2 clade TCP genes imply that the ventral stamen abortion might have evolved by convergent genetic mechanisms in different lineages of Lamiales s.l. It merits further research in function and upstream regulatory pathway to determine how the expression divergence is caused among paralogues of OpdCYC in Opithandra. In addition, since the diverse variations of zygomorphy in Lamiales s.l. might have involved independent shifts in stamen number [3,21], further investigation of expression pattern and functional analyses of CYC-like genes with identification of their upstream cis- or trans-regulators as well as research in finding other factors possibly coopted to this regulatory pathway in more zygomorphic groups would shed new lights on the mechanisms that underlie the vast morphological diversity of zygomorphy in Lamiales s.l.

Conclusion

As the first to document the expression domain of ECE-CYC clade genes in the ventral stamens, we here report that the expressions of OpdCYC genes are correlated with abortion of both dorsal and ventral stamens in Opithandra, strengthened by the negatively correlated expression of their putative direct target OpdcyclinD3 genes. The complement of OpdcyclinD3 to OpdCYC gene expressions further indicates that OpdCYC expressions are related to the dorsal and ventral stamen abortion through the negative effect on OpdcyclinD3 genes. The novel patterns of CYC-like gene expressions in Opithandra, along with previous reports, suggest that ECE-CYC2 clade TCP genes are not only functionally conserved in the dorsal stamen repression, but also involved in arresting ventral stamens, a genetic mechanism underlying the establishment of zygomorphy with abortion of both the dorsal and ventral stamens evolved within Lamiales s.l.. It would be important to further find whether ECE-CYC2 clade TCP genes are recruited repeatedly for arresting ventral stamens and (or) whether there is any other genetic pathways underlying the ventral stamen abortion, independently or interacting with ECE-CYC2 clade TCP genes, in different lineages of Lamiales s.l.

Methods

Plant Materials

All materials used in this study, including gene cloning, in situ hybridization and RT-PCR, were collected from the wild fields, i.e. Dinghu Mountains, Guangdong province, China, where plants of Opithandra dinghushanensis W. T. Wang are mainly distributed.

Gene Cloning and Sequence Analyses

CYC-like genes were isolated from O.dinghushanensis using degenerate oligonucleotide primers in 3' and 5'-RACE according to described methods [47] and the manufacture's protocol (INVITROGEN), respectively. Two D3-type cyclin genes were also isolated from O. dinghushanensis using the above methods. Total RNA was extracted from the floral buds of O. dinghushanensis using the Plant RNA Purification Reagent (INVITROGEN) according to the manufacture's protocol. First-strand cDNAs were synthesized from total RNA with the Supertranscript™ III RNase H- Reverse Transcriptase (INVITROGEN). To examine the intron/exon structures we isolated and sequenced the corresponding genomic DNA of OpdCYC and OpdcyclinD3 genes from leaves.

According to the known sequence information, phylogenetic analyses of CYC-like and D-type cyclin genes were conducted to identify the position of CYC-like and D3-type cyclin genes isolated herein in their gene families, respectively. AmCYC and AmDICH are from A. majus [4,5], and LjCYC1/2/3, TCP1/12/18 and TB1 are from Lotus japonicus [13], Arabidopsis [12] and maize [34]. GCYC1A/1B are from Saintpaulia ionantha and Streptocarpus primulifolius [43-45]. GCYC1C/1D (designated as GCYC1 respectively in some taxa) are from Chirita heterotricha [25], Jankaea heldeichii [43], Conandron ramondioides, Haberlea ferdinandi-coburgii, Primulina tabacum, Ramonda myconi [44], Cyrtandra apiculata, Didymocarpus citrinus and Loxostigma sp. [46]. GCYC2 (2A/2B) are from Chirita heterotricha [25], Conandron ramondioides, Haberlea ferdinandi-coburgii, Ramonda myconi [44], Cyrtandra apiculata [46] and Didymocarpus citrinus (Yin-Zheng Wang, unpublished). Amino acid sequences of D-type cyclin genes are from A. majus (Amcyclin D1/D3a/D3b) [6], Populus trichocarpa (Ptcyclin D1/D2/D3) [48], Nicotiana tabacum (Ntcyclin D/D3) [49], Solanum tuberosum (Stcyclin D) (accession nos.EU325650), Lycopersicon esculentum (Lecyclin D3-1) [50] and Arabidopsis (Atcyclin D1/D2/D3) [32]. Phylogenetic analyses with the neighbor joining method and p-distance were carried out using PAUP*4.0b4a [51] and bootstrap was estimated with 1,000 resampling replicates. The DNA sequences of genes reported in this paper, i.e. OpdCYC1C, OpdCYC1D, OpdCYC2A, OpdCYC2B, OpdcyclinD3a and OpdcyclinD3b have been deposited in the GenBank database (accession nos. FJ710518, FJ710519, FJ710520, FJ710521, FJ710522 and FJ644637).

RNA In Situ Hybridization

Floral tissue for in situ hybridization was fixed, sectioned and hybridized to digoxygenin-labeled probes of OpdCYC1C, OpdCYC1D, OpdCYC2A and OpdCYC2B with reference to described methods [52]. Four gene-specific fragments of OpdCYC1C, OpdCYC1D, OpdCYC2A and OpdCYC2B in the coding region were amplified, respectively, and then were purified and cloned into pGEM®-T Easy vectors. Digoxygenin-labeled probes of OpdCYC1C, OpdCYC1D, OpdCYC2A and OpdCYC2B were prepared from linearized templates amplified using primer Yt7 and Ysp6 from pGEM®-T plasmids [53].

Gene-Specific Semiquantitative RT-PCR

Flowers of different stages were collected as follows: Flower buds of middle-stage (less than 1 cm long) and flowers of late-stage (3-4 cm long) were collected separately. Sepals were removed from the outer whorl. The petals with corresponding corolla-tube were dissected into dorsal (including the attached dorsal staminode), lateral, and ventral regions. Lateral stamens and ventral staminodes were dissected from the corolla-tube and collected each for RT-PCR. All materials were frozen in liquid nitrogen immediately after collection for ribonucleic acid (RNA) isolation. The extraction of total RNAs, purification of poly (A) mRNAs, and synthesis of the first-strand cDNAs were performed according to the methods described above. The template quantity was regulated to be uniform using the ACTIN gene [54]. PCR was performed by using gene-specific primers of OpdCYC1C, OpdCYC1D, OpdCYC2A, OpdCYC2B, OpdcyclinD3a and OpdcyclinD3b. To make sure that each pair of primers was suitable, we first used them to amplify genomic DNA of O. dinghushanensis. The PCR products were then cloned. At least 20 clones of each PCR product were sequenced, and all the primers used could amplify the specific copies of OpdCYC and OpdcyclinD3 genes. The following thermocycling conditions were employed: initial denaturation at 96°C for 3 min, 30 cycles of 96°C for 30 s, 55-60°C (depending on the Tm value of primer pairs) for 30s, and 72°C for 1 min, and a final extension at 72°C for 10 min. The amplified products were separated on a 1.5% agarose gel, and the density of ethidium bromide-stained bands was determined using a Bioimaging System (Gene Tools Program, Syngene, UK). We repeated the RT-PCR experiments five times independently with a new RNA extraction each time. In addition, all RT-PCR products were cloned into pGEM-T Easy-vector, and at least 20 clones from each product were sequenced to test the gene specificity of RT-PCR. The OpdCYC/ACTIN and Opdcyclin/ACTIN ratios represented the relative level of OpdCYC and OpdcyclinD3 mRNA expression. Data are presented as the mean ± SD of independent RT-PCR experiments, and one-way analysis of variance was

used to analyze the expression difference of these transcripts in floral tissue from
O. dinghushanensis. A P value less than 0.05 was taken to indicate statistical
significance.

Authors' Contributions

CFS and QBL isolated OpdCYC and OpdcyclinD3 genes and performed the
laboratory work of in situ hybridization and gene-specific RT-PCR. RHL carried
out the phylogenetic analyses of OpdCYC and OpdcyclinD3 genes. YZW conceived of and designed the studies, and CFS and QBL participated in the design.
YZW drafted the manuscript and all the authors participated in the editing of the
manuscript. All the authors read and approved the final manuscript.

Acknowledgements

We thank Dr. James Smith for his critical comments and language improvement
on the manuscript. This work was supported by CAS Grant KSCX2-YW-R-135
and National Natural Science Foundation of China Grant, no.30770147.

References

1. Cubas P: Floral zygomorphy, the recurring evolution of a successful trait. BioEs-
    says 2004, 26:1175–1184.

2. Endress PK: Antirrhinum and Asteridae - evolutionary changes of floral sym-
    metry. Symposium Series of the Society of Experimental Biology 1998, 51:133–
    140.

3. Preston JC, Hileman LC: Developmental genetics of floral symmetry evolution.
    Trends in Plant Science 2009, 14(3):147–154.

4. Luo D, Carpenter R, Vincent C, Copsey L, Coen E: Origin of floral asymmetry
    in Antirrhinum Nature 1996, 383:794–799.

5. Luo D, Carpenter R, Copsey L, Vincent C, Clark J, Coen E: Control of organ
    asymmetry in flowers of Antirrhinum Cell 1999, 99:367–376.

6. Gaudin V, Lunness PA, Fobert PR, Towers M, Riou-Khamlichi C, Murray JA,
    Coen E, Doonan JH: The expression of D-cyclin genes defines distinct devel-
    opmental zones in snapdragon apical meristems and is locally regulated by the
    Cycloidea gene. Plant Physiology 2000, 122:1137–1148.

7. Doebley J, Stec A, Gustus C: teosinte branched1 and the origin of maize: evidence for epistasis and the evolution of dominance. Genetics 1995, 141:333–346.

8. Doebley J, Stec A, Hubbard L: The evolution of apical dominance in maize. Nature 1997, 386:485–488.

9. Kosugi S, Ohashi Y: PCF1 and PCF2 specifically bind to cis-elements in the rice proliferating cell nuclear antigen gene. Plant Cell 1997, 9:1607–1619.

10. Kosugi S, Ohashi Y: DNA binding and dimerization specificity and potential targets for the TCP protein family. Plant Journal 2002, 30:337–348.

11. Cubas P, Lauter N, Doebley J, Coen E: The TCP domain: a motif found in proteins regulating plant growth and development. Plant Journal 1999, 18:215–222.

12. Howarth DG, Donoghue MJ: Phylogenetic analysis of the "ECE" (CYC/TB1) clade reveals duplications predating the core eudicots. Proceedings of the National Academy of Sciences, USA 2006, 103:9101–9106.

13. Feng X, Zhao Z, Tian Z, Xu S, Luo Y, Cai Z, Wang Y, Yang J, Wang Z, Weng L, Chen J, Zheng L, Guo X, Luo J, Sato S, Tabata S, Ma W, Cao X, Hu X, Sun C, Luo D: Control of petal shape and floral zygomorphy in Lotus japonicus. Proceedings of the National Academy of Sciences, USA 2006, 103:4970–4975.

14. Wang Z, Luo YH, Li X, Wang LP, Xu SL, Yang J, Weng L, Sato S, Tabata S, Ambrose M, Rameau C, Feng XZ, Hu XH, Luo D: Genetic control of floral zygomorphy in pea (Pisum sativum L.). Proceedings of the National Academy of Sciences, USA 2008, 105:10414–10419.

15. Cubas P, Coen E, Zapater JM: Ancient asymmetries in the evolution of flowers. Current Biology 2001, 11:1050–1052.

16. Busch A, Zachgo S: Control of corolla monosymmetry in the Brassicaceae Iberis amara . Proceedings of the National Academy of Sciences, USA 2007, 104:16714–16719.

17. Broholm SK, Tähtiharju S, Laitinen RAE, Albert VA, Teeri TH, Elomaa P: A TCP domain transcription factor controls flower type specification along the radial axis of the Gerbera (Asteraceae) inflorescence. Proceedings of the National Academy of Sciences, USA 2008, 105:9117–9122.

18. Kim M, Cui M-L, Cubas P, Gillies A, Lee K, Chapman MA, Abbott RJ, Coen E: Regulatory genes control a key morphological and ecological trait transferred between Species. Science 2008, 322:1116–1119.

19. Donoghue MJ, Ree RH, Baum DA: Phylogeny and the evolution of flower symmetry in the Asteridae. Trends in Plant Science 1998, 3:311–317.

20. Endress PK: Symmetry in flowers: diversity and evolution. International Journal of Plant Sciences (Suppl.) 1999, 160:3–23.

21. Preston JC, Kost MA, Hileman C: Conservation and diversification of the symmetry developmental program among close relatives of snapdragon with divergent floral morphologies. New Phytologist 2009, 182(3):751–762.

22. Xia Z, Wang Y-Z, Smith JF: Familial placement and relations of Rehmannia and Triaenophora (Scrophulariaceae s.l.) inferred from five gene regions. American Journal of Botany 2009, 96(2):519–530.

23. Hileman LC, Kramer EM, Baum DA: Differential regulation of symmetry genes and the evolution of floral morphologies. Proceedings of the National Academy of Sciences, USA 2003, 100:12814–12819.

24. Li ZY, Wang YZ: Plants of Gesneriaceae in China. Zhengzhou, China, Henan Science & Technology Publishing House; 2004.

25. Gao Q, Tao JH, Yan D, Wang YZ: Expression differentiation of floral symmetry CYC-like genes correlated with their protein sequence divergence in Chirita heterotricha (Gesneriaceae). Development Genes and Evolution 2008, 218:341–351.

26. Kalisz S, Ree RH, Sargent RD: Linking floral symmetry genes to breeding system evolution. Trends in Plant Science 2006, 11:568–573.

27. Citerne HL, Pennington RT, Cronk QCB: An apparent reversal in floral symmetry in the legume Cadia is a homeotic transformation. Proceedings of the National Academy of Sciences, USA 2006, 103:12017–12020.

28. Wortley AH, Rudall PJ, Harris DJ, Scotland RW: How much data are needed to resolve a difficult phylogeny? case study in Lamiales. Systematic Biology 2005, 54:697–709.

29. Nugent JH, Alfa CE, Young T, Hyams JS: Conserved structural motifs in cyclins identified by sequence analysis. Journal of Cell Science 1991, 99:669–674.

30. Dowdy SF, Hinds PW, Louie K, Reed SI, Arnold A, Weinberg RA: Physical interaction of the retinoblastoma protein with human D cyclins. Cell 1993, 73:499–511.

31. Kato J, Matsushime H, Hiebert SW, Ewen ME, Sherr CJ: Direct binding of cyclin D to the retinoblastoma gene product (pRb) and pRb phosphorylation by the cyclin D-dependent kinase CDK4. Genes & Development 1993, 7:331–342.

32. Soni R, Carmichael JP, Shah ZH, Murray JA: A family of cyclin D homologs from plants differentially controlled by growth regulators and containing the

conserved retinoblastoma protein interaction motif. Plant Cell 1995, 7:85–103.

33. Meijer M, Murray JA: The role and regulation of D-type cyclins in the plant cell cycle. Plant Molecular Biology 2000, 43:621–633.

34. Hubbard L, McSteen P, Doebley J, Hake S: Expression patterns and mutant phenotype of teosinte branched1 correlate with growth suppression in maize and teosinte. Genetics 2002, 162:1927–1935.

35. Zhou X-R, Wang Y-Z, Smith JF, Chen RJ: Altered expression patterns of TCP and MYB genes relating to the floral developmental transition from initial zygomorphy to actinomorphy in Bournea (Gesneriaceae). New Phytologist 2008, 178:532–543.

36. Perez-Rodriguez M, Jaffe FW, Butelli E, Glover BJ, Martin C: Development of three different cell types is associated with the activity of a specific MYB transcription factor in the ventral petal of Antirrhinum majus flowers. Development 2005, 132:359–370.

37. Nath U, Crawford BC, Carpenter R, Coen E: Genetic control of surface curvature. Science 2003, 299:1404–1407.

38. Palatnik JF, Allen E, Wu X, Schommer C, Schwab R, Carrington JC, Weigel D: Control of leaf morphogenesis by microRNAs. Nature 2003, 425:257–263.

39. Du ZY, Wang YZ: Significance of RT-PCR expression patterns of CYC-like genes in Oreocharis benthamii (Gesneriaceae). Journal of Systematics and Evolution 2008, 46(1):23–31.

40. Burtt BL: Studies in the Gesneriaceae of the old world XXXI: some aspects of functional evolution. Notes from the Royal Botanic Garden Edinburgh 1970, 15(1):1–10.

41. Wang WT, Pan KY, Li ZY: Key to the Gesneriaceae of China. Edinburgh Journal of Botany 1992, 49(1):5–74.

42. Weber A: Gesneriaceae. In The families and genera of vascular plants, Dicodyledons. Lamiales (except Acanthaceae including Avicenniaceae). Volume 7. Edited by: Kubitzki K, Kadereit JW. Berlin, Germany, Springer; 2004:63–158.

43. Möller M, Clokie M, Cubas P, Cronk QCB: Integrating molecular phylogenies and developmental genetics: a Gesneriaceae case study. In Molecular systematics and plant evolution. Edited by: Hollingsworth PM, Bateman RJ, Gornal RJ. London, UK, Taylor & Francis; 1999:375–402.

44. Citerne H, Möller M, Cronk QCB: Diversity of cycloidea-like genes in Gesneriaceae in relation to floral symmetry. Annals of Botany 2000, 86:167–176.

45. Wang L, Gao Q, Wang YZ, Lin QB: Isolation and sequence analysis of two CYC-like genes, SiCYC1A and SiCYC1B, from zygomorphic and actinomorphic cultivars of Saintpaulia ionantha (Gesneriaceae). Acta Phytotaxonomica Sinica 2006, 44:353–361.

46. Wang CN, Möller M, Cronk QCB: Phylogenetic position of Titanotrichum oldhamii (Gesneriaceae) inferred from four different gene regions. Systematic Botany 2004, 29:407–418.

47. Sambrook J, Russell DW: Molecular Cloning: a laboratory manual. Volume 2. Third edition. New York, USA, Cold Spring Harbor Laboratory Press; 2001:8.61–8.65.

48. Menges M, Pavesi G, Morandini P, Bogre L, Murray JA: Genomic organization and evolutionary conservation of plant D-type cyclins. Plant Physiology 2007, 145:1558–1576.

49. Sorrell DA, Combettes B, Chaubet-gigot N, Gigot C, Murray JA: Distinct cyclin D genes show mitotic accumulation or constant levels of transcripts in tobacco bright yellow-2 cells. Plant Physiology 1999, 119:343–352.

50. Joubes J, Walsh D, Raymond P, Chevalier C: Molecular characterization of the expression of distinct classes of cyclins during the early development of tomato fruit. Planta 2000, 211:430–439.

51. Swofford DL: PAUP*. Phylogenetic Analysis Using Parsimony (* and other methods). Version 4.0b10. Sinauer, Associates, Sunderland, MA; 2002.

52. Bradley D, Carpenter R, Sommer H, Hartley N, Coen E: Complementary floral homeotic phenotypes result from opposite orientations of a transposon at the plena locus of Antirrhinum. Cell 1993, 72:85–95.

53. Divjak M, Glare EM, Walters EH: Improvement of non-radioactive in situ hybridization in human airway tissues: use of PCR-generated templates for synthesis of probes and an antibody sandwich technique for detection of hybridization. Journal of Histochemistry and Cytochemistry 2002, 50:541–548.

54. Prasad K, Sriram P, Kumar CS, Kushalappa K, Vijayraghavan U: Ectopic expression of rice OsMADS1 reveals a role in specifying the lemma and palea, grass floral organs analogous to sepals. Development Genes and Evolution 2001, 211(6):281–290.

CITATION

Originally published under the Creative Commons Attribution License. Song C-F, Lin Q-B, Liang R-H, Wang Y-Z. Expressions of ECE-CYC2 Clade Genes Relating to Abortion of Both Dorsal and Ventral Stamens in Opithandra (Gesneriaceae). BMC Evolutionary Biology 2009, 9:244 doi:10.1186/1471-2148-9-244.

A Comparative Analysis of Pollinator Type and Pollen Ornamentation in the Araceae and the Arecaceae, Two Unrelated Families of the Monocots

Julie Sannier, William J. Baker, Marie-Charlotte Anstett and
Sophie Nadot

ABSTRACT

Background

The high diversity of ornamentation type in pollen grains of angiosperms has often been suggested to be linked to diversity in pollination systems. It is commonly stated that smooth pollen grains are associated with wind or water

pollination while sculptured pollen grains are associated with biotic polli-
nation. We tested the statistical significance of an association between pollen
ornamentation and pollination system in two families of the monocotyledons,
the Araceae and the Arecaceae, taking into account the phylogenetic frame-
work.

Findings

Character optimization was carried out with the Maximum Parsimony
method and two different methods of comparative analysis were used: the
Concentrated-Change test and the Discrete method. The ancestral ornamen-
tation in Araceae is foveolate/reticulate. It is probably the same in Arecaceae.
The ancestral flowers of Araceae were pollinated by beetles while ancestral pol-
lination in Arecaceae is equivocal. A correlation between ornamentation type
and pollination was highlighted in Araceae although the results slightly differ
depending on the method and the options chosen for performing the analyses.
No correlation was found in palms.

Conclusion

In this study, we show that the relationships between the ornamentation type
and the pollination system depend on the family and hence vary among tax-
onomic groups. We also show that the method chosen may strongly influence
the results.

Findings

The exine wall of the pollen grains of flowering plants displays patterns of ornamentation (the external aspect of pollen grains, also called sculpturing) that are
highly diversified. The reasons accounting for such variation in the ornamentation of pollen grains in flowering plants still remain unclear. Among the different
types of relationship implying pollen ornamentation that have been suggested,
the existence of a link between exine sculpturing and pollinator type has often
been proposed and was even evidenced in certain situations. It is often considered
that smooth pollen grains are associated with abiotic pollination (wind or water)
while echinulate or reticulate pollen grains are associated with biotic pollination,
particularly entomophily [1,2]. These results show that the adaptiveness of this
character still remains largely debated.

The study presented here aims to test the hypothesis suggested by Grayum [3]
concerning a relationship between pollen ornamentation and pollinator type in
the Araceae, using Phylogenetic Comparative Methods. He established a correlation between (a) psilate and verrucate pollen and pollination by beetles and (b)

echinulate pollen and pollination by flies. We think that the flaw of this study is inherent to the fact that correlations were established without statistical analysis and without taking into account the phylogenetic background of the family, making it impossible to know whether the correlations observed between the pollen and pollinator types result from adaptation or from common ancestry.

The processes underlying a relationship between two characters remain generally extremely difficult to determine [4,5]. A correlation may be the result of adaptation, but also of developmental constraints. It may also be simply the result of phylogenetic inertia i.e., that related species resemble each other more than they resemble species drawn at random [6]. Various mathematical approaches, called Phylogenetic Comparative Methods or PCM [4,7], have been proposed over the last twenty years [8-10] and take into account the phylogenetic background of the organisms studied.

Here we re-examine the correlation between pollen sculpturing and pollinator type proposed by Grayum [3], in light of the phylogenetic framework available for the Araceae family [11] using two PCM applied to discrete characters. In the conclusion of his paper, Grayum suggested to investigate other groups of mono-cotyledons, palms in particular. In this family a large amount of pollen data has been recorded but rarely studied from an evolutionary point of view, except for the number of apertures [12]. Moreover data on pollinators are available and a detailed and well resolved phylogeny including almost all of the genera [13] now exists. Consequently we also examine the correlation between pollen and pollinator types in the palm family (Arecaceae).

Methods

Character optimization was carried out with the Maximum Parsimony method implemented in the Mesquite software [14].

Two PCMs were used: the Concentrated Changes Test or CCT [9] and Discrete [10].

Results and Discussion

Character Evolution in the Araceae

To our knowledge, there is little data in the literature concerning the evolution of ornamentation of pollen grains in monocots [15]. Concerning the angiosperms, a recent study showed that the ancestral exine structure had a continuous or microperforate surface [16]. However, foveolate-reticulate tectum would have arisen soon after [16]. The work of Grayum [17] that is re-examined here, provides hypotheses about the ancestral and derived states of pollen wall sculpturing within the Araceae (monocots). His proposition that the most primitive aroid pollen had foveolate to reticulate exine is not in contradiction with our phylogenetic analysis of the character. Indeed, our results suggested that the hypothetical aroid pollen was either 'Foveolate/Reticulate' or 'Psilate' for pollen ornamentation (Figure 1A). The equivocal ancestral state is probably due to the polytomies, coded as soft (uncertainty in resolution), that are present in the tree. From this equivocal type, different types of sculpturing evolved [17]. However, no type of ornamentation is restricted to one clade and no particular trend in the evolution of the character emerged clearly from our analysis. It can be noted that several ornamentations originated several times independently and most of all, from different character states. This indicates that some transitions may occur indifferently among the different states.

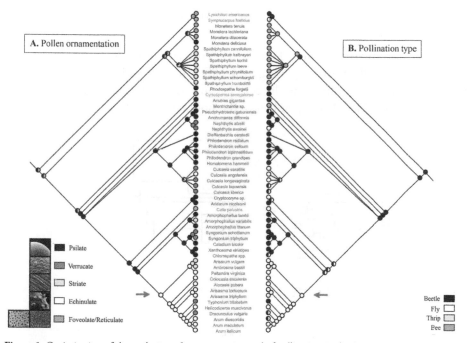

Figure 1. Optimization of the evolution of ornamentation and of pollination in the Araceae.
Composite phylogenetic tree of the family where each mirror tree presents the optimization of one character. A. Optimization of ornamentation type (five character states: Psilate, Verrucate, Striate, Echinulate and Foveolate/Reticulate). Pictures are given as illustration for each of these types, they do not correspond to a particular species of the family. They were obtained from http://www.paldat.org. B. Optimization of pollination type (four character states: Beetle, Fly, Thrip and Bee pollination). Species names are coloured according to the subfamilies (Orontioideae in pink, Monsteroideae in blue, Lasioideae in orange, Calloideae in green and Aroideae in red). The last common ancestor of the group Areae+Arisaemateae is indicated by a red arrow.

Concerning the pollination type, optimization of character evolution suggested that the last common ancestor of the family was pollinated by 'Beetle' (Figure 1B), according to outgroup comparison with other Alismatales [17]. From this condition, the other types of pollination each evolved several times within the family. In particular, fly pollination is clearly derived from beetle pollination in Aroideae, where it evolved in several unrelated genera and is synapomorphic for the Arisaemateae+Areae clade (indicated by a red arrow in Figure 1).

After the transformation of the coding from multistate to binary characters (Figure 2), we sought to test a correlation between pollen ornamentation and pollinators using the concentrated-change test [9] and the discrete method [10]. When polymorphic species were removed, both methods found a correlation between the ornamentation and the pollination (Tables 1 and 2). The Echinulate type was found as contingent upon Fly pollination (changes towards Echinulate pollen happened more often in Fly pollinated taxa; Table 1) with the CCT only. However, with the Discrete method, the flow diagram showed that transitions towards Fly pollination were probably followed by transitions towards Echinulate ornamentation (Figure 3).

When polymorphic species were duplicated, a correlation was found between ornamentation and pollination only using the CCT. With the coding 'Echinulate' ornamentation and 'Fly' pollination, the correlation was significant when the ornamentation was considered as dependent of the pollinator (Table 2).

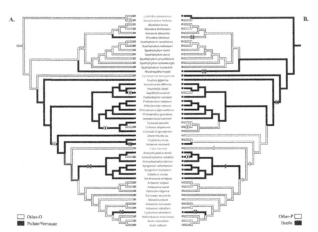

Figure 2. Evolution of ornamentation and pollination in Araceae with polymorphic species removed.
A. Optimization of ornamentation type coded as 'Other-O' (white) and 'Psilate/Verrucate' (black). B. Optimization of pollination type coded as 'Other-P' (white) and 'Beetle' (black). The bicoloured branches indicate an equivocal inference of the ancestral character state. The transitions towards 'Beetle' pollination and 'Psilate/Verrucate' ornamentation are indicated by full crossbars and the reversals towards 'Other-P' pollination and 'Other-O' ornamentation are indicated by open crossbars (red and blue crossbars correspond respectively to ACCTRAN and DELTRAN optimizations). Species names are coloured according to the subfamilies (Orontioideae in pink, Monsteroideae in blue, Lasioideae in orange, Calloideae in green and Aroideae in red).

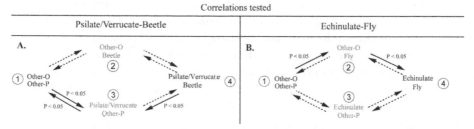

Figure 3. Flow diagrams of correlated evolution between ornamentation and pollination in Araceae and Arecaceae.

Flow diagrams of correlated evolution between 'Psilate/Verrucate' ornamentation and 'Beetle' pollination (A-B) and between 'Echinulate' ornamentation and 'Fly' pollination (C-D) in the Araceae. Solid arrows indicate significant transitions; dotted arrows indicate non-significant transitions. The larger the arrow, the greater the level of significance. The numbers correspond to the different situations. The situations in gray correspond to transitional intermediate states. A-C: polymorphic species duplicated. B-D: polymorphic species removed.

Table 1. Comparative analyses conducted with the Concentrated-Changes Test [9] in Araceae.

		P-value
Coding 1	P/V → B	P < 0.05
	B → P/V	P < 0.05
Coding 2	E → F	NS
	F → E	P < 0.05

Table 2. Comparative analyses performed using Discrete.

Family	Polymorphic species	Type of test	|LR|	df	P
Araceae	Removed	• Omnibus test (P/V correlated with B)	12.80	4	< 0.05
		○ Contingent change test:			
		P/V → B	2.90	1	NS
		B → P/V	0.77	1	NS
		◈ Temporal order test	1.78	1	NS
		• Omnibus test (E correlated with F)	15.34	4	< 0.01
		○ Contingent change test:			
		E → F	0	1	NS
		F → E	0.28	1	NS
		◈ Temporal order test	0.33	1	NS
	Duplicated	• Omnibus test (P/V correlated with B)	0.86	4	NS
		• Omnibus test (E correlated with F)	0.29	4	NS
Arecaceae	Duplicated	• Omnibus test (P/V correlated with B)	2.61	4	NS
		• Omnibus test (E correlated with F)	0.3	4	NS

These results indicate that the impact of adding/removing information on the detection of a correlation varies according to the method used (CCT or Discrete). The interpretation of the results may then be strongly influenced by the method chosen, as already shown [18]. In our case, duplicating polymorphic species leads to increase the number of opposing associations (associations for which we do not seek correlation). However, treat polymorphic species as pairs of species (duplicate) with contrasting character states [19] is the most conservative option and avoids loss of information. In conclusion, it is important to be aware of this problematic when a choice has to be made.

Character Evolution in the Arecaceae

In spite of the important literature on pollen ornamentation available for the Arecaceae [20], no steady hypotheses have been proposed about the ancestral ornamentation, apart from a suggestion that psilate exine could be the primitive condition [21]. The present study is the first that makes hypotheses about the ancestral features of pollen grain in palms using a phylogenetic approach. In our analysis, the ancestral character state for the family was inferred as 'Echinulate' (Figure 4A). However, the reconstruction of the ornamentation character on a phylogeny of the family including all of the genera inferred a 'Foveolate/Reticulate' ancestral character state (personal information). This conflict is explained by the presence of only polymorphic genera in Calamoideae in our study.

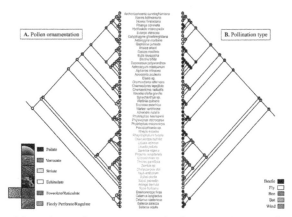

Figure 4. Optimization of the evolution of ornamentation and pollination in Arecaceae. Supertree of Arecaceae where each mirror tree presents the optimization of one character.

A. Optimization of ornamentation type (five character states: Psilate, Verrucate, Striate, Echinulate and Foveolate/Reticulate). Pictures are given as illustration for each of these types, they do not correspond to a particular species of the family. They were obtained from http://www.paldat.org. B. Optimization of pollination type (five character states: Beetle, Fly, Bee, Bat and Wind pollination). Species names are coloured according to the subfamilies (Calamoideae in blue, Nypoideae in orange, Coryphoideae in green, Ceroxyloideae in pink and Arecoideae in red).

The Pollination type appeared as a very variable character and consequently, no clear evolutionary trend emerged from the character optimisation (Figure 4B). The ancestral pollinators either were bees, beetles or flies, and the pollination type for each subfamily (except Nypoideae) was ambiguous. In this family, even with the binary coding (Figure 5), polymorphic species were so numerous that when they were removed, according to one of the option chose, there were not enough data left to perform any test. When polymorphic species were duplicated, as a result of the high variability in characters, the comparative analyses failed to detect a correlation between ornamentation type and pollination type in Arecaceae, whatever the method used (Tables 2 and 3).

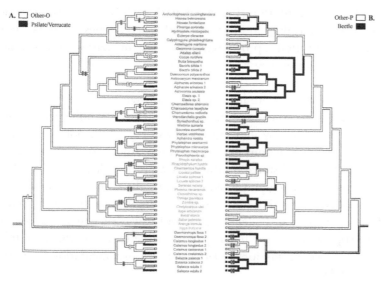

Figure 5. Evolution of ornamentation and pollination in Arecaceae with polymorphic species duplicated.
A. Optimization of ornamentation type coded as 'Other-O' (white) and 'Psilate/Verrucate' (black). B. Optimization of pollination type coded as 'Other-P' (white) and 'Beetle' (black). The bicoloured branches indicate an equivocal inference of the ancestral character state. The transitions towards 'Beetle' pollination and 'Psilate/Verrucate' ornamentation are indicated by full crossbars and the reversals towards 'Other-P' pollination and 'Other-O' ornamentation are indicated by open crossbars (red and blue crossbars correspond respectively to the ACCTRAN and DELTRAN optimizations). Species names are coloured according to the subfamilies (Calamoideae in blue, Nypoideae in orange, Coryphoideae in green, Ceroxyloideae in pink and Arecoideae in red).

Table 3. Comparative analyses conducted with the Concentrated-Changes Test [9] in Arecaceae.

		P-value
Coding I	P/V → B	NS
	B → P/V	NS
Coding 2	E → F	NS
	F → E	NS

Relationships between Ornamentation Type and Pollination

The reason why Angiosperms display such a large diversity of pollen ornamentation remains to date rather unclear. Several studies addressing this question have been produced, often leading to conflicting, or at least different, conclusions. Most studies however examine the hypothesis of a link between pollen ornamentation and the pollination system. The underlying idea is that since pollen grains need a vector (biotic or abiotic) to reach the female parts, the pollen surface may play a role in the efficiency of the interaction either with the pollination agent or with the receptive area of the female organs. A relationship between abiotic pollination (wind and water) and smooth (or nearly) pollen grains has been proposed in many studies [1,22] but concerning biotic pollination, the results are more controversial. [2,3,23,24]. According to previously published studies [3,20,24], our results suggest that the relationship between pollen ornamentation and pollinators may actually depend on the taxon. The association between psilate (=smooth) pollen and beetles seems to be rather specific to the Araceae, since entomogamous species are generally thought to produce pollen grains with a deeply sculptured exine [1,23]. The idea is that the sculptures would enhance the adherence of the pollen grains to the insect body by allowing a better storage of the pollenkitt. This sticky substance, of which functions are not yet quite understood, is produced by entomogamous species and stored on the surface of the pollen wall [25]. The pollenkitt would enable pollen grains to adhere on the hairs of insects or on the feathers of birds in case of ornithophily [23,24]. In the Araceae however, pollen grains were depicted as poor in pollenkitt [26] and it was suggested that sticky secretions on the stigma and/or the inner spathe surface may play the same role as the pollenkitt [27], accounting for the lack of pollen ornamentation in beetle pollinated species of Araceae.

The fact that no correlations could be detected in the palm family may be due to various reasons. First, it has to be stressed that the sampling was sparser for palms than for the Araceae. In particular, there was a poor overlap at the species level between the pollen and the pollinator datasets. This led us to combine information from different species for the ornamentation type, which we are aware may be questionable considering the high lability of the character even at the intraspecific level. We tried to overcome this problem by attributing polymorphic character states whenever intrageneric diversity was recorded, and by applying two different treatments to polymorphic species in the comparative analyses but we cannot exclude that the choices made here (due to the scarcity of data concerning pollination systems in palms) may have biased the results. However, there is a possibility that our results indeed reflect the reality and that pollen ornamentation is not involved in the pollination syndrome in palms. In this family, the traits linked to pollinator identity still remain almost unknown. Palm flowers are relatively poorly diversified in morphology when compared to the spectacular flowers of other groups [20]. In many species, flowers are visited by many insect families and species (often more than 50 species), although maybe among these insects only a single or a small number of species effectively act as pollinators [28]. The lack of correlation between pollen ornamentation and pollinators may be accounted for by a weak degree of specialization in the pollination system, or it may be that some other factors (like pollenkitt or scents for example) may be dominant in the pollen-pollinator interaction.

In conclusion, it is our feeling that there is little to be gained from seeking a general tendency concerning the relationship between the type of pollen ornamentation and the pollination system across the angiosperms. The ornamentation of the pollen wall is only one of the numerous elements that constitute the pollination syndrome and certainly not the most important [29]. Like the other factors [30], the relative importance of its role in the plant-pollinator interaction may indeed vary among plant taxa or according to geography, and it may have an adaptive value in some groups but evolve in a neutral way in other groups.

Competing Interests

The authors declare that they have no competing interests.

Authors' Contributions

JS performed the coding and the optimization of the reconstruction of the characters, carried out all the comparative analyses, interpreted the data and contributed to writing the manuscript. WJB provided the phylogenetic framework for the Arecaceae. MCA assembled the dataset on the pollination type in Arecaceae. SN supervised the study and contributed to writing the manuscript. All authors read and approved the final manuscript.

Acknowledgements

This work was supported by the Laboratoire d'Écologie, Systématique et Évolution of the Université Paris-Sud 11. Thanks to Florian Jabbour for all the constructive discussions on comparative analyses and to Madeline M. Harley for providing data on the ornamentation type in Arecaceae.

References

1. Lumaga MR, Cozzolino S, Kocyan A: Exine Micromorphology of Orchidinae (Orchidoideae, Orchidaceae): Phylogenetic Constraints or Ecological Influences? Annals of Botany 2006, 98:237–244.

2. Tanaka N, Uehara K, Murata J: Correlation between pollen morphology and pollination mechanisms in the Hydrocharitaceae. Journal of Plant Research 2004, 117:265–276.

3. Grayum MH: Correlations between pollination biology and pollen morphology in the Araceae, with some implications for angiosperm evolution. In Pollen and Spores: form and function. Edited by: Blackmore S, Ferguson IK. NY London: Academic Press; 1986:313–327.

4. Harvey PH, Pagel MD: The comparative method in evolutionary biology. New York: Oxford University Press; 1991.

5. Leroi AM, Rose MR, Lauder GV: What does the comparative method reveal about adaptation? The American Naturalist 1994, 143(3):381–402.

6. Blomberg SP, Garland TJ: Tempo and mode in evolution: phylogenetic inertia, adaptation and comparative methods. Journal of Evolutionary Biology 2002, 15:899–910.

7. Morand S: Comparative analyses of continuous data: the need to be phylogenetically correct. In The origin of biodiversity in insects: phylogenetic tests of evolutionary scenarios. Mém Mus natn Hist nat Edited by: Grandcolas P. 1997, 173:73–90.

8. Felsenstein J: Phylogenies and the comparative method. The American Naturalist 1985, 125(1):1–15.

9. Maddison WP: A method for testing the correlated evolution of txo binary characters: are gains or losses cencentrated on certain branches of a phylogenetic tree? Evolution 1990, 44(3):539–557.

10. Pagel M: Detecting correlated evolution on phylogenies: a general method for the comparative analysis of discrete characters. Proceedings of the Royal Society of London Series B: Biological Sciences 1994, 255:37–45.

11. Cabrera LI, Salazar GA, Chase MW, Mayo SJ, Bogner J, Davila P: Phylogenetic relationships of Aroids and Duckweeds (Araceae) inferred from coding and noncoding plastid DNA. American Journal of Botany 2008, 95(9):1153–1165.

12. Harley MM, Baker WJ: Pollen aperture morphology in Arecaceae: application within phylogenetic analyses, and a summary of the fossil record of palm like pollen. Grana 2001, 40:45–77.

13. Baker WJ, Savolainen V, Asmussen-Lange CB, Chase MW, Dransfield J, Forest F, Harley MM, Uhl NW, Wilkinson M: Complete generic level phylogenetic analyses of palms (Arecaceae) with comparisons of Supertree and Supermatrix approaches. Systematic Biology 2009, in press.

14. Maddison WP, Maddison DR: Mesquite: a modular system for evolutionary analysis. [http://mesquiteproject.org] Version 2.0 2006.

15. Ferguson IK, Havard AJ, Dransfield J: The pollen morphology of the tribe Borasseae (Palmae: Coryphoideae). Kew Bulletin 1987, 42(1):404–422.

16. Doyle JA: Early evolution of angiosperm pollen as inferred from molecular and morphological phylogenetic analyses. Grana 2005, 44:227–251.

17. Grayum MH: Evolution and phylogeny of the Araceae. Annals of the Missouri Botanical Garden 1990, 77:628–697.

18. Ward D, Seely MK: Adaptation and constraint in the evolution of the physiology and behavior of the namib desert tenebrionid beetle genus Onymacris. Evolution 1996, 50(3):1231–1240.

19. Donoghue MJ, Ackerly DD: Phylogenetic uncertainties and sensitivity analyses in comparative biology. Philosophical transactions of the Royal Society of London, serie B 1996, 351:1241–1249.

20. Dransfield J, Uhl NW, Asmussen CB, Baker WJ, Harley MM, Lewis CE: Genera Palmarum. The Evolution and Classification of Palms. Kew, GB: Kew Publishing; 2008.

21. Harley MM: Occurence of simple, tectate, monosulcate or trichotomosulcate pollen grains within the Palmae. Review of Palaeobotany and Palynology 1990, 64:137–147.

22. Osborn JM, El-Ghazaly G, Cooper RL: Development of the exineless pollen wall in Callitriche truncata (Callitrichaceae) and the evolution of underwater pollination. Plant Systematics and Evolution 2001, 228:81–87.

23. Ferguson IK, Skvarla JJ: Pollen morphology in relation to pollinators in Papilionoideae (Leguminosae). Botanical Journal of the Linnean Society 1982, 84:183–193.

24. Hesse M: Pollen wall stratification and pollination. Plant Systematics and Evolution 2000, 222:1–17.

25. Heslop-Harrison J: The adaptative significance of the exine. In The evolutionary significance of the exine. Edited by: Ferguson IK, Muller M. London: Academic press; 1976:27–37.

26. Hesse M: Entwicklungsgeschichte und ultrastruktur von polenkitt und exine bei nahe verwandten entomophilen und anemophilen angiospermensippen der

Alismataceae, Liliaceae, Juncaceae, Cyperaceae, Poaceae und Araceae. Plant Systematics and Evolution 1980, 134:229–267.

27. Gibernau M, Barabé D, Labat D: Flowering and pollination of Philodendron melinonii (Araceae) in French Guiana. Plant Biology 2000, 2:331–334.

28. Barfod AS, Burholt T, Borchsenius F: Contrasting pollination modes in three species of Licuala (Arecaceae: Coryphoideae). Telopea 2003, 10(1):207–223.

29. Fenster CB, Armbruster WS, Wilson P, Dudash MR, Thomson JD: Pollination syndromes and floral specialization. Annual Review of Ecology, Evolution and Systematic 2004, 35:375–403.

30. Ollerton J, Alarcón R, Waser NM, Price MV, Watts S, Cranmer L, Hingston A, Peter CI, Rotenberry J: A global test of the pollination syndrome hypothesis. Annals of Botany 2009, 103:1471–1480.

CITATION

Originally published under the Creative Commons Attribution License. Sannier J, Baker WJ, Anstett M-C, Nadot S. A Comparative Analysis of Pollinator Type and Pollen Ornamentation in the Araceae and the Arecaceae, Two Unrelated Families of the Monocots. BMC Research Notes 2009, 2:145. doi:10.1186/1756-0500-2-145.

Life History Traits in Selfing Versus Outcrossing Annuals: Exploring the 'Time-Limitation' Hypothesis for the Fitness Benefit of Self-Pollination

Rebecca Snell and Lonnie W. Aarssen

ABSTRACT

Background

Most self-pollinating plants are annuals. According to the 'time-limitation' hypothesis, this association between selfing and the annual life cycle has evolved as a consequence of strong r-selection, involving severe time-limitation for completing the life cycle. Under this model, selection from frequent density-independent mortality in ephemeral habitats minimizes time to flower maturation, with selfing as a trade-off, and / or selection minimizes the time between flower maturation and ovule fertilization, in which case selfing has a direct fitness benefit. Predictions arising from this hypothesis were evaluated using phylogenetically-independent contrasts of several life history traits in predominantly selfing versus outcrossing annuals from a data base of 118 species distributed across 14 families. Data for life history traits specifically related to maturation and pollination times were obtained by monitoring the start and completion of different stages of reproductive development in a greenhouse study of selfing and outcrossing annuals from an unbiased sample of 25 species involving five pair-wise family comparisons and four pairwise genus comparisons.

Results

Selfing annuals in general had significantly shorter plant heights, smaller flowers, shorter bud development times, shorter flower longevity and smaller seed sizes compared with their outcrossing annual relatives. Age at first flower did not differ significantly between selfing and outcrossing annuals.

Conclusions

This is the first multi-species study to report these general life-history differences between selfers and outcrossers among annuals exclusively. The results are all explained more parsimoniously by selection associated with time-limitation than by selection associated with pollinator/mate limitation. The shorter bud development time reported here for selfing annuals is predicted explicitly by the time-limitation hypothesis for the fitness benefit of selfing (and not by the alternative 'reproductive assurance' hypothesis associated with pollinator/mate limitation). Support for the time-limitation hypothesis is also evident from published surveys: whereas selfers and outcrossers are about equally represented among annual species as a whole, selfers occur in much higher frequencies among the annual species found in two of the most severely time-limited habitats where flowering plants grow – deserts and cultivated habitats.

Background

Most flowering plants that are predominantly self-pollinating have an annual life history [1-3]. Interpretations of this association usually involve one of two main hypotheses. (i) Compared with perennials, annuals may generally accrue greater fitness benefits from selfing through 'reproductive assurance', i.e., because ovules may be generally more outcross-pollen-limited and/or pollen grains may be more outcross-ovule-limited [2,4-8]. (ii) Perennials may incur a higher fitness cost of selfing through seed discounting and inbreeding depression; hence, possibly most selfers are annuals simply because relatively few perennials can be selfers [9,10].

A recent third hypothesis, the 'time-limitation' hypothesis, predicts that both selfing and the annual life cycle are concurrent products of strong 'r-selection' associated with high density-independent mortality risk in ephemeral habitats with a severely limited period of time available to complete the life cycle [11]. Both the traditional reproductive assurance hypothesis and the time-limitation hypothesis involve a fitness advantage for selfing through ensuring that at least some reproduction occurs, but they involve very different selection mechanisms – pollinator/mate-limitation (where outcross pollen is not available at all due to a lack of pollinators or mates), versus time-limitation (where outcross pollen is available but arrives too late to allow sufficient time for development of viable seeds). Accordingly, these two hypotheses for selfing involve very different assumptions and predictions.

The time-limitation hypothesis has direct and indirect components. The indirect component predicts higher selfing rates in annuals as a trade-off of selection for earlier reproductive maturity in annuals [12,13] (Figure 1a). More rapid floral maturation is expected to result in smaller flowers with increased overlap of anther dehiscence and stigma receptivity in both space (reduced herkogamy) and time (reduced dichogamy) thus, increasing the frequency of selfing as an incidental consequence [12] (Figure 1a). If selfing also shortens the time between flower maturation and ovule fertilization, then higher selfing rates for annuals in time-limited habitats may also be predicted as a direct fitness benefit; abbreviating the time between anthesis and ovule fertilization may ensure that there is enough remaining time in the growing season (after ovule fertilization) to allow complete seed and fruit maturation [11] (Fig 1b). Selection favors selfing here by favoring increased overlap in anther dehiscence and stigma receptivity in both space and time, which are in turn facilitated by smaller flower size and shorter flower development time, respectively (Figure 1b).

However, the two components of time-limitation cannot be separated clearly, as they operate simultaneously; i.e., earlier onset of flowering, shorter flower development time, smaller flowers and selfing can all be interpreted to have direct fitness benefits because they may all contribute directly to accelerating the life cycle [11]. Indeed, time-limitation associated with strong r-selection would be expected also to favor an acceleration of the final stage in the life cycle – seed/fruit development time (Figure 1) – resulting, as a trade-off, in smaller seeds and/or fruits [11].

The time-limitation hypothesis remains untested. Some recent studies have explored the rapid growth and maturation time of annuals in terms of bud development rates and ontogeny [13-15]. However, these studies have compared growth and development rates between selfing and outcrossing populations of only a single species. Since their effective sample size is only one, this makes it difficult to extrapolate the predominant selection pressures that may have promoted the general association of selfing with the annual life cycle.

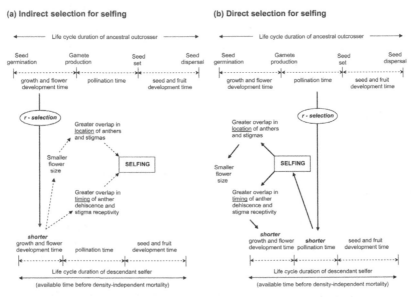

Figure 1. Two components of the 'time-limitation' hypothesis for the evolution selfing in annuals. In (a), selfing is a trade-off of selection favoring a shorter time to reproductive maturity (fully developed flowers) under strong r-selection. As a tradeoff (dashed arrows), flowers become smaller with greater overlap in location and timing of anther dehiscence and stigma receptivity, thus increasing the rate of selfing as an incidental consequence. In (b) strong r-selection favors a shorter pollination time directly; i.e., selfing is selected for directly because it shortens the amount of time between flower maturation and ovule fertilization, thus leaving sufficient remaining time for seed and fruit maturation before the inevitable early mortality of the maternal plant under strong r-selection. In this case, smaller flower size and shorter flower development time are favored by selection because they facilitate selfing (see text).

The objective of the present study was to compare, for annuals exclusively, life history traits associated with selfing versus outcrossing using several species from a wide range of plant families. Phylogenetically-independent contrasts (PIC) were used to control for confounding effects due to common ancestry among species [16]. Using a database of 118 species involving 14 families, plant size, flower size, and seed size were compared between selfing and outcrossing annuals. The time-limitation hypothesis predicts that all of these traits should be smaller in selfing annuals because the severely time-limited growing season that promotes selfing also imposes an upper limit on the maximum sizes that can be attained for plant traits [11] (Figure 1). The trend for outcrossers to be taller, and have larger flowers and larger seeds has often been noted [1,17-19]. We used a multi-species, across-family comparison, however, to investigate whether this trend also holds true within annuals exclusively.

Data on the timing of life history stages (i.e. age at first flower, bud development time, and flower longevity) were also obtained from a greenhouse study of 25 annual species involving 5 families. The time-limitation hypothesis predicts that selfers should produce mature flowers more quickly and should have shorter flowering times.

Results

Data Base Analyses

Based on phylogenetically-independent contrasts, selfing annuals had significantly shorter plant heights (Wilcoxon test for matched pairs, n = 12, T = 15.5, one-tailed P = 0.032, Figure 2a), significantly smaller flowers (Wilcoxon test for matched pairs, n = 14, T = 13, one-tailed P = 0.0054, Figure 2b), and significantly smaller seeds (Wilcoxon test for matched pairs, n = 13, T = 13, one-tailed P < 0.01, Figure 2c).

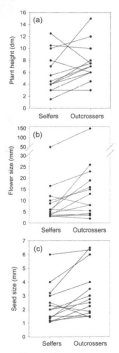

Figure 2. (a) Plant height contrasts for 13 selfing and outcrossing pairs (some points overlap), where each pair consists of the median value of the selfing and outcrossing species within one family. (b) Flower size contrasts for 14 selfing and outcrossing pairs. (c) Seed size contrasts between 14 selfing and outcrossing pairs.

Greenhouse Study

Bud development time (Figure 3) and flower longevity (Figure 4) were significantly (P < 0.05) shorter in selfing annuals in all of the families except the Fabaceae (P = 0.123 and P = 0.056 respectively). Selfing annuals also had significantly shorter bud development times (Figure 5), and floral longevities (Figure 6) in three of the four genus pairs. Selfing and outcrossing annuals of the genus Ipomoea did not differ significantly in either bud development time (P = 0.402) or flower longevity (P = 0.328). Age at first flower was not significantly related to mating system for any of the family or genus comparisons (P > 0.05; data not shown).

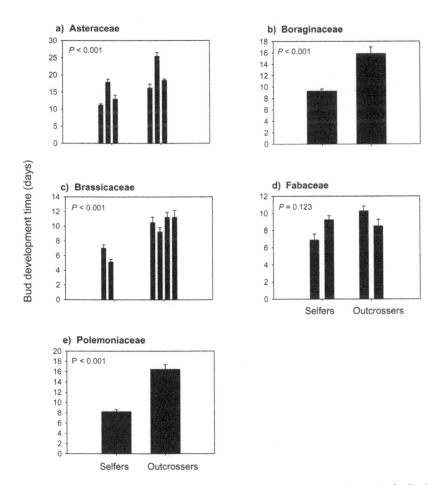

Figure 3. Mean (SE) bud development time for selfing and outcrossing species within each of 5 families. For each species, n = 4 or 5. P – values are from ANOVA.

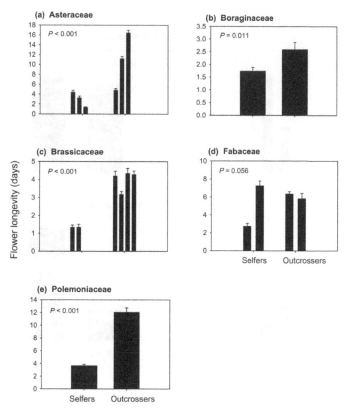

Figure 4. Mean (SE) flower longevity for selfing and outcrossing species within each of 5 families. For each species, n = 4 or 5. P – values are from ANOVA.

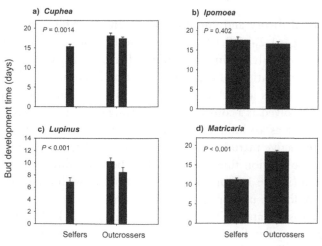

Figure 5. Mean (SE) bud development time for selfing and outcrossing species within each of 4 genus pairs. For each species, n = 4 or 5. P – values are from ANOVA.

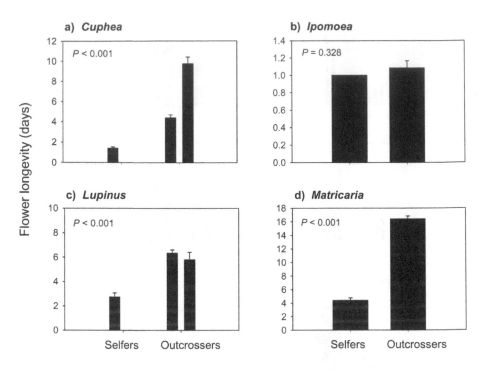

Figure 6. Mean (SE) flower longevity for selfing and outcrossing species within each of 4 genus pairs. For each species, n = 4 or 5. P – values are from ANOVA.

Discussion

There is a rich body of theory and empirical work on the evolution of selfing in flowering plants [e.g. [1,2,4-7,9,10]], but practically none of it involves an explicit role of selection involving time-limitation. The present paper is only the second to explore the implications of the time-limitation hypothesis and contribute to the maturation of this idea. According to the time-limitation hypothesis, selfing in annuals has evolved as a consequence of strong r-selection in ephemeral habitats, resulting either as an indirect consequence (trade-off) of selection for shorter time to reproductive maturity (Figure 1a), or as a direct consequence of selection for shorter pollination time, i.e., the time between flower maturation and ovule fertilization (Figure 1b), or both [11]. Consistent with the predictions of this hypothesis, we found, using phylogenetically-independent contrasts, that (compared with outcrossing annuals) selfing annuals in general had significantly shorter plant heights, smaller flowers, shorter bud development time, shorter flower longevity and smaller seed sizes.

At the same time, these results are not inconsistent with the predictions of selection resulting from pollinator/mate-limitation associated with the traditional reproductive assurance hypothesis. Just as with many situations where two different mechanisms can potentially produce the same outcome/pattern, it is not easy here to clearly distinguish between the roles of "pollinator/mate-limitation" and "time-limitation". Nevertheless there are two important contributions from our study: First, in reporting significant life history differences between selfers and outcrossers, our multi-species study is unique in its comparison of monocarpic annual species exclusively. All previous multi-species studies of trait comparisons between selfers and outcrossers have involved variable mixes of monocarpic and longer-lived polycarpic species. Second, by comparing annuals exclusively, our results provide indirect support for the time-limitation hypothesis, not by rejecting the role of pollinator/mate-limitation, but rather by representing a system in which it is more plausible to argue for the role of time-limitation; i.e., compared with pollinator/mate-limitation, time-limitation as a selection factor favoring selfing is likely to have been much stronger, more persistent and more widespread. The strength of this argument lies in the fact that the annual life history is unequivocally a product of some type of time-limitation favoring an abbreviated life cycle, which is promoted by (among other things) selfing (as opposed to outcrossing) (Fig. 1). It is much less plausible to suspect that selection associated with pollinator/mate-limitation has been sufficiently strong and persistent to favor selfing in such a wide range of annual taxa across the many genera and families considered here. We emphasize, therefore, that for annuals the time-limitation hypothesis provides a more parsimonious explanation for the differences in traits between selfers and outcrossers. We consider each of these traits in turn below.

Plant Height and Time to Anthesis

Taller plants may attract more pollinators and, hence, experience greater outcrossing rates [20,21]. The pollination benefit of being relatively tall, therefore, is presumably experienced only by outcrossers. If, however, selfers have evolved from outcrossers [3], then why should selfers be shorter than their outcrossing ancestors? The relatively small size, including short height of selfers can be predicted as an indirect consequence of selection, from time-limitation, favoring precocious maturation time [22,23] (Figure 1a). In the present study, however, selfers and outcrossers did not differ significantly in age at first flower. Andersson [18] found similar results between selfing and outcrossing populations of Crepis tectorum. Arroyo [24], however, reported that selfing individuals of Limnanthes floccosa flowered earlier than the outcrossing L. alba, as predicted by the time-limitation hypothesis. The results for flowering times in the present study may be confounded by the controlled greenhouse environment of constant day-length, temperature and moisture regime. In the field, flowering times may be triggered by environmental cues. L. floccosa, for example, uses soil moisture to trigger the early onset of flowering, thus escaping the detrimental effects of soil desiccation during seed development [24]. Note also that age at first flower is only a crude estimate of time to reproductive maturity. Future studies may employ more detailed measures such as rate of mature flower production.

Flower Size

One of the most well established trends of predominantly self-fertilizing species is their reduced flower sizes compared with outcrossing species [1,17]. The present results indicate that this trend is also evident even within annuals exclusively. In all but three of the 14 PICs, selfing annuals had smaller flowers than the outcrossing annuals (Figure 2b). Outcrossers and selfers had similar flower sizes in the Fabaceae and Plantaginaceae. In the Poaceae, outcrossing annuals had smaller flowers than selfers.

Under the time-limitation hypothesis, smaller flowers and selfing may be tradeoffs of selection for precocity (Figure 1a), or smaller flowers may be favored by selection because they promote selfing and hence, direct fitness benefits by abbreviating pollination time (Figure 1b). Also, if selfing evolves from outcrossing (by whatever mechanism), then selection may subsequently favour a reduction in flower size since relatively large flowers are no longer needed to attract pollinators. Hence, higher fitness may result if the resources required to construct and support these larger flowers are invested instead in other functions (e.g. seed and fruit development) [17].

Bud Development Time

Selfers had significantly shorter bud development times in all but one of the independent family contrasts (Figure 3) and all but one of the genus comparisons (Figure 5). Results from previous studies, however, are inconsistent. Shorter bud development times were found in selfing populations of Mimulus guttatus [25] and in Clarkia xantiana [14]. However, no significant differences in bud growth rates were found between the selfing and outcrossing populations of C. tembloriensis [15]. Hill, Lord and Shaw [13] reported that flowers from selfing populations of Arenaria uniflora develop over a longer period of time than observed in outcrossing populations. In the field, selfing populations of A. uniflora were also observed to flower at the same time or even later than outcrossing populations [13], suggesting that time-limitation is not currently a strong selection pressure. Self-fertilization in A. uniflora may have arisen through reproductive assurance in response to competition for pollinators [7]. The evolution of self-fertilizing species from outcrossing progenitors has occurred repeatedly and independently in several lineages [1,3,14], each of which may have been associated with different contexts of natural selection vis-à-vis the fitness benefits of selfing.

Flower Longevity

The families and genera in which selfers had shorter bud development times also had significantly shorter flower longevities (Figure 4). In fact, all of the selfers had flowers that remained open for less than four days (except in Trifolium hirtum; Fabaceae), with a large proportion of flowers open for only one day, which is common amongst self-fertilizing species [17]. The present data again indicate that this generalization apparently holds true even within annuals exclusively. By having flowers that remain open longer, outcrossers increase the probability of visitation by pollinators and successful cross-pollination [17]. This fitness benefit is realized, however, only if there is sufficient time remaining after cross-pollination to complete seed and fruit development before the maternal plant succumbs to density-independent mortality in strongly r-selecting habitats [11]. If time is limiting in this context, selection should favor selfing (Figure 1b) with no advantage in having long-lived flowers.

It is important to note that our data measure maximum flower longevity, since there were no pollinators in the greenhouse, nor was hand pollination conducted. Pollination has been shown to induce floral senescence in numerous species [26]. This effect was not tested on any of the study species, which means that our observed flower longevities in outcrossing species may be longer than would normally be seen in the wild. Nevertheless, since selfing may have evolved as a method of shortening pollination time, and flower longevity was used as a measure of pollination time, the maximum floral longevity gives an indication of how long outcrossers can delay flower abscission or self-pollination (i.e. through delayed selfing).

Seed Size

Strong r-selection associated with the annual life form presumably favors wide dispersal mechanisms (for colonizing new and distant sites) which may be conferred by small seed sizes [19]. The reproductive assurance hypothesis would predict, therefore, that most selfers are annuals because annuals are more likely than perennials to disperse further, or colonize new habitats where conditions are unsuitable for successful outcrossing (because of a shortage of mates or pollinators) and where selfing, therefore, provides reproductive assurance. The present study indicates that even among annuals only, selfers have smaller seeds than outcrossers (Figure 2c). Future studies are required to test whether smaller-seeded selfing annuals are more likely than their outcrossing annual relatives to disperse further or colonize new habitats and thereby incur potential reproductive assurance benefits of selfing.

An alternative explanation, however, is offered by an extension of the time-limitation hypothesis: strong r-selection favors an acceleration of all stages of the life cycle (Figure 1), including not only earlier reproductive maturity (Figure 1a) and a shorter pollination time (facilitated through selfing) (Figure 1b), but also a shorter seed and fruit maturation time, which, on a per-seed basis, is facilitated in turn through the production of smaller seeds. Andersson [18] found that self-fertilizing individuals of Crepis tectorum took an average of 16 days for fruit maturation, whereas outcrossing individuals of the same species required 43.3 days. Small seed size may also be simply a trade-off of selection for high fecundity, also favored by strong r-selection [11].

Habitat Selection and Time-Limitation

While most selfers are annuals, it is not the case that most annuals are selfers. An unbiased literature survey [27] suggests that roughly half of all annual species are selfers and half are outcrossers. If, however, selfing annuals evolved in habitats with a short window of time for completing the life cycle (Figure 1), then selfing annuals should be significantly more common than expected (i.e. comprising greater than 50% of resident annuals) within habitats associated with historically regular, early-season disturbances (e.g. cultivated fields, gardens), or in habitats where severe droughts follow quickly after a wet season (i.e. deserts, Mediterranean climates, vernal pools). Hence, we should expect to find more selfers than outcrossers among annual weeds of cultivated habitats and among desert annuals in particular. Similarly, for annuals with both selfing and outcrossing ecotypes or races, we should expect selfers (or a higher selfing rate) to be more commonly associated with these severely time-limited habitats [11].

While rigorous tests of these predictions have yet to be explored, some preliminary support is available from published surveys. From a representative sample of Mediterranean annuals [28], we find a much greater representation of selfers: i.e. 34 selfers versus 11 outcrossers. Selfing and outcrossing desert annuals have been shown to be distributed along a moisture gradient. Outcrossing annuals are found generally in the wetter areas and selfers in the more arid zones, as seen in Clarkia xantiana [29] and between outcrossing populations of Limnanthes alba and its selfing relative L. floccosa [24]. Since the length of the growing season is limited by the amount of moisture in the soil, selfers have a much narrower window of time to complete their life cycle before desiccation. During a severe drought, seed production in L. alba was reduced by one sixth, whereas the seed set of L. floccosa found in the same area was virtually unaffected by the identical drought [24].

The association between 'weediness' and self-fertilization has also been noted [2,30]. An extensive survey of colonizing herbaceous plants of Canada showed that agricultural weeds of row crops and grain fields are almost exclusively annuals, and most of these are self-compatible [31]. A published list of the world's worst weeds of agricultural crops [32] includes 76 species, 41 of which are annuals. Based on previous literature, we were able to identify the breeding system for 24 of these annuals, and, as predicted, the majority (20 out of 24) are selfers.

Conclusions

Botanists have long known that selfing is particularly associated with the annual life cycle in flowering plants [2]. The present study shows further that, among annuals exclusively, selfing is particularly associated with shorter plant heights, smaller flowers, shorter bud development time, shorter flower longevity and smaller seed sizes compared with annuals that are outcrossing. Also, in spite of the null prediction that selfing and outcrossing annuals should be equally represented if there is no bias associated with time-limitation, we found instead that two of the most time-limited habitats on earth that support flowering plants have a significantly higher percentage of selfers among the resident species that are annuals. Because we focused on annual species only, all of these results are explained more parsimoniously by selection associated with time-limitation than by selection associated with pollinator/mate limitation. The role of pollinator/mate-limitation (as traditionally associated with the reproductive assurance hypothesis for the evolution of selfing) is likely to be of greater importance in longer-lived polycarpic species (not considered here), simply because by comparison, there is no convincing basis to argue that selection associated with time-limitation is likely to have been important in species with longer life cycles. We suggest therefore, that most selfers, because most of them are annuals, are likely to have evolved not because of fitness benefits through reproductive assurance associated with selection from pollinator/mate limitation, but rather because of fitness benefits associated with selection from time limitation.

The effect of time-limitation under strong r-selection is to minimize the duration of the life cycle, with selfing favored directly (Figure 1b) and/or indirectly (Figure 1a). There is no basis for predicting that either mechanism is more probable than the other; both are likely to operate simultaneously and perhaps indistinguishably. Indeed, the predicted effects under direct and indirect selection involve the same phenotypic outcome for the same suite of traits (Figure 1). It is particularly significant that the shorter bud development time reported here for selfing annuals is predicted explicitly by the time-limitation hypothesis but not by selection associated with pollinator/mate limitation. Although, we cannot of course rule out the possibility that shorter bud development time may be a pleiotropic consequence of the evolution of autonomous selfing through other mechanisms.

Designing empirical studies that clearly distinguish between mechanisms involving time-limitation versus pollinator/mate limitation remain a challenge but we anticipate that our results and our discussion of these issues may help to inspire further research along these lines. Future studies may be designed to test more directly the role of limited pollination time (vis-à-vis Figure 1b) by comparing the time required for effective pollination under selfing versus outcrossing for closely related species or ecotypes within natural habitats, taking care of course to control for other aspects of the pollination environment (such as mate and pollinator availability) that might affect time-to-effective pollination.

Methods

Data Base Analyses

The literature was surveyed to obtain breeding information (i.e. selfing versus outcrossing) for as many annuals species as possible. For each species, data on plant height, flower size, and seed size were obtained where possible from standard floras and other published literature. A complete database was assembled for 118 species from both Europe and North America, involving 14 families (Table 1). For each species, the maximum published value for each trait was used. Plant height was the maximum recorded vertical extent of the plant. The measure used for flower size depended on the usual convention specific for each family, e.g. maximum petal length, corolla width, lemma length (in the Poaceae). Seed size was measured as the length of the longest axis. Within each family, for each trait, the median value across selfing species and the median value across outcrossing species was calculated and used in the phylogenetically-independent contrasts.

The contrasts were based on 14 phylogenetically-independent pairs, where each pair consisted of median values of the selfing and outcrossing species within one family, which by definition are species that are more closely related to each other than to any other species in the data set [19]. For plant height, only 13 pairs were used due to missing information. The data were analyzed using a Wilcoxon matched pairs test.

Table 1. Species list, with breeding system (O – outcrosser; S – selfer), for database from published literature.

Family
    Species
Asteraceae
    Anthemis cotula (O)
    Cosmos bipinnatus (O)
    Centaurea cyanus (O)
    Centaurea montana (O)
    Crepis capillaris (O)
    Crepis tectorum (O)
    Helianthus annuus (O)
    Lapsana communis (O)
    Matricaria maritima (O)
    Matricaria matricarioides (S)
    Senecio viscosus (S)
    Senecio vulgaris (S)
    Silybium marianum (S)
    Sonchus oleraceus (S)
    Xanthium strumarium (S)

Boraginaceae
    Anchusa arvensis (O)
    Borago officinalis (O)
    Lappula squarrosa (S)
    Myosotis arvensis (S)
    Myosotis ramosissima (S)
    Myosotis stricta (S)
    Plagiobothrys calandrinioides (S)

Brassicaceae
    Arabidopsis thaliana (S)
    Berteroa incana (O)
    Brassica juncea (O)
    Brassica nigra (O)
    Brassica rapa (O)
    Cakile edentula (S)
    Capsella bursa-pastoris (S)
    Cardamine hirsute (S)
    Descurainia pinnata (S)
    Diplotaxis muralis (O)
    Erucastrum gallicum (S)
    Erysimum cheiranthoides (O)
    Erysimum repandum (S)
    Lepidium sativum (O)
    Lepidium campestre (S)
    Lepidum ruderale (S)
    Rorippa palustris (S)
    Sinapis alba (O)
    Sinapis arvensis (O)
    Sisymbrium officinale (S)
    Thlaspi arvense (S)
    Thlaspi perfoliatum (S)

Caryophyllaceae
    Agrostemma githago (O)
    Arenaria serpyllifolia (S)
    Cerastium nutans (O)
    Silene dichotoma (O)
    Silene noctiflora (S)
    Spergula arvensis (S)
    Stellaria media (S)

Fabaceae

Family
    Species
Malvaceae
    Abutilon theophrasti (S)
    Hibiscus trionum (O)
    Malva neglecta (O)
    Malva rotundiflora (S)

Plantaginaceae
    Plantago arenaria (O)
    Plantago virginica (S)

Poaceae
    Aira praecox (O)
    Avena fatua (S)
    Avena sativa (S)
    Bromus hordeaceus (S)
    Bromus secalinus (O)
    Bromus sterilis (S)
    Bromus tectorum (S)
    Desmazeria rigida (S)
    Echinochloa crus-galli (S)
    Hordeum vulgare (S)
    Lolium temulentum (S)
    Panicum miliaceum (S)
    Phalaris canariensis (O)
    Poa annua (S)
    Secale cereale (O)
    Setaria italica (S)
    Setaria verticillata (S)
    Setaria virdis (S)
    Triticum aestivum (S)
    Zea mays (O)

Polemoniaceae
    Allophyllum gilioides (S)
    Allophyllum integrifolium (S)
    Collomia grandiflora (O)
    Collomia linearis (S)
    Gilia australis (S)
    Gilia capitata (O)
    Gilia caruifolia (O)
    Gilia clivorum (S)
    Gilia inconspicua (S)
    Gilia millefoliata (S)
    Gilia sinuata (S)
    Gilia tenuiflora (O)
    Gilia transmontana (S)
    Gilia tricolor (O)
    Navarretia atrictyloides (O)
    Navarretia squarrosa (S)

Polygonaceae
    Fagopyrum esculentum (O)
    Polygonum aviculare (S)
    Polygonum convolvulus (S)
    Polygonum hydropiper (S)
    Polygonum lapathifolium (S)
    Polygonum persicaria (S)

Ranunculaceae
    Myosurus minimus (S)

Table 1. *(Continued)*

Medicago lupulina (O)	*Ranunculus reptans* (O)
Trifolium arvense (S)	*Ranunculus sceleratus* (O)
Trifolium aureum (S)	
Trifolium campestre (S)	Scrophulariaceae
Vicia sativa (S)	*Chaenorrhinum minus* (S)
Vicia tetrasperma (O)	*Veronica agrestis* (O)
	Veronica arvensis (S)
Lamiaceae	*Veronica peregrina* (S)
Galeopsis tetrahit (O)	*Veronica persica* (S)
Lamium amplesicaule (S)	
Lamium purpureum (S)	Apiaceae
	Aethusa cynapium (S)
	Anethum graveolens (O)

Greenhouse Study

The species included in this study were chosen if there was a known breeding system, if germinable seeds were available, and if a complementary species (i.e. in the same family with the opposite breeding system) was known and could also be obtained as germinable seeds. Seeds were obtained from a variety of sources; Herbiseed, Rancho Santa Ana Botanic Gardens, Chiltern Seeds, S&S Seeds, and the National Plant Germplasm System. Our search lead to an unbiased sample of 25 candidate species, allowing five pair-wise family comparisons and four pair-wise genus comparisons (Table 2).

Table 2. List of species, with breeding system (O – outcrosser; S – selfer), used in the greenhouse study.

Family	Family
Species	Species
Asteraceae	Convolvulaceae
Crepis capillaris (O)	*Ipomoea hederacea* (S)
Helianthus annuus (O)	*Ipomoea purpurea* (O)
Matricaria maritime (O)	
Matricaria matricarioides (S)	Fabaceae
Senecio viscosus (S)	*Lupinus bicolor* (O)
Senecio vulgaris (S)	*Lupinus nanus* (S)
	Lupinus succulentus (O)
Boraginaceae	*Trifolium hirtum* (S)
Borago officinalis (O)	
Myosotis arvensis (S)	Lythraceae
	Cuphea laminuligera (O)
Brassicaceae	*Cuphea lanceolata* (O)
Brassica juncea (O)	*Cuphea lutea* (S)
Brassica nigra (O)	
Capsella bursa-pastoris (S)	Polemoniaceae
Cardamine hirsuta (S)	*Navarretia squarrosa* (S)
Sinapis alba (O)	*Phlox drummondii* (O)
Sinapis arvensis (O)	

Most species were germinated in 15 cm pots filled to 3 cm below the top with standard potting soil (Promix BX©). Pots were placed in a greenhouse and watered daily until the appearance of their first true leaves. Subsequently, they were watered uniformly every second day to ensure that the soil was kept moist. Some species were germinated in a petri-dish in a growth chamber (23°C, 12 hour cycles of light and dark), after which they were transplanted into pots and placed in the greenhouse. Each species was replicated five times, with one plant per pot. Pots were arranged randomly on benches at a density of 1 pot per 0.093 m2.

The plants were exposed to 16 hours of daylight each day, with maximal natural light levels at ca 1200 μE. Before sunrise and after sunset, artificial lights (250–300 μE) were used to supplement the light exposure to 16 hours of light per day. The greenhouse was kept at an average temperature of 23.1°C during the day and dropped to 20.0°C at night. The plants were fertilized every 2 weeks with 200 ml per pot of a 2g/L concentration of 20–20–20 N-P-K fertilizer.

For each plant, age at first flower, bud development time, and flower longevity were measured. Emergence of the first pair of true leaves, after the cotyledons, was considered day 1 of the plant's life. Age at first flower was measured in days from day 1 to when the first flower opened on each plant. Bud development time (n = 3 buds per plant) was calculated as the number of days from the first appearance of a new bud until the flower opened. The same three buds on each plant were then monitored every day after opening, and the number of days until the flower senesced (flower longevity) was recorded for each. A flower was considered to be senesced when the corolla wilted, fell apart, or became discolored, as designated by Primack [17]. Any flower that was open for only one day was considered a one-day flower, regardless of whether it was open for the whole day or only part. Flowers in the Asteraceae were considered withered when the whole inflorescence had senesced, rather than the individual florets [17].

For bud development time and flower longevity, the three replicate measurements for each plant were averaged, and then these values for the five replicate plants were averaged to obtain a mean value for the species. The data were analyzed with a standard least squares one-way analysis of variance (ANOVA) model, with a post-hoc contrast between selfers and outcrossers. These analyses were done for each family and genus separately in order to control for phylogeny at these levels. In cases where the data were non-normal, a log-transformation was applied which corrected the distribution.

Authors' Contributions

RS collected the data, performed most of the analyses, participated in the design of the study, and wrote the first draft as a B.Sc.(Hons) thesis. LWA conceived of the study, participated in its design and coordination, and wrote the final draft for submission to BMC. Both authors read and approved the final manuscript.

Acknowledgements

This research was supported by the Natural Sciences and Engineering Research Council of Canada through a research grant to LWA and a USRA to RS. Thanks to Spencer Barrett for provided breeding information for several species, Dale Kristensen and Richard Gold for assistance in the greenhouse, and Christopher Eckert and two anonymous reviewers for comments on an earlier version of the text.

References

1. Stebbins GL: Adaptive radiation of reproductive characteristics in angiosperms, I. Pollination mechanisms. Annu Rev Ecol Syst 1970, 1:307–326.

2. Lloyd DG: Demographic factors and mating patterns in angiosperms. In Demography and Evolution in Plant Populations. Edited by: Solbrig OT. London: Blackwell; 1980:67–88.

3. Barrett SCH, Harder LD, Worley AC: The comparative biology of pollination and mating in flowering plants. In Plant Life Histories. Edited by: Silvertown J, Franco M, Harper JL. London: Cambridge University Press; 1997:57–76.

4. Stebbins GL: Self-fertilization and population variability in the higher plants. Am Nat 1957, 91:337–354.

5. Wyatt R: Pollinator-plant interactions and the evolution of breeding systems. In Pollination biology. Edited by: Real L. London: Academic Press; 1983:51–95.

6. Wyatt R: Phylogenetic aspects of the evolution of self-pollination. In Plant evolutionary biology. Edited by: Gottlieb LD, Jain SK. London: Chapman and Hall; 1988:109–131.

7. Fishman L, Wyatt R: Pollinator-mediated competition, reproductive character displacement, and the evolution of selfing in Arenaria uniflora (Caryophyllaceae). Evolution 1999, 53:1723–1733.

8. Fausto JA, Eckhart VM, Geber MA: Reproductive assurance and the evolutionary ecology of self-pollination in Clarkia xantiana (Onagraceae). Am J Bot 2001, 88:1794–1800.

9. Lloyd DG: Self and cross-fertilization in plants. I. The selection of self-fertilization. Int J Plant Sci 1992, 153:370–380.

10. Morgan MT, Schoen DJ, Bataillon TM: The evolution of self-fertilization in perennials. Am Nat 1997, 150:618–638.

11. Aarssen LW: Why are most selfers annuals? A new hypothesis for the fitness benefit of selfing. Oikos 2000, 89:606–612.

12. Guerrant EO: Early maturity, small flowers and autogamy: a developmental connection? In The Evolutionary Ecology of Plants. Edited by: Brock JH, Linhart YB. Boulder: Westview Press; 1989:61–84.

13. Hill JP, Lord EM, Shaw RG: Morphological and growth rate differences among outcrossing and self-pollinating races of Arenaria uniflora (Caryophyllaceae). J Evol Biol 1992, 5:559–573.

14. Runions JC, Geber MA: Evolution of the self-pollinating flower in Clarkia xantiana (Onagraceae). I. Size and development of floral organs. Am J Bot 2000, 87:1439–1451.

15. Sherry RA, Lord EM: A comparative developmental study of the selfing and outcrossing flowers of Clarkia tembloriensis (Onagraceae). Int J Plant Sci 2000, 161:563–574.

16. Silvertown J, Dodd M: Comparing plants and connecting traits. In Plant Life Histories. Edited by: Silvertown J, Franco M, Harper JL. London: Cambridge University Press; 1997:3–16.

17. Primack RB: Longevity of individual flowers. Annu Rev Ecol Syst 1985, 16:15–37.

18. Andersson S: Floral reduction in Crepis tectorum (Asteraceae): tradeoffs and dominance relationships. Biol J Linn Soc Lond 1996, 57:59–68.

19. Eriksson O, Jakobsson A: Abundance, distribution and life histories of grassland plants: a comparative study of 81 species. J Ecol 1998, 86:922–933.

20. Donnelly SE, Lortie CJ, Aarssen LW: Pollination in Verbascum thapsus (Scrophulariaceae): the advantage of being tall. Am J Bot 1998, 85:1618–1625.

21. Lortie CJ, Aarssen LW: The advantage of being tall: higher flowers receive more pollen in Verbascum thapsus L. (Scrophulariaceae). Ecoscience 1999, 6:68–71.

22. Briggs D, Hodkinson H, Block M: Precociously developing individuals in populations of chickweed [Stellaria media (L.) Vill.] from different habitat types, with special reference to the effects of weed control measures. New Phytol 1991, 117:153–164.

23. Theaker AJ, Briggs D: Genecological studies of groundsel (Senecio vulgaris L.). Rate of development in plants from different habitat types. New Phytol 1993, 123:185–194.

24. Arroyo MTK: Chiasma frequency evidence on the evolution of autogamy in Limnanthes floccosa (Limnanthaceae). Evolution 1973, 27:679–688.

25. Fenster CB, Diggle PK, Barrett SCH, Ritland K: The genetics of floral development differentiating two species of Mimulus (Scrophulariaceae). Heredity 1995, 74:258–266.

26. Van Doorn WG: Effects of pollination on floral attraction and longevity. J Exp Bot 1997, 48:1615–1622.

27. Hamrick JL, Godt MJ: Effects of life history traits on genetic diversity in plant species. In Plant Life Histories. Edited by: Silvertown J, Franco M, Harper JL. London: Cambridge University Press; 1997:102–118.

28. Kunin WE, Shmida A: Plant reproductive traits as a function of local, regional and global abundance. Cons Biol 1997, 11:183–192.

29. Eckhart VM, Geber MA: Character variation and geographic range in Clarkia xantiana (Onagraceae): breeding system and phenology distinguish two common subspecies. Madrono 2000, 46:117–125.

30. Baker HG: Self-compatibility and establishment after "long distance" dispersal. Evolution 1955, 9:347–348.

31. Mulligan GA: Recent colonization by herbaceous plants in Canada. In The Genetics of Colonizing Species. Edited by: Baker HG, Stebbins GL. New York: Academic Press; 1965:127–143.

32. Holm LG, Plucknett DL, Pancho JV, Herberger JP: The Worlds Worst Weeds. Honolulu: The University Press of Hawaii; 1977.

CITATION

Originally published under the Creative Commons Attribution License. Snell R, Aarssen LW. Life History Traits in Selfing Versus Outcrossing Annuals: Exploring the 'Time-Limitation' Hypothesis for the Fitness Benefit of Self-Pollination. BMC Ecology 2005, 5:2. doi:10.1186/1472-6785-5-2.

Functional Diversity of Plant–Pollinator Interaction Webs Enhances the Persistence of Plant Communities

Colin Fontaine, Isabelle Dajoz, Jacques Meriguet and
Michel Loreau

ABSTRACT

Pollination is exclusively or mainly animal mediated for 70% to 90% of angiosperm species. Thus, pollinators provide an essential ecosystem service to humankind. However, the impact of human-induced biodiversity loss on the functioning of plant–pollinator interactions has not been tested experimentally. To understand how plant communities respond to diversity changes in their pollinating fauna, we manipulated the functional diversity of both plants and pollinators under natural conditions. Increasing the functional diversity of both plants and pollinators led to the recruitment of more diverse plant communities. After two years the plant communities pollinated by the most functionally diverse pollinator assemblage contained about 50% more plant species than did plant communities pollinated by less-diverse pollinator assemblages. Moreover, the positive effect of functional diversity was explained by a complementarity between functional groups of pollinators and plants. Thus, the functional diversity of pollination networks may be critical to ecosystem sustainability.

Introduction

Understanding the consequences of biodiversity loss for ecosystem functioning and services is currently a major aim of ecology [1,2]. Animal-mediated pollination is one of the essential ecosystem services provided to humankind [3,4]. The negative impact of pollinator decline on the reproductive success of flowering plants has been documented at the species level [5–7], but little information is available at the community level [8]. Increasing the scale of study to the community level is essential to account for potential competitive or facilitative effects among species that belong to the plant–pollinator network. Such effects, which are often linked to diversity [9,10], are known to have large influences on ecological processes such as community productivity and stability [11,12].

Experimental evidence for diversity effects on the functioning of terrestrial ecosystems is mainly available for plants. As primary producers, plants play a central role in the flow of energy within ecosystems [13,14]. Animal-pollinated angiosperms represent up to 70% of plant species in numerous communities and ecosystems [15]. Mutualistic interactions between animals and plants form several intricate interaction webs [16]. Recent analysis of plant–pollinator and plant–frugivore interaction webs demonstrates that these contain a continuum from fully specialist to fully generalist species [17,18]. However, these networks are structured in a nested way [19,20], with specialists mainly interacting with generalists. Such a pattern might have important consequences for ecosystem functioning, because it might confer resilience to perturbations such as the extinction of species [21] if, for example, generalist pollinators buffer the loss of specialist pollinators [18,22–24]. Furthermore, this hypothesis does not take into account the dynamical properties of these networks. In a plant–pollinator community, variations in species diversity at different trophic levels may lead to an adaptation of interaction strengths [25], which may in turn affect the total effectiveness of pollination. We conclude that more information is urgently needed concerning the impacts of biodiversity loss on multispecies and multitrophic interactions.

To experimentally test the effect of functional diversity on the functioning and persistence of plant–pollinator communities, we defined functional groups of plants and pollinators based on morphological traits. For plants, two functional groups with three species each were defined according to accessibility of floral rewards (pollen and nectar; see Figure 1). The first group (group 1) included Matricaria officinalis, Erodium cicutarium, and Raphanus raphanistrum, which have easily accessible floral rewards and will be called "open flowers." The second group (group 2), called "tubular flowers," included Mimulus guttatus, Medicago sativa, and Lotus corniculatus, all of which present floral rewards hidden at the bottom of a tubular corolla. For pollinators, two functional groups were defined according to mouthparts length (Figure 1). The first group (group A) included three species of syrphid flies (Diptera) with short mouthparts: Saephoria sp., Episyrphus balteatus, and Eristalis tenax. The second group (group B) included three species of bumble bees with longer mouthparts: Bombus terrestris, B, pascuorum, and B, lapidarius. Note that in this case a functional trait (long mouthparts) and a phylogenetic group are confounded. Preliminary observations showed that these six insect species contribute up to 70% of all pollinating visits to flowers in our study area in France. Constructing a plant–pollinator network with these four functional groups leads to a nested structure with specialists interacting with generalists (Figure 1, third column). In principle, syrphid flies cannot efficiently pollinate tubular flowers because their mouthparts are too short.

Pollinators species and groups	Mouthpart length (mm ± S.E.)	Theoretical pollination network	Plants species and groups	Accessibility	
				pollen	nectar
Sphaerophoria sp.	2.66 ± 0.35		M. officinalis	easy	easy
E. balteatus	2.3 ± 0.20	Syrphid-flie Open flower	E. cicutarium	easy	easy
E. tenax	5.47 ± 0.29		R. raphanistrum	easy	difficult
B. terrestris	9.02 ± 0.19		M. guttatus	easy	difficult
B. hortorum	9.21 ± 1.02	Bumble-bee Tubular flower	M. sativa	difficult	difficult
B. lapidarius	8.10 ± 0.86		L. corniculatus	difficult	difficult

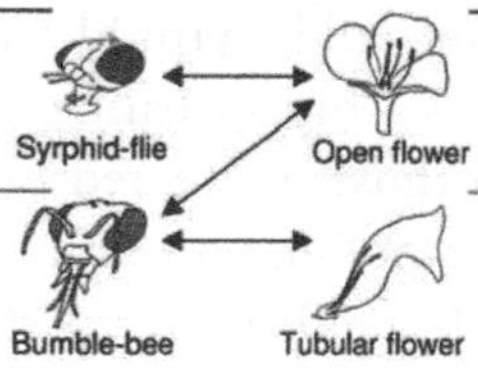

Figure 1. Experimental Pollination Web.
Summary of the characteristics upon which functional groups of pollinators (left) and plants (right) were based. In the middle, the arrows linking insect heads to flower types show the theoretical pollination network when all functional groups are present.

At the beginning of spring 2003, we set up 36 4-m2 caged experimental plant communities. There were three plant treatments following a "substitutive" design [26]. Two of them contained one of the two plant functional groups alone (group 1 or 2), whereas the third contained both plant functional groups in combination (group 3). We applied three different pollination treatments to each plant treatment, by introducing each pollinator functional group alone (group A or B), or both groups together (group C). This full factorial design led to nine experimental treatments, which were replicated four times each, making a total of 36 experimental units. The pollination treatments were applied in two consecutive years (June–July 2003 and 2004). We controlled for the total number of pollinator visits received by each plot during the two pollination seasons (1,000 visits in 2003 and 1,200 visits in 2004) to allow an unbiased comparison of pollination efficiency among the various experimental treatments.

In August and September 2003, we counted the number of fruits on each plant in every plot. We also counted the number of seeds per fruit on five collected fruits per plant. Lastly, in April 2004 and 2005, we measured both the number of plant species present at the seedling stage (recruitment richness) and the total number of seedlings (recruitment density) to determine the effects of the experimental treatments on the natural recruitment of the next plant generation.

Results

Effects on Plant Reproductive Success

The reproductive success of the two plant functional groups after the first season is analysed in Table 1. There was a significant effect of pollination treatment on the number of fruits per plant (Table 1, left; standardized means ± standard error [SE]: syrphid -0.278 ± 0.061, bumble bee 0.221 ± 0.065, and both 0.063 ± 0.068). Orthogonal contrasts on pollination treatment indicate that the identity of the pollinator guild (syrphid [A] versus bumble bee [B]) had a significant effect. There was a higher fruit production in bumble bee–pollinated communities than in those pollinated by syrphids. Moreover, the breakdown of the interaction of pollination and plant treatments into the orthogonal contrasts A1 versus B1 and A2 versus B2 indicates that the two plant functional groups responded differently to the identity of the pollinator functional group. Tubular 3 flowers (group 2) produced significantly fewer fruits in the syrphid treatment, whereas open flowers (group 1) produced the same amount of fruits whatever the identity of the pollinator functional group (Figure 2A). This supports our hypothesis that bumble bees were able to pollinate both plant functional groups whereas syrphids could only efficiently pollinate open flowers. Although the functional diversity of plant or pollinator treatment alone had no significant effect, fruit production tended to increase with both plant and pollinator functional diversity (contrast [A1 + A2 + B1 + B2] versus C3; Figure 2B).

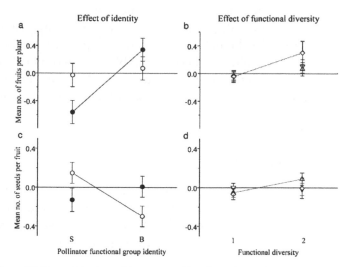

Figure 2. Effects of Pollinator Identity and Diversity on Plant Reproductive Success

The left panels show the effects of pollinator guild identity (S indicates syrphid flies, B indicates bumble bees) on the reproductive success of the two plant guilds (open circle indicates open-flowers [group 1], closed circle indicates tubular-flowers [group 2]). Reproductive success was measured by (A) the standardized number of fruits per plant and (B) the standardized number of seeds per fruit. The right panels show the effects of the functional diversity of pollination treatments (triangle), plant treatment (inverted triangle) and both (diamond) on the standardized numbers of fruits per plant (C) and seeds per fruit (D). Lines connecting symbols indicate significant effects (solid indicates p < 0.001, dashed indicates p < 0.08). Error bars represent one standard error. See Table 1 for statistical analysis.

Table 1. Analysis of Plant Reproductive Success.

Effects and Contrasts	Effect of	df	Fruits per Plant		Seeds per Fruit	
			F Value	Pr > F	F Value	Pr > F
Poll trt		2,27	6.83	0.0040	1.96	0.1603
A vs. B	Identity of poll guild	1,27	13.04	0.0012	0.52	0.4757
(A + B) vs. C	Functional diversity of poll trt	1,27	0.60	0.4446	3.34	0.0785
Plant trt		2,27	1.17	0.3252	0.03	0.9694
1 vs. 2	Identity of plant guild	1,27	0.68	0.4159	0.02	0.8924
(1 + 2) vs. 3	Functional diversity of plant trt	1,27	1.64	0.2111	0.04	0.8401
Poll trt × Plant trt		4,27	1.80	0.1588	2.65	0.0552
A1 vs. B1	Identity of poll guild in open plant trt	1,27	0.17	0.6813	8.60	0.0068
A2 vs. B2	Identity of poll guild in tubular plant trt	1,27	14.70	0.0007	0.71	0.4076
(A1 + A2 + B1 + B2) vs. C3	Functional diversity of both plant and poll trt	1,27	3.37	0.0774	0.28	0.5984

With respect to seed set per fruit, the interaction between plant and pollination treatment was marginally significant (Table 1, right). As with fruit production, the contrasts A1 versus B1 and A2 versus B2 indicate that the two plant functional groups responded differently to pollinator functional group identity. The pattern, however, was different: Open flowers produced significantly fewer seeds per fruit in the bumble bee treatment than in the syrphid treatment (Figure 2C). This means that bumblebees were less-efficient pollinators than syrphids for open flowers. This could be due to the higher rate of geitonogamous visits (i.e., consecutive visits to different flowers of the same plant, resulting in self-fertilization) by bumblebees. Indeed, preliminary observations using a similar experimental design showed that bumble bees perform a higher percentage of geitonogamous visits than do syrphids (I. Dajoz, unpublished data). Finally, the mean number of seeds per fruit in the plant communities tended to increase with functional diversity of pollination treatments (contrast [A + B] versus C; Figure 2D).

Effects on Natural Recruitment

We analysed the long-term effects of our pollination treatments on the natural recruitment of our experimental plant treatments after the first and second pollination seasons. The results are presented in Table 2. There was a significant effect of year on recruitment richness with a higher richness after the second pollination season (mean ± SE: 1.916 ± 0.075 in 2004, and 2.291 ± 0.0856 in 2005). Among the possible causes was a severe drought in 2003 [27], which likely affected both plant and insect populations. Such a drought did not occur in 2004. This difference in climate between years may account for a large part of the year effect.

Table 2. Analysis of Plant Recruitment Richness and Density.

Effects	df	Recruitment Richness		Recruitment Density	
		F Value	Pr > F	F Value	Pr > F
Pollination trt	2,27	6.77	0.0041	3.51	0.0442
Plant trt	2,27	35.88	<0.0001	6.22	0.0060
Year	1,27	19.97	0.0001	12.69	0.0014
Pollination trt × plant trt	4,27	3.63	0.0171	4.74	0.0050
Pollination trt × year	2,27	0.43	0.6539	0.42	0.6585
Plant trt × year	2,27	0.64	0.5366	2.54	0.0978
Pollination trt × plant trt × year	4,27	0.91	0.4698	1.62	0.1978

Recruitment richness was significantly different among plant treatments, with fewer species recruiting in tubular communities (Figure 3). This is very likely due to two perennial species (whereas all species are annuals in the other group) which may have different reproductive traits and create differences in competitive intensity among the plant treatments. There was a significant effect of pollination treatment, with a higher recruitment richness when both groups of pollinators were present (means ± SE: syrphid 1.854 ± 0.973, bumble bee 2.052 ± 0.826, and both

2.406 ± 1.062). However, as suggested by the significant interaction between plant and pollination treatments, the pattern was more complex (Figure 3A). In fact, pollination treatments had no effect on recruitment richness in open-flower plant treatment (Figure 3A, left). In the tubular-flower plant treatment, recruitment in the syrphid fly treatment tended to be lower than in the other pollination treatments (Figure 3A, centre). But the positive effect of pollinator functional diversity was obvious in the plant treatment that contained both plant functional groups (Figure 3A, right). In the mixed plant treatment, recruitment richness under the most functionally diverse pollination treatment was substantially above that in the two other treatments.

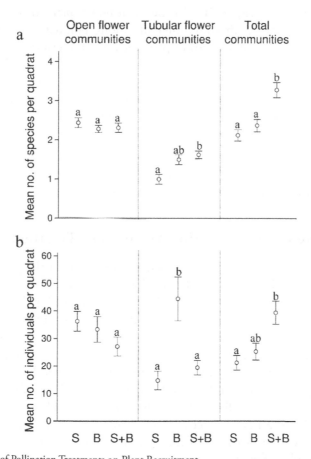

Figure 3. Effects of Pollination Treatments on Plant Recruitment.
Effects of pollination by syrphid flies (S), bumble bees (B), or both (S + B) on (A) recruitment richness (mean number of plant species present as seedlings in a quadrat) and (B) recruitment density (mean number of plant individuals present as seedlings in a quadrat) in the various plant treatments. Error bars represent one standard error. Lower-case letters indicate statistically significant differences among pollination treatments within a plant treatment (Bonferroni-adjusted t-test, p < 0.05).

Considering recruitment density, there was also a significant effect of year, with a higher density after the second pollination season (mean ± SE: 26.784 ± 2.324 in 2004 and 31.319 ± 1.937 in 2005), and a significant effect of plant treatment, with fewer individuals recruiting in tubular communities (Figure 3B, centre). These year and plant-treatment effects can be explained in the same way as for recruitment richness (see above). There was also a significant effect of pollination treatment, with a lower recruitment density when plant communities were pollinated by syrphid flies alone (means ± SE: syrphids: 24.104 ± 20.464, bumble bees: 34.364 ± 32.781, and both 28.688 ± 21.459). This is congruent with our results on the number of fruits produced per plant (see Table 1, contrast A versus B). As for recruitment richness, there was a significant interaction between plant and pollination treatments (Figure 3B). In the open-flower plant treatment, recruitment density was not significantly different among pollination treatments (Figure 3B, left). But in the tubular-flower plant treatment, recruitment density was significantly higher in the bumble bee treatment than in the other pollination treatments (Figure 3B, centre). Finally, in the mixed plant treatment, the same pattern as for recruitment richness was observed: There was a higher density in the mixed pollination treatment than in single-guild pollination treatments (Figure 3B, right).

Note that these results on natural recruitment are not an artefact caused by sampling small quadrats in heterogeneous experimental plots since the same patterns were observed when data from all quadrats in a plot were pooled.

Pollination Visitation Web in the Mixed Plant Treatment

To explain the strong effect of pollinator functional diversity on the persistence of mixed plant communities, we carried out a log-linear analysis on the visitation rate of each insect species in a given pollination treatment, for the six plant species of the mixed plant treatment. Data from the year 2003 are illustrated in Figure 4, and the results of the analysis on both years are presented in Table 3. In the second year, there was a significant effect of plant functional group identity: Tubular flowers received a higher number of visits than did open flowers (mean visitation frequency ± SE: for open flowers 0.236 ± 0.097 and for tubular flowers 0.763 ± 0.097). This is very likely due to the two well-established perennial species, which produced a more attractive floral display during the second year of the experiment. For the two years of the experiment, there was a significant interaction between plant functional group and pollinator functional group. This indicates that the two pollinator functional groups were specialised on different plant functional groups (mean visitation frequency ± SE on open flowers and tubular flowers, respectively: in 2003, for bumble bees 0.128 ± 0.058 and 0.433

± 0.075; for syrphids 0.327 ± 0.043 and 0.113 ± 0.052; in 2004, for bumble bees 0.01 ± 0.005 and 0.58 ± 0.075; for syrphids 0.23 ± 0.055 and 0.18 ± 0.087). Syrphids mainly visited open flowers whereas bumble bees preferentially visited tubular flowers (Figure 4). Even though bumble bees can pollinate open flowers quite efficiently when this is the only plant functional group present (as shown by the reproductive success, recruitment diversity, and recruitment density of the open-flower plant treatment in the bumble bee treatment, Figures 2 and 3), they focus on the tubular-flower group in the mixed plant treatment. In the mixed pollination treatment, the match between plant and pollinator functional groups leads to a more homogenous distribution of pollinator visits among plant groups than in the other pollination treatments. Ultimately, this significantly increases the reproductive success of plants, most likely through the homogenisation of pollinator visits and the minimization of inefficient pollinator visits.

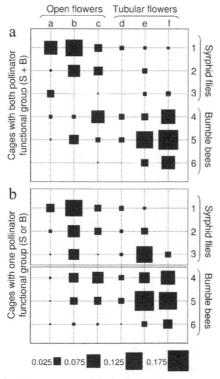

Figure 4. Visitation Web in the Communities with Both Plant Types.

Distribution of pollinator visits for the year 2003, among the six plant species in the plant treatment containing the two plant functional groups, (A) for the mixed pollination treatment (S + B) and (B) for the single functional group pollination treatments (S or B). The length of the side of the black squares shows the proportion of visits by a given pollinator species on each plant species. Lower-case letters represent plant species: a, Ma. officinalis; b, E. cicutarium; c, R. raphanistrum; d, Mi.guttatus; e, Me. sativa; f, L. corniculatus. Numbers represent pollinator species: 1, Saephoria sp.; 2, Ep. balteatus; 3, Er. tenax; 4, B. terrestris; 5, B. pascuorum; 6, B. lapidarius.

Table 3. Analysis of Visitation Rates.

Effects	df	Year 2003		Year 2004	
		F Value	Pr > F	F Value	Pr > F
Poll species (poll guild)	4,248	6.95	<0.0001	2.41	0.0499
Plant species (plant guild)	4,248	9.16	<0.0001	36.22	<0.0001
Poll func diversity	1,4	0.00	0.9661	0.00	0.9911
Poll guild identity	1,4	1.66	0.2676	2.28	0.2059
Plant guild id	1,4	0.95	0.3859	28.60	0.0059
Poll func diversity × poll guild id	1,4	2.03	0.2271	3.42	0.1382
Poll func diversity × plant guild id	1,4	0.12	0.7496	0.13	0.7338
Poll guild id × plant guild id	1,4	34.50	0.0042	44.47	0.0026
Poll func diversity × poll guild id × plant guild id	1,4	2.29	0.2050	1.82	0.2487

Discussion

Previous studies on the diversity of plant–pollinator interaction webs were either descriptive [16], carried out on a single plant species [6,7,28–30], or based on simulation [21] and theoretical approaches [22,31]. To our knowledge, this is the first experimental evidence that the persistence of a plant community can be affected by a loss of diversity of its pollinating fauna. Of course, our experimental communities differed from natural ones in several respects. Among other things, the interaction networks we studied were much simpler than those occurring in nature; in particular, they contained fewer species in each trophic level. But such simplifications from natural situations are often necessary to carry out controlled experiments.

In plant communities that contained only open flowers, plants produced fewer seeds per fruit in the bumblebee treatment than in the syrphid treatment (Figure 2C), but this was compensated by a sufficiently high fruit production, leading to a richness and density of natural recruitment that was similar to the other pollination treatments (Figure 3A and 3B left). Thus, in these communities, all pollination treatments were equally effective in the long term.

In plant communities that contained only tubular flowers, syrphids were inefficient pollinators; fruit production was very low (Figure 2A) and insufficient to allow a good natural recruitment. Bumble bees were the most effective pollinators (Figure 3A and 3C, centre). Note that in the bumble bee treatment, the very high value of average recruitment density was due to three measurements in two replicates, in which only M. guttatus seedlings were recorded at a very high density (more than 150 seedlings per quadrat). To test the effect of these outliers, we removed them and repeated our analysis. The same significant effects were observed, except for the effect of pollination treatment, which became marginally significant (p = 0.0645). The new mean number of seedlings per quadrat for this experimental treatment was 32.17 ± 4.55 (SE), which is still slightly above the value for the pollination treatment with both pollinator groups. For plant communities that contained only tubular flowers, recruitment richness in the two pollination treatments that contained bumblebees was similar.

These results are in agreement with our theoretical pollination network presented in Figure 1. In our experimental system, syrphids can be considered as specialist pollinators since they efficiently pollinate only open flowers. Bumble bees were potentially generalists as they induced an important fruit production of the two plant types and a good recruitment in the open- and tubular-flower plant treatments. Our results on the reproductive success and recruitment of single-guild plant treatments indicate that there are strong functional group identity effects since our plant functional groups responded differently to our pollinator functional groups.

However, the functional diversity of both the plant and pollination treatments was also important. Plant reproductive success tended to increase with pollinator functional diversity when the number of seeds per fruit was considered, and with both plant and pollinator functional diversity when the number of fruits per plant was considered (Figures 2B and 2D). Although recruitment in single-guild plant treatments was mainly affected by the identity of functional groups, the effect of functional diversity was dramatic in the mixed plant treatment. Natural recruitment of plant communities visited by mixed pollinator guilds was largely above that in other pollination treatments.

Pollination by syrphids alone allowed the reproduction of open flowers but not tubular flowers, as expected from the specialisation of syrphids. More surprisingly, however, bumble bees failed to be efficient generalist pollinators. Most of their visits occurred on tubular flowers (Figure 4), resulting in a relatively poor recruitment of open flowers. The only pollination treatment that achieved a high recruitment of both open and tubular flowers when they were mixed, was the one containing the two insect functional groups (Figure 3, right). When syrphids and bumble bees simultaneously pollinated mixed plant communities, they each focused on their target plant functional group, leading to more efficient visits and a better distribution of visits among plant functional groups (Figure 4). Ultimately, it was the pollination treatment with both pollinator functional groups that produced the highest richness and density of natural recruitment. Consequently, since most natural plant communities contain both open and tubular flowers, pollinator functional diversity should strongly enhance the persistence of these communities.

Although our experimental system differed from natural communities, and information about the reciprocal effects of the functional diversity of plant communities on the diversity of pollinator communities would be useful, our study indicates that the functional diversity of plant–pollinator interaction webs may be critical for the persistence and functioning of ecosystems and should be carefully monitored and protected. The loss of pollinator functional diversity is likely to trigger plant population decline or extinctions [4], which in turn are likely to affect the structure and composition of natural plant communities and the productivity of many agroecosytems that rely on insect pollination [8]. Ultimately, higher trophic levels may be affected since the diversity and biomass of consumers depend on primary production. Our results strongly suggest that the functional diversity of complex interaction webs plays a crucial role in the sustainability of ecosystems.

Materials and Methods

Experimental Plant Communities

At the beginning of spring 2003, plant communities were set up in a meadow that remained almost undisturbed for 10 years at the Station Biologique de Foljuif, France, 80 km southwest of Paris. Prior to the establishment of the communities, soil was sterilized by injecting 120 °C steam (30 min) to destroy the seed bank and soil pathogens. In each of the 36 4-m² plots, a total of 30 adult plants were planted on a grid, spaced 25 cm from each other, to minimize competition and homogenise spatial distribution. Thus, plant density was the same in all experimental plots. We selected a moderate density to maintain within- and among-species competition to a low level, and to allow enough space for future recruitment in the plots. Each of these plant communities was enclosed in a 2-m–high nylon mesh cage in order to eliminate natural pollinator visitation.

Pollination Rounds

During the flowering seasons (June–July 2003 and 2004), pollinators were captured around the study area and introduced into the cages. The relative abundance of pollinator species in the various pollination treatments reflects their natural abundances. From preliminary observations, we had noticed that, in order to have no more than three insects active at the same moment in a 4-m² plot, it was necessary to put about eight syrphid flies, or six bumble bees, or a mixture of six syrphids and four bumble bees in each pollination cage. Each pollination round in a given plot included 200 visits in the year 2003 and 300 in the year 2004. In total, each plot received either four (in 2004) or five (in 2003) pollination rounds, leading to a total of 1,000 visits per plot in 2003 and 1,200 in 2004.

Pollination Activity

Bumble bees needed approximately 30 min after introduction in the cages to calm down and start to pollinate. In the pollination treatment with both pollinator guilds, we then introduced syrphids, which started to pollinate immediately. Mean visitation time was not significantly different between insects in the cages and in nature. This was true both for bumble bees (mean visitation time in cages: 3.25 ± 0.92 s, mean visitation time in nature: 2.91 ± 1.33 s, t = 1.51, df = 96, p = 0.133) and for syrphids (mean visitation time in cages: 40.21 ± 8.89 s, mean visitation time in nature: 35.38 ± 14.75 s, t = 0.77, df = 12, p = 0.45).

Measurement of Reproductive Success

One month after the first pollination treatments, we counted the total number of fruits on each plant, except for M. guttatus and M. officinalis in which fruits cannot be counted without collecting them. We randomly took five fruits per plant of each species to estimate the number of seeds per fruit.

Measurement of Recruitment Richness and Density

Recruitment richness and density were estimated during the second (April 2004) and third (April 2005) year of the experiment by counting the number of seedlings of each species in four 1,600-cm² quadrats in each plot.

Statistical Analysis

Statistical analyses were performed using SAS 8.2 software.

For the analysis of plant reproductive success, we log-transformed the data to ensure normality. We standardized the data by species using the formula: $x - \mu/\sigma$ (where μ = the mean and σ = the standard deviation of number of fruits or number of seeds per fruit for a given plant species) in order to make the data comparable among the various species and functional groups. We used a mixed analysis of variance (ANOVA) model (SAS proc mixed), in which the fixed effects were plant treatment, pollination treatment, and their interaction term. To investigate the effects of the various plant and pollination treatments, we subdivided a priori each main effect into two components using orthogonal contrasts. The first contrast tested the effect of the identity of the plant or pollinator functional group, i.e. one group versus the other. The second tested the effect of the functional diversity of the plant or pollination treatment, i.e. single-guild versus mixed-guild plant or pollination treatments. Similarly, we subdivided the interaction into three orthogonal contrasts testing the effects of pollinator functional group identity on each plant guild, and the effect of the functional diversity of both plant and pollination treatments. See Table 1 for the construction of the contrasts.

For the analysis of plant recruitment, we used a repeated measure ANOVA model (SAS proc mixed). The fixed effects were pollination treatment, plant treatment, year, and all the interaction terms. The repeated effect was year, and the subject effect was replicate. For recruitment density, data were log transformed.

For each year of the experiment, the visitation rate of pollinators on each plant species in the communities with both plant functional groups was analysed using a mixed log-linear model (glimix macro, SAS). We subdivided the pollination treatment into two effects: pollinator functional diversity (one or two pollinator functional groups) and identity of the pollinator functional group (bumble bees or syrphids). The model included pollinator species nested within identity of pollinator functional groups, plant species nested within identity of plant functional group, identity of pollinator functional groups, identity of plant functional groups, pollinator functional diversity, and all interaction terms. The replicate was a random effect.

Acknowledgements

We thank Carine Collin, Romain Gallet, Jean-Francois le Galliard, Jacques Gignoux, Andy Gonzalez, Gérard Lacroix, Gaelle Lahoreau, Louis Lambrecht, Manuel Massot, Naoise Nunan, Virginie Tavernier, and Elisa Thebault for useful discussions; and Marco Banchi, Yves Bas, Mathilde Baude, Alix Boulouis, Marion Decoust, Patricia Genet, Alexandra Kabadajic, Mohsen Kayal, Fanny Marlin, and Emilie Patural for great help in the field and in the lab. We also thank Andy Gonzalez, Andy Hector, Marcel van der Heijden, Claire Kremen, Jane Memmott, Nick Waser, and three anonymous reviewers for constructive and useful comments on the manuscript. We acknowledge the financial support of the Quantitative Ecology Coordinated Incentive Action (ACI Ecologie Quantitative) of the Ministry of Research (France).

Authors' Contributions

CF and ID designed the experiment. CF, ID, and JM performed the experiment. CF analysed the data. CF, ID, and ML conceived the work and wrote the paper.

References

1. Loreau M, Naeem S, Inchausti P, Bengtsson J, Grime JP, et al. (2001) Biodiversity and ecosystem functioning: Current knowledge and future challenges. Science 294: 804–808.

2. Naeem S, Thompson LJ, Lawler SP, Lawton JH, Woodfin RM (1994) Declining biodiversity can alter the performance of ecosystems. Nature 368: 734–737.

3. Costanza R, Arge R, Groot R, Farber S, Grasso M, et al. (1997) The value of the world's ecosystem services and natural capital. Nature 387: 253–260.

4. Kearns CA, Inouye DW, Waser NM (1998) Endangered mutualisms: The conservation of plant-pollinator interactions. Annu Rev Ecol Syst 29: 83–112.

5. Kremen C, Williams NM, Thorp RW (2002) Crop pollination from native bees at risk from agricultural intensification. Proc Natl Acad Sci USA 99: 16812–16816.

6. Klein AM, Steffan-Dewenter I, Tscharntke T (2003) Fruit set of highland coffee increases with the diversity of pollinating bees. Proc R Soc Lond B Biol Sci 270: 955–961.

7. Herrera CM (1987) Components of pollinator 'quality': Comparative analysis of a diverse insect assemblage. Oikos 50: 79–90.

8. Tepedino VJ (1979) The importance of bees and other pollinators in maintaining floral species composition. Great Basin Nat Mem 3: 139–150.

9. Van der Heijden MGA, Klironomos JN, Ursic M, Moutoglis P, Streitwolf-Engel R, et al. (1998) Mycorrhizal fungal diversity determines plant biodiversity, ecosystem variability and productivity. Nature 396: 69–72.

10. Cardinale BJ, Palmer MA, Collins SL (2002) Species diversity enhances ecosystem functioning through interspecific facilitation. Nature 415: 426–429.

11. Callaway RM, Brooker RW, Choler P, Kikvidze Z, Lortie CJ, et al. (2002) Positive interactions among alpine plants increase with stress. Nature 417: 844–848.

12. Thebault E, Loreau M (2003) Food-web constraints on biodiversity-ecosystem functioning relationships. Proc Natl Acad Sci USA 100: 14949–14954.

13. Hector A, Schmid B, Beierkuhnlein C, Caldeira MC, Diemer M, et al. (1999) Plant diversity and productivity experiments in European grasslands. Science 286: 1123–1127.

14. Tilman D, Reich PB, Knops J, Wedin D, Mielke T, et al. (2001) Diversity and productivity in a long-term grassland experiment. Science 294: 843–845.

15. Axelrod DI (1960) The evolution of flowering plants. In: Tax S, editor. Evolution after Darwin. Volume 1, The evolution of life. Chicago: University of Chicago Press. pp. 227–305.

16. Memmott J (1999) The structure of a plant-pollinator food web. Ecol Lett 2: 276–280.

17. Olesen JM, Jordano P (2002) Geographic patterns in plant-pollinator mutualistic networks. Ecology 83: 2416–2424.

18. Jordano P, Bascompte J, Olesen JM (2003) Invariant properties in coevolutionary networks of plant-animal interactions. Ecol Lett 6: 69–81.
19. Bascompte J, Jordano P, Melian CJ, Olesen JM (2003) The nested assembly of plant-animal mutualistic networks. Proc Natl Acad Sci USA 100: 9383–9387.
20. Vázquez DP, Aizen MA (2004) Asymmetric specialization: A pervasive feature of plant-pollinator interactions. Ecology 85: 1251–1257.
21. Memmott J, Waser NM, Price MV (2004) Tolerance of pollination networks to species extinctions. Proc R Soc Lond B Biol Sci 271: 2605–2611.
22. Ashworth L, Aguilar R, Galetto L, Aizen MA (2004) Why do pollination generalist and specialist plant species show similar reproductive susceptibility to habitat fragmentation? J Ecol 92: 717–719.
23. Aizen MA, Feinsinger P (1994) Forest fragmentation, pollination, and plant reproduction in a chaco dry forest, Argentina. Ecology 75: 330–351.
24. Vázquez DP, Simberloff D (2002) Ecological specialization and susceptibility to disturbance: Conjectures and refutations. Am Nat 159: 606–623.
25. Kondoh M (2003) Foraging adaptation and the relationship between food-web complexity and stability. Science 299: 1388–1391.
26. Austin MP, Fresco LFM, Nicholls AO, Groves RH, Kaye PE (1998) Competition and relative yield: Estimation and interpretation at different densities and under various nutrient concentrations using Silybum marianum and Cirsium vulgare. J Ecol 76: 157–171.
27. Ciais P, Reichstein M, Viovy N, Granier A, Ogee J, et al. (2005) Europe-wide reduction in primary productivity caused by the heat and drought in 2003. Nature 437: 529–533.
28. Motten AF, Campbell DR, Alexander DE, Miller HL (1981) Pollination effectiveness of specialist and generalist visitors to a North Carolina population of Claytonia virginica. Ecology 62: 1278–1287.
29. Thostesen AM, Olesen JM (1996) Pollen removal and deposition by specialist and generalist bumblebees in Aconitum septentrionale. Oikos 77: 77–84.
30. Kremen C, Williams NM, Bugg RL, Fay JP, Thorp RW (2004) The area requirements of an ecosystem service: Crop pollination by native bee communities in California. Ecol Lett 7: 1109–1119.
31. Vázquez DP, Aizen MA (2003) Null model analyses of specialization in plant-pollinator interactions. Ecology 84: 2493–2501.

CITATION

Originally published under the Creative Commons Attribution License. Fontaine C, Dajoz I, Meriguet J, Loreau M (2006) Functional Diversity of Plant–Pollinator Interaction Webs Enhances the Persistence of Plant Communities. PLoSBiol 4(1): e1. doi:10.1371/journal.pbio.0040001.

How to be an Attractive Male: Floral Dimorphism and Attractiveness to Pollinators in a Dioecious Plant

Marc O. Waelti, Paul A. Page, Alex Widmer and
Florian P. Schiestl

ABSTRACT

Background

Sexual selection theory predicts that males are limited in their reproductive success by access to mates, whereas females are more limited by resources. In animal-pollinated plants, attraction of pollinators and successful pollination is crucial for reproductive success. In dioecious plant species, males should thus be selected to increase their attractiveness to pollinators by investing more than females in floral traits that enhance pollinator visitation. We tested the prediction of higher attractiveness of male flowers in the dioecious, moth-pol-

linated herb Silene latifolia, by investigating floral signals (floral display and fragrance) and conducting behavioral experiments with the pollinator-moth, Hadena bicruris.

Results

As found in previous studies, male plants produced more but smaller flowers. Male flowers, however, emitted significantly larger amounts of scent than female flowers, especially of the pollinator-attracting compounds. In behavioral tests we showed that naïve pollinator-moths preferred male over female flowers, but this preference was only significant for male moths.

Conclusion

Our data suggest the evolution of dimorphic floral signals is shaped by sexual selection and pollinator preferences, causing sexual conflict in both plants and pollinators.

Background

According to sexual selection theory, males compete with each other over access to females since the reproductive success of a male is limited by the number of females he can fertilize, whereas female reproductive success is limited by resources available for producing offspring [1,2]. In plants, access to pollinators should therefore limit the reproductive success of males to a greater extent than it restricts the reproductive success of females [3-5]. Consequently, different selection pressures are expected to act on males and females, resulting in male-male competition over mates [1].

The majority of plants rely on pollinators for successful pollen transfer [6]. Prefertilization-competition among male gametophytes has been described as pollen competition within the female organs [7], which can be influenced through physiological interactions with the pistils [8]. However, pollinator attraction is the first step in the reproductive cycle of animal-pollinated plants and in dioecious species, sexual reproduction is impossible without the transfer of pollen from male to female flowers. In contrast to most other plants that have hermaphroditic flowers, males and females can respond differently to pollinator-mediated selection. In this situation, selection may hence favor traits that improve pollination and fertilization success, which may lead to sexual dimorphisms in pollinator attracting traits rendering male flowers more attractive than females since access to mates is a function of access to pollinators [4,9,10]. In addition, natural selection on females should reduce attractiveness, since besides pollinators, floral signals also attract granivores that can drastically reduce fitness [11,12]. Even for male flowers, however, pollinator attraction is risky, as it may lead to infection by anther smut fungus that sterilizes the flowers [13].

It has been known for a long time that floral traits like color, shape, size and odor influence the behavior of flower visitors, but less is known about the relative importance of these traits in different pollination systems [14,15]. Colorful floral displays are often accompanied by fragrance, and both visual and olfactory signals attract pollinators and serve as learning cues [16]. In many nocturnal pollination systems, however, scent is thought to be of primary importance for pollinator attraction [16-18]. Floral signals in plants have a similar role in sexual reproduction as mating signals in animals, although they act indirectly, through the behavior of their pollinators. Thus, sexual dimorphisms in floral signals should evolve as a result of sexual selection, acting in concert with the neuronal and behavioral purge of the pollinators. Therefore, the signature of sexual selection should be especially pronounced in the signals that play a particular role in attraction of a given (guild of) pollinator(s).

We tested the prediction of higher attractiveness in male flowers by investigating floral signals and pollinator behavior in the perennial dioecious herb Silene latifolia that is primarily pollinated by nocturnal moths [11,19-21]. This species is an example of nursery pollination, as one of the main pollinators, the noctuid moth Hadena bicruris, oviposits in female flowers where larvae feed on developing seeds. We investigated flower size, flower number and floral odor emission in male and female S. latifolia plants. Further, we assessed the attractiveness of individual flowers of both sexes using male and female Hadena bicruris moths in a wind tunnel bioassay.

Results

Floral Odor

In both populations investigated, male flowers produced significantly more odor than female flowers (Figure 1; Mean ± SE: Switzerland: males: 422.05 ± 47.34 ng h^{-1}, females: 202.56 ± 25.57 ng h^{-1}; Spain: males: 134.19 ± 16.88 ng h^{-1}, females: 81.56 ± 7.90 ng h^{-1}). In the GLM, both sex and population showed a significant effect and there was a significant interaction between sex and population (Table 1; GLM: sex*population: $F_{1,555}$ = 11.252, P = 0.001), indicating that the Swiss population differed more strongly among sexes than did the Spanish population. The covariate flower diameter was not significant (Table 1).

Table 1. GLM of the effects of sex and population of Silene latifolia plants on log transformed total amount of odor per flower.

Source	Type III sum of squares	df	mean square	F	P
population	33.571	1	33.571	98.623	< 0.001
sex	12.560	1	12.560	36.899	< 0.001
population*sex	3.830	1	3.830	11.251	= 0.001
flower diameter	0.104	1	0.104	0.304	= 0.581
error	187.217	550	0.337		

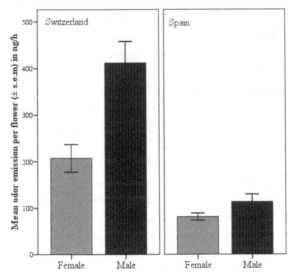

Figure 1. Absolute amounts of odor emitted by female (grey bars) and male (black bars) flowers of two S. latifolia populations. In both populations, male flowers emit significantly more odor than female flowers (GLM, P < 0.001).

In the analysis of individual compounds, more active compounds (compounds triggering electrophysiological responses or affecting behavioral responses in pollinators) [17] were significantly different between the sexes than non-active compounds (Switzerland: active 93%, non-active: 50%; Spain active: 86%, non-active: 62%). The emission of most compounds behaviorally active in Hadena bicruris [17] were found to be significantly higher in male flowers than in female flowers in both populations (Figure 2a, b, Table 2). In Switzerland, 2-methoxy phenol, the lilac aldehydes A, B and C, and veratrole were found in significantly higher amounts in male flowers. The amounts of phenylacetaldehyde and linalool were not significantly different between the sexes. In Spain, phenylacetaldehyde, lilac aldehyde A, and veratrole were found in significantly higher amounts in male flowers. 2-methoxy phenol, and the lilac aldehydes B and C were not significantly different in males and females, but showed a trend to higher emission in males. Only linalool was found in significantly higher amounts in females.

Table 2. Mean absolute amounts of odor compounds (± SEM; ng h^{-1}) in headspace samples of Silene latifolia flowers (asterisks (*) indicate significant differences between female and male amounts within populations).

Compounds[1]	Switzerland		Spain	
	Females (N = 123) Mean ± SE	Males (N = 79) Mean ± SE	Females (N = 217) Mean ± SE	Males (N = 136) Mean ± SE
Fatty Acid Derivates[2]	**0.33%**	**0.13%**	**0.37%**	**0.17%**
Octanal	0.43 ± 0.14	0.33 ± 0.10	0.11 ± 0.01 *	0.10 ± 0.03
Nonanal[3]	0.24 ± 0.02	0.20 ± 0.03	0.19 ± 0.02 *	0.13 ± 0.02
Aromatics[2]	**41.41%**	**43.82%**	**57.58%**	**59.56%**
Benzaldehyde[3]	0.59 ± 0.08	0.48 ± 0.08	1.14 ± 0.18	0.94 ± 0.11
Phenylacetaldehyde[4]	3.49 ± 0.66	3.85 ± 1.53	34.80 ± 4.02 *	45.93 ± 6.07
2-Methoxy phenol[4]	0.68 ± 0.11 *	1.15 ± 0.21	0.08 ± 0.01	0.19 ± 0.04
Methyl benzoate[3]	0.04 ± 0.01 *	0.80 ± 0.46	0.06 ± 0.01 *	0.06 ± 0.01
2-Phenylethanol[3]	0.28 ± 0.13	0.31 ± 0.13	3.02 ± 0.54 *	2.26 ± 0.30
Veratrole[4]	72.94 ± 14.95 *	173.27 ± 25.43	1.88 ± 1.41 *	7.28 ± 3.53
Methyl salicylate[3]	0.85 ± 0.10 *	1.14 ± 0.34	5.41 ± 0.81 *	4.01 ± 0.84
Benzyl benzoate[3]	5.22 ± 3.20 *	3.95 ± 1.01	0.58 ± 0.11 *	0.34 ± 0.08
Monoterpenes[2]	**54.55%**	**52.67%**	**33.73%**	**37.24%**
α-Pinene	0.16 ± 0.01 *	0.12 ± 0.02	0.13 ± 0.01 *	0.08 ± 0.01
Camphene	0.16 ± 0.01	0.14 ± 0.02	0.12 ± 0.01 *	0.08 ± 0.01
β-Pinene	0.10 ± 0.01 *	0.06 ± 0.01	0.10 ± 0.01 *	0.06 ± 0.01
Limonene	0.46 ± 0.04	0.36 ± 0.03	0.36 ± 0.02 *	0.23 ± 0.02
Eucalyptol	0.52 ± 0.11 *	0.86 ± 0.17	0.76 ± 0.10	0.72 ± 0.11
Trans-β-Ocimene[3]	1.25 ± 0.53	1.95 ± 0.84	1.73 ± 0.25	1.23 ± 0.27
Linalool[4]	0.14 ± 0.01	0.11 ± 0.01	0.15 ± 0.01 *	0.08 ± 0.01
Lilac aldehyde A[4]	36.63 ± 4.34 *	80.64 ± 8.02	8.41 ± 1.21	19.34 ± 3.83
Lilac aldehyde B[4]	61.92 ± 7.02 *	122.04 ± 12.12	13.81 ± 1.88	24.89 ± 5.13
Lilac aldehyde C[4]/Benzyl acetate	7.29 ± 2.09 *	11.99 ± 2.43	1.37 ± 0.21	2.20 ± 0.40
Lilac alcohol[3]	2.15 ± 0.25 *	4.05 ± 0.38	0.58 ± 0.07	1.01 ± 0.18
Sesquiterpenes[2]	**0.05%**	**0.01%**	**0.09%**	**0.02%**
B-Farnesene	0.11 ± 0.02 *	0.05 ± 0.01	0.07 ± 0.02	0.03 ± 0.00
Irregular terpenes[2]	**0.05%**	**0.02%**	**0.12%**	**0.04%**
6-Methyl-5-hepten-2-one	0.11 ± 0.02	0.10 ± 0.02	0.10 ± 0.01	0.06 ± 0.01
Unidentified with Kovat's retention index (R_i)[2]	**3.61%**	**3.35%**	**8.11%**	**2.95%**
Unknown 1 (978)	0.99 ± 0.10	0.95 ± 0.12	0.90 ± 0.08 *	0.65 ± 0.08
Unknown 2 (992)	3.63 ± 0.41 *	2.47 ± 0.55	4.98 ± 0.66 *	1.86 ± 0.23
Unknown 3 (1009)	0.09 ± 0.01	0.05 ± 0.01	0.09 ± 0.02 *	0.02 ± 0.01
unknown 4 (1112)	0.29 ± 0.06 *	0.82 ± 0.11	0.14 ± 0.02 *	0.26 ± 0.04
unknown 5 (1191)	2.33 ± 0.36 *	9.85 ± 1.20	0.51 ± 0.06 *	1.17 ± 0.19

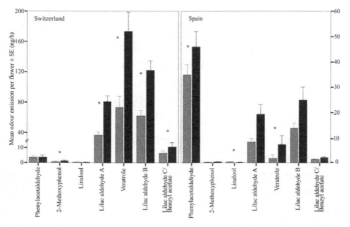

Figure 2. Mean emission of behaviorally active odor compounds by female (grey bars) and male (black bars) S. latifolia flowers in the Swiss (a) and the Spanish (b) population (Mann-Whitney U-test, * = p < 0.05). Note the difference in scale in the y-axes.

Morphology

Flower number was higher in male plants compared to female plants in both populations (flowers plant^{-1} ± SE: Switzerland males 7.74 ± 0.53, Switzerland females 5.26 ± 0.50, Mann-Whitney U-test: U = 2887, P < 0.001; Spain males 8.63 ± 0.54, Spain females 4.67 ± 0.19, Mann-Whitney U-test: U = 8908, P < 0.001). Flower diameter was significantly smaller in male flowers than in female flowers in both populations (mean flower diameter ± SE: Switzerland males 2.51 ± 0.03, Switzerland females 2.60 ± 0.03, t-test: t = 2.017, df = 193, P < 0.05; Spain males 2.67 ± 0.02, Spain females 2.84 ± 0.02, t = 5.405, df = 351, P < 0.001).

Moth Behavior

Most flower-naïve pollinator moths chose male flowers in their first approach to S. latifolia, suggesting higher attractiveness of male flowers compared to female flowers, however the preference was only significant for male moths. Of the 25 male moths tested, 6 chose female and 19 male flowers (Chi^2 = 6.76, df = 1, P = 0.009). Of the 31 unmated females tested, 11 chose female and 20 male flowers (Chi^2 = 2.61, df = 1, P = 0.11).

Discussion

Floral traits that increase pollinator attraction are expected to evolve under stronger pollinator-mediated selection in male plants because males compete for pollinator visitation whereas females are usually limited by resources other than pollen [1,3-5]. For female plants, and especially in nursery pollination systems, attractiveness to pollinators is risky because it is linked to seed predation [11,12]. Consistent with theoretical expectations of sexual selection on floral attractiveness, we found that individual male flowers were more attractive to naïve pollinators, especially to male moths. The likely reason for this is the significantly higher emission of floral scent per flower in male versus female flowers.

In many dioecious plant species, males produce larger or more flowers than females [3,10,22-24]. Plants with increased floral display usually receive higher numbers of visits by pollinators [25-28]. In S. latifolia, male flowers are smaller than female flowers, but male plants produce more flowers than females, suggesting a trade-off between flower size and number [29]. Silene latifolia pollinators prefer plants with larger floral displays [13]. Therefore, the increased number of flowers found on male S. latifolia plants enhances the attractiveness to pollinators [29,30]. However, as yet it was not clear whether the increased attractiveness of male plants is simply a function of the higher number of flowers, or whether individual flowers have evolved traits of higher attractiveness to pollinators. We showed in dual choice experiments in the wind-tunnel that individual male flowers are indeed more attractive to the males of the main pollinator moth. Male flowers, despite being smaller than female flowers, produced significantly higher amounts of odor, both in the Swiss and Spanish population. The significant interaction between sex and population indicates that the sex differences in the amount of scent produced is different among the two populations. Indeed, in the Spanish population, the sex difference was less pronounced, and these plants emitted an overall lower amount of scent. Population specific differences in sex specific traits may be related to differences in pollinator composition, or differences in Swiss and Spanish Hadena moths, however, data are not available.

Interestingly, we found that in both populations more of the compounds involved in pollinator attraction, so-called active compounds, were significantly different among sexes, suggesting that selection for higher attractiveness is mediated by the sensory ecology of the pollinator. Overall, our results strongly suggest that in S. latifolia, scent is more important for the attractiveness of individual flowers than size, as we used flowers of similar size in our behavioral assays. Other studies on S. latifolia floral scent emission failed to detect a statistically significant higher odor production by males [20], but similar trends were found [31]. Both studies, however, were not designed to investigate the effect of gender on floral fragrance and analyzed fewer plants than our study; given the usually high variation in floral scent, large sample size is an important factor in detecting significant effects. Sex differences in amounts of scent were also detected in other plant species, with male plants releasing more attractive volatiles or higher amounts than conspecific females [32,33]. Ashmann et al. (2005) showed that in the gynodioecious wild strawberry Fragaria virginiana the smaller hermaphroditic flowers emitted significantly more odor, which resulted in more visits by pollinators compared to conspecific females [34]. However, these authors suggest that the odor of hermaphroditic flowers is preferred due to the production of unique compounds produced by the anthers, rather than due to quantitative differences in odor production.

A positive association between scent concentration and attractiveness was also found in earlier experiments, where higher odor concentration amplified the response of the pollinators in wind tunnel bioassays with Hadena bicruris [17]. Schiestl (2004) showed that larger amounts of a biologically active floral odor compound attracted more pollinators of the orchid Chiligottis trapeziformis [35]. In natural populations of S. latifolia, preferences of pollinators for male flowers could be the result of learning, since male flowers produce higher sugar concentrations and thus higher quality rewards [13]. As we used naïve pollinators for our experiments, learning should not have influence the preferred choice of male flowers. Alternatively, preference of stronger odor sources may be due to the stronger excitement of olfactory receptors, a form of supernormal stimulation found in floral mimicry systems [35].

Interestingly, the higher attractiveness of male flowers was less pronounced in female moths, which showed no significant preference for male flowers. Female moths need to find flowers for oviposition, besides nectar consumption for energy supply. Because only female flowers are optimal sites for larval development, female moths are expected to evolve preference for female flowers, and especially so after mating. It would be interesting to test in future experiments behavioral preferences of unmated vs. mated females. This situation represents an interesting example of sexual conflict, both in the plant and its pollinator [36]. In the moths, male preference of stronger odor emission acts against female interests, i.e. preference of female flowers, at least after mating. In the plant, the stronger odor emission for increased pollen export should be selected against in female plants that are probably not limited in reproductive success by pollen income and should avoid the attraction of seed predating female moths.

Conclusion

In conclusion our study confirms the predictions by sexual selection theory that male flowers should be more attractive to pollinators than female flowers. This result was not obvious in this plant species, as male are smaller than female flowers, however, the floral scent is decisive for attractiveness in this pollination system. We suggest that taking into account the pollinators' sensory and behavioral ecology will likely give a better picture on sexual and natural selection acting on floral signals, and thus lead to a better understanding how flowers evolve under pollinator-mediated selection.

Methods

(a) Plant Material and Odor Collection

For volatile collection, S. latifolia plants from seeds collected in two populations (Leuk, Switzerland, n = 25; Ribes de Fraser, Spain, n = 4) were grown in a common garden setup in a green house. Floral odor of 555 S. latifolia plants (Switzerland: 79 males and 123 females, Spain: 217 males and 136 females) was collected in May at night, between 9 p.m. and 7 a.m. by headspace sorption as described in [18]. We used only newly opened flowers for odor collection. All floral odor samples were stored in sealed glass vials at -20°C for subsequent gas chromatograph (GC) analysis.

(b) Chemical Analysis

The samples were analyzed with a gas chromatograph (GC, Agilent 6890 N) fitted with a HP5 column (30 m 0.32 mm internal diameter x 0.25 μm film thickness) and a flame ionization detector (FID); hydrogen served as carrier gas. We injected one micro-liter of each odor sample splitless at 50°C (1 min) followed by heating to 150°C at a rate of 5°C min^{-1}, and then to 300°C at a rate of 10°C per minute before keeping the oven at 300°C for ten minutes. For odor compound identification peak retention times were compared with those of authentic standard compounds and confirmed by comparison of spectra obtained by gas chromatography-mass spectrometry (GC-MS). One micro liter aliquots of the odor samples were injected into a GC (HP G1800A) with a mass selective detector using the oven and column parameters described above. We discriminated between active and non-active compounds in the attraction of Handena bicruris on the basis of electrophysiological recordings and behavioral assays done in another study [20].

(c) Morphology

We counted all newly opened flowers per plant before odor sampling. The next morning, two flowers per plant were collected and images taken with a digital camera. We measured floral diameter from these images using the software ImageJ http://rsb.info.nih.gov/ij and calculated average values.

(d) Moth Rearing

Hadena bicruris (Lepidoptera: Noctuidae) moths were bred in the lab, starting with adult moths (n = 10, 50% males and 50% females) collected in the surroundings of Zürich, Switzerland, during summer 2005. Wild moths (approx 10 per year) were added to the colony each summer in order to ensure outbreeding. All insects were reared in controlled conditions under a L16:D8 photoperiod and temperature of 20 ± 2°C, at 65 – 80% relative humidity (RH). After hatching of the eggs, larvae were kept in separate plastic containers to avoid cannibalism and were individually fed with fresh capsules of Silene latifolia until pupation. The pupae were then sexed and placed in separate rearing cages until emergence. These emerged adults were used for behavioral experiments before they had any contact to flowers. Over three successive seasons, a total of 56 flower-naïve moths (31 females and 25 males) were used for dual choice experiments with flowers of S. latifolia.

(e) Behavioral Assays

A 200 x 80 x 80 cm wind tunnel was used for behavioral tests. A Fischbach speed-controller fan (D340/E1, FDR32, Neunkirchen, Germany) pushed air through the tunnel with an air velocity of 0.35 m s^{-1}. Four charcoal filters (145.457 mm, carbon thickness 16 mm, Camfil Farr, Reinfeld, Germany) cleaned the incoming air. The experiments were performed at night with red light illumination (< 0.01 µE) 1–3 h after the start of the dark period. An hour prior to testing, the moths were exposed to ambient room temperature (20°C), RH (65%) and experimental light conditions. Naïve moths were tested individually in the wind tunnel. Each moth was placed in an open glass tube mounted on a stand at the downwind side of the tunnel. At the upwind end were set two freshly collected flowers of S. latifolia (from the Swiss population) of different sexes and approximately equal corolla size (to avoid an effect of flower size). A distance of 20 cm separated the two flowers in order to ensure that the moth's response entirely depended on scent source. After take-off, each moth was followed visually until it landed on one of the two flowers, which always resulted in proboscis extension and nectar drinking.

(f) Statistical Analysis

All data were tested for homogeneity of variances (Levene's test) and for normality (Kolmogorov-Smirnov test). We used a GLM approach to examine the effects of sex and population on mean absolute odor emission. Log-transformed total amount of odor was used as dependent variable and sex and population as factors. Floral diameter was used as covariate to correct for flower size. Nonparametric Mann-Whitney U-tests were used to analyze the differences in absolute amounts of individual compounds and of flower numbers between males and females in each population separately, since no transformation allowed analysis with a parametric test. Flower diameter was analyzed by a t-test. Frequency of moth choices were compared using a Chi^2 test. All analyses were carried out using SPSS 11.0.4 for Mac OS X (SPSS Inc., Chicago, USA).

Authors' Contributions

MOW collected and analyzed data, and wrote part of the manuscript, PAP performed bioassays with moths, AW designed the experiments, FPS designed experiments, analyzed data, and wrote part of the manuscript. All authors contributed to the discussion, read and approved the final manuscript.

Acknowledgements

We thank S. Dötterl and R. Kaiser for kindly providing reference compounds used in this study. K. Förster provided valuable comments on the manuscript. Funding was provided by an ETH Zurich TH-grant to AW and FPS (Grant-N° TH – 32/03-2) and by an SNF grant to FPS (SNF grant No. 2-77843-06).

References

1. Bateman AJ: Intra-sexual selection in Drosophila. Heredity 1948, 2(3):349–368.

2. Trivers RL: Parental investment and sexual selection. In Sexual selection and the descent of man, 1871–1971. Edited by: Campbell B. London: Heinemann; 1972:136–179.

3. Bell G: On the functions of flowers. Proceedings of the Royal Society of London Series B-Biological Sciences 1985, 224(1235):223–265.

4. Queller D: Sexual selection in flowering plants. In Sexual selection: testing the alternative. Edited by: Bradbury JW, Andersson, MB. Chichester: Wiley; 1987:165–179.

5. Willson MF: Sexual selection in plants. American Naturalist 1979, 113(6):777–790.

6. Faegri K, Pijl L: The principles of pollination ecology. Oxford: Pergamon Press; 1979.

7. Howard DJ: Conspecific sperm and pollen precedence and speciation. Annual Review of Ecology and Systematics 1999, 30:109–132.

8. Heslop-Harrison J: Incompatibility and pollen-stigma interactions. Annual Review of Plant Physiology 1975, 26:403–425.

9. Stanton ML: Male-male competition during pollination in plant populations. American Naturalist 1994, 144:S40–S68.

10. Stephenson AG, Bertin RI: Male competition, female choice, and sexual selection in plants. In Pollination Biology. Edited by: Real L. New York: Academic Press; 1983:109–149.

11. Brantjes NBM: Riddles around pollination of Melandrium album (Mill) Garcke (Caryophyllaceae) during oviposition by Hadena bicruris Hufn (Noctuidae, Lepidoptera) .2. Proceedings of the Koninklijke Nederlandse Akademie Van Wetenschappen Series C-Biological and Medical Sciences 1976, 79(2):127–141.

12. Bopp S, Gottsberger G: Importance of Silene latifolia ssp. alba and S. dioica (Caryophyllaceae) as host plants of the parasitic pollinator Hadena bicruris (Lepidoptera, Noctuidae). Oikos 2004, 105(2):221–228.

13. Shykoff JA, Bucheli E: Pollinator visitation patterns, floral rewards and the probability of transmission of Microbotryum violaceum, a venereal disease of plants. Journal of Ecology 1995, 83(2):189–198.

14. Waser NM, Price MV: Optimal and actual outcrossing in plants, and the nature of plant-pollinator interactions. New York: Van Nostrand Reinhold; 1983.

15. Dobson HEM: Floral volatiles in insect biology. In Insect-Plant Interactions. Edited by: Bernays EA. CRC Press; USA; 1994:47–81.

16. Raguso RA: Floral scent, olfaction, and scent-driven foraging behavior. In Cognitive Ecology of Pollination. Edited by: Chittka L, Thomson, JD. Cambridge: Cambridge University Press; 2001:83–105.

17. Dötterl S, Jurgens A, Seifert K, Laube T, Weissbecker B, Schutz S: Nursery pollination by a moth in Silene latifolia : the role of odours in eliciting antennal and behavioural responses. New Phytologist 2006, 169(4):707–718.

18. Huber FK, Kaiser R, Sauter W, Schiestl FP: Floral scent emission and pollinator attraction in two species of Gymnadenia (Orchidaceae). Oecologia 2005, 142(4):564–575.

19. Brantjes NBM: Sensory responses to flowers in night-flying moths. In The pollination of flowers by insects. Edited by: Richards A. London: Academic Press; 1978:13–19.

20. Dötterl S, Wolfe LM, Jurgens A: Qualitative and quantitative analyses of flower scent in Silene latifolia. Phytochemistry 2005, 66(2):203–213.

21. Wälti MO, Mühlemann JK, Widmer A, Schiestl FP: Floral odour and reproductive isolation in two species of Silene. Journal of Evolutionary Biology 2008, 21:111–121.

22. Delph LF, Galloway LF, Stanton ML: Sexual dimorphism in flower size. American Naturalist 1996, 148(2):299–320.

23. Lloyd DG, Webb CJ: Secondary sex characters in plants. Botanical Review 1977, 43(2):177–216.

24. Costich DE, Meagher TR: Impacts of floral gender and whole-plant gender on floral evolution in Ecballium elaterium (Cucurbitaceae). Biological Journal of the Linnean Society 2001, 74(4):475–487.

25. Vaughton G, Ramsey M: Floral display, pollinator visitation and reproductive success in the dioecious perennial herb Wurmbea dioica (Liliaceae). Oecologia 1998, 115(1–2):93–101.

26. Johnson SG, Delph LF, Elderkin CL: The effect of petal-size manipulation on pollen removal, seed set, and insect-visitor behavior in Campanula americana. Oecologia 1995, 102(2):174–179.

27. Young HJ, Stanton ML: Influences of flroal variation on pollen removal and seed production in wild radish. Ecology 1990, 71(2):536–547.

28. Ashman TL, Diefenderfer C: Sex ratio represents a unique context for selection on attractive traits: Consequences for the evolution of sexual dimorphism. American Naturalist 2001, 157(3):334–347.

29. Meagher TR: The quantitative genetics of sexual dimporphism in Silene latifolia (Caryophyllaceae) .1. Genetic variation. Evolution 1992, 46(2):445–457.

30. Carroll SB, Delph LF: The effects of gender and plant architecture on allocation to flowers in dioecious Silene latifolia (Caryophyllaceae). International Journal of Plant Sciences 1996, 157(4):493–500.

31. Dötterl S, Jurgens A: Spatial fragrance patterns in flowers of Silene latifolia : Lilac compounds as olfactory nectar guides? Plant Systematics and Evolution 2005, 255(1–2):99–109.

32. Dufay M, Hossaert-McKey M, Anstett MC: Temporal and sexual variation of leaf-produced pollinator-attracting odours in the dwarf palm (vol 139, 392, 2004). Oecologia 2004, 140(2):379–379.

33. Miyake T, Yafuso M: Floral scents affect reproductive success in fly-pollinated Alocasia odora (Araceae). American Journal of Botany 2003, 90(3):370–376.

34. Ashman TL, Bradburn M, Cole DH, Blaney BH, Raguso RA: The scent of a male: the role of floral volatiles in pollination of a gender dimorphic plant. Ecology 2005, 86(8):2099–2105.

35. Schiestl FP: Floral evolution and pollinator mate choice in a sexually deceptive orchid. Journal of Evolutionary Biology 2004, 17(1):67–75.

36. Chapman T, Arnqvist G, Bangham J, Rowe L: Sexual conflict. Trends in Ecology & Evolution 2003, 18(1):41–47.

CITATION

Originally published under the Creative Commons Attribution License. Waelti MO, Page PA, Widmer A, Schiestl FP. How to be an Attractive Male: Floral Dimorphism and Attractiveness to Pollinators in a Dioecious Plant. BMC Evol Biol. 2009 Aug 6;9:190. doi:10.1186/1471-2148-9-190.

Pollen Development in *Annona cherimola mill.* (Annonaceae). Implications for the Evolution of Aggregated Pollen

Jorge Lora, Pilar S. Testillano, Maria C. Risueño,
Jose I. Hormaza and Maria Herrero

ABSTRACT

Background

In most flowering plants, pollen is dispersed as monads. However, aggregated pollen shedding in groups of four or more pollen grains has arisen independently several times during angiosperm evolution. The reasons behind this phenomenon are largely unknown. In this study, we followed pollen development in Annona cherimola, a basal angiosperm species that releases pollen in groups of four, to investigate how pollen ontogeny may explain the rise and establishment of this character. We followed pollen development using immunolocalization and cytochemical characterization of changes occurring from anther differentiation to pollen dehiscence.

Results

Our results show that, following tetrad formation, a delay in the dissolution of the pollen mother cell wall and tapetal chamber is a key event that holds the four microspores together in a confined tapetal chamber, allowing them to rotate and then bind through the aperture sites through small pectin bridges, followed by joint sporopollenin deposition.

Conclusion

Pollen grouping could be the result of relatively minor ontogenetic changes beneficial for pollen transfer or/and protection from desiccation. Comparison of these events with those recorded in the recent pollen developmental mutants in Arabidopsis indicates that several failures during tetrad dissolution may convert to a common recurring phenotype that has evolved independently several times, whenever this grouping conferred advantages for pollen transfer.

Background

Pollen development is a well characterized and highly conserved process in flowering plants [1-3]. Typically, following anther differentiation, a sporogenous tissue develops within the anthers producing microsporocytes or pollen mother cells. Prior to meiosis, pollen mother cells become isolated by a wall with the deposition of a callose layer. Each pollen mother cell, as the result of the two meiotic divisions, generates four haploid cells forming a tetrad and, for a short time, these four sibling microspores are held together in a persistent pollen mother cell wall that is surrounded by callose. The tapetum then produces an enzyme cocktail that dissolves the pollen mother cell wall and the microspores are shed free and become independent [2]. The unicellular microspores go through an asymmetric mitotic division (pollen mitosis I) to produce a pollen grain with two cells, a larger vegetative cell that hosts a smaller generative cell; the latter will divide once more to produce two sperm cells (Pollen mitosis II). Pollen mitosis II can take place before or after pollen release and, depending on when it occurs, the pollen will be bicellular or tricellular at the time of anther dehiscence. Throughout the manuscript we will use the term "pollen tetrads" for mature pollen to avoid confusion with the tetrads of early developmental stages ("microspore tetrads").

Angiosperms pollen is most commonly released as single pollen grains or monads [4] which represent the basic angiosperm pollen-unit. Dehiscence of aggregated pollen (mostly in groups of four) is considered a recent apomorphic characteristic [5,6] that has arisen independently several times during evolution primarily in animal-pollinated taxa although, in some cases, monads may have evolved secondarily from groups of four grains [6]. Pollen release as tetrads has been reported in some or all members of 55 different angiosperm families and also in some pteridophytes [7]. Blackmore and Crane (1988) [8] put forward that the maintenance of pollen tetrads could be the result of relatively minor ontogenetic changes and, consequently, this could be an excellent example of convergence in situations where the release of pollen as tetrads is an effective reproductive strategy. Interestingly, the dissemination of pollen as tetrads has also been reported in the quartet mutants of Arabidopsis [9,10].

Annonaceae, included in the order Magnoliales, is the largest family within the basal angiosperm Magnoliid clade [11,12]. Due to its phylogenetic position among the basal angiosperms, the family has been the object of considerable interest from a taxonomic and phylogenetic point of view [13-15] and a number of studies have focused on pollen morphology [16-20]. Although most genera of the Annonaceae produce solitary pollen at maturity, in several species of the family pollen is released aggregated in groups of four or in polyads [17]. Recent studies on the mechanism of pollen cohesion in this family have been performed in species of the genera Pseuduvaria [21], Annona and Cymbopetalum [22,23]. Pollen cohesion in these species is generally acalymmate (four pollen grains are grouped only by partial fusion) with simple cohesion [21]. But these studies show differences in cohesion mechanisms; thus, while pollen grains in Pseuduvaria are connected by wall bridges (crosswall cohesion), involving both the exine and the intine, in A. glabra, A. montana and Cymbopetalum cohesion is achieved through a mass of callose-cellulose. Evolutionary transitions in flowering plant reproduction are proving to have a clear potential in plant evolutionary biology [24], and the need for more detailed ontogenetic studies in the family has been put forward [22]. Indeed the fact of being the largest family among basal angiosperms, together with the puzzling connection mechanisms so far described in the different species examined, provide an excellent opportunity to investigate the ontogeny of pollen development and its evolutionary implications.

In this work, pollen development is characterized in A. cherimola, one of the species in the Annonaceae where pollen is shed aggregated in groups of four, paying special attention to the events close to pollen formation and retention of the individual pollen grains together, observed by immunolocalization of different wall components. Results are discussed in relation to the shedding of pollen in groups of four in other species and how this event may have occurred and settled during evolution.

Results

The mature A. cherimola flower is a syncarpous gynoecium with a conic shape composed of about 100 fused carpels surrounded at its base by several rows of anthers with up to 200 stamens, encircled by two whorls of three petals. The flower cycle from opening to anther dehiscence lasts two days: the flower opens on the morning of the first day in the female stage and remains in this stage until the afternoon of the following day when the flower enters the male stage. Anther dehiscence occurs concomitantly in all stamens of a flower and, as the anthers dehisce, they detach from the flower and fall over the open petals.

Flower buds of A. cherimola develop in the leaf axes following leaf expansion; the basal nodes are differentiated in the year preceding anthesis. The uppermost distal buds differentiate in synchrony with shoot growth [25]. Flower bud growth begins 39 days prior to anthesis. Anther differentiation proceeded centripetally, with the most developmentally advanced anthers placed in the outermost rows and the different stages of anther and pollen development present within the same flower. This fact was helpful for establishing successive stages of anther development. The anther becomes septate with pollen mother cells positioned between rows of interstitial tapetum similar to the anthers described in a sister species, Annona squamosa [26].

To determine if pollen development followed a standard pattern and whether pollen tetrads at anther dehiscence corresponds with the cytological and morphological features of mature pollen, anther and pollen development were examined from microsporogenesis to maturity. Special attention was given to the events responsible for pollen cohesion. Microgametogenesis and tapetum degeneration were also examined sequentially.

Microsporogenesis

Initial hypodermal archesporial cells were apparent 24 days before anthesis (Figure 1A). From them, anther septa initials and pollen mother cells (PMC) developed in 9 cm long flower buds 19 days before anthesis (Figure 1B). Each anther contained a uniseriate row of pollen mother cells with a conspicuous common wall. The PMC began to accumulate starch grains (Figure 1C) and increased in size (Figure 1C, 1D). Starch grains vanished concomitantly with the beginning of meiosis, some 15 days before anthesis, as a translucent cell wall layer was apparent surrounding the PMC (Figure 1D). Meiosis proceeded rapidly and was followed by a new accumulation of starch grains in the young microspores (Figure 1E) 14 days before anthesis. The microspore tetrads remained together in isolated tapetal chambers surrounded by the PMC wall that stained positively with periodic acid-Shiff's reagent (PAS) for carbohydrates (Figure 1F, 1G).

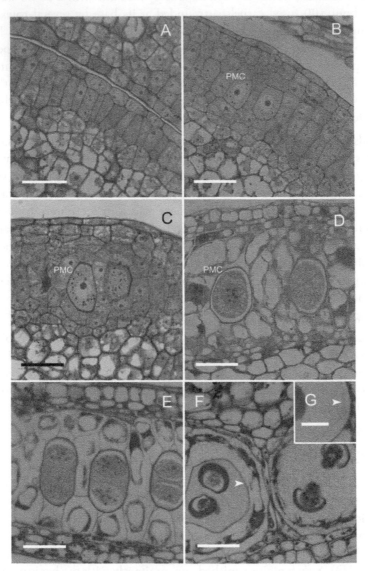

Figure 1. Microsporogenesis of Annona cherimola.
(A) Uniseriate row of arquesporial cells. (B) Septal and pollen mother cells (PMC), showing a conspicuous wall, alternate in the sporogenous tissue. (C) PMC increase in size and starch grains are visible. (D) Starch grains vanish, a translucent layer appears in the PMC wall, and PMC starts meiosis. The tapetum vacuolates and the tapetal chambers are apparent. (E) Following meiosis, starch grains accumulate again in the young microspores, which are surrounded by callose. (F) The young microspores, with an incipient exine, appear to float and turn within the still remaining PMC wall (arrow) that holds the four microspores together. (G) Detail of PMC wall (arrow). Longitudinal sections of the anthers stained with PAS and Toluidine blue. Bar = 20 μm.

Immunocytochemical essays revealed the localization of various cell wall components (Figure 2). Callose surrounded the PMC wall and, following meiosis I, an additional furrow of callose developed inwards (Figure 2A) forming a callose positive band between the dyad cells (Figure 2B, 2C). Successive cytokinesis followed (Figure 2C), resulting in a tetrad (2D), each separated by callose. Dyad and tetrad stages coexist in the flower as centripetal maturation progresses. The PMC wall also reacted positively to JIM7 (Figure 2E) and JIM5 (Figure 2F) staining, indicating the presence of methyl-esterified and unesterified pectins respectively. However, while the walls of the anther somatic and tapetal cells also reacted positively to the JIM7 for methyl-esterified pectins (Figure 2E), they showed only a faint staining for the presence of unesterified pectins (Figure 2F).

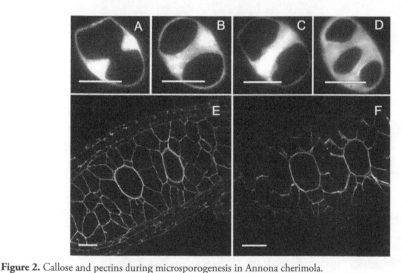

Figure 2. Callose and pectins during microsporogenesis in Annona cherimola.
(A-D) Anticallose in dyad/tetrad phases. (A) A furrow of callose developed inwards, forming a wall between the dyad cells. (B) Dyad phase, showing (B) an incipient, and (C) a well developed callose wall. (D) Tetrad microspore showing in the section plane three of the microspores separated by callose walls. (E) PMC and other anther tissue walls showing methyl-esterified pectins. (F) PMC wall also shows unesterified pectins. Specific cell components were localized using antibodies against callose (A, B, C, D), methyl-esterified pectin (JIM7) (E), and unesterified pectin (JIM5) (F). A-D Bar = 10 μm. E-F: Bar = 20 μm.

Pollen Cohesion

Following tetrad formation, callose disappeared but the microspores remained within the PMC calcofluor-positive cellulosic wall (Figure 3A). At this developmental stage, an interesting event was detected: the microspores within each tapetal chamber, which initially had their pollen aperture sites facing outward towards the PMC wall (Figure 3A), appeared to float and rotate within their individual chambers (Figure 3B). This movement was not random, but the microspores turned 180° until the pollen aperture sites faced each other (Figure 3B). The remaining PMC cellulosic wall, which persisted for some time, together with the confined space provided by the tapetal chamber, appear to contribute towards keeping the microspore tetrad together. Subsequently the PMC cellulosic wall disappeared completely (Figure 3C) and the tapetum degenerated as the microspores increased in size. They remained in their new orientation attached by their apparently sticky aperture sites that now faced each other (Figure 3D).

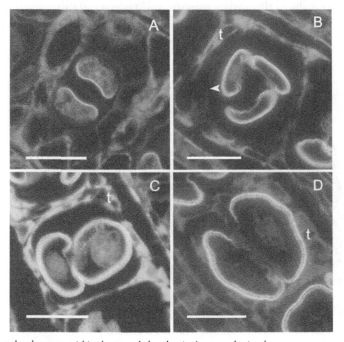

Figure 3. Pollen development within the tapetal chamber in Annona cherimola.
(A) Two young microspores in a tetrad which still keeps the pollen mother cell wall. Aperture sites are located towards the outside facing the pollen mother cell wall. (B) Pollen is shed free, within the PMC wall, in the tapetal chamber. Within this confined space the young microspores turn (C) with their aperture sites facing now each other as the PMC cellulosic wall is digested. (D) The pollen grains regroup sticking through the aperture sites, and enlarge as the tapetum degenerates. Longitudinal anther 2 μm resin sections stained with calcofluor and auramine. Bar = 20 μm.

At this stage, both the cell walls of the somatic cells of the anther and the inner wall of microspores (intine) reacted similarly for methyl-esterified pectins (Figure 4A), while unesterified pectins were present just in the microspore intine (Figure 4B). The exine showed a low unspecific autofluorescence but in a different fluorescence wavelength (yellowish color) than the fluorescent marker of the antibodies, AlexaFluor 488, which emitted green fluorescence. As a consequence, exine autofluorescence was clearly differentiated from the immunofluorescence signals.

Anti-callose immunofluorescence revealed remnants of callose at the pollen aperture sites where the thick external layer of the microspore wall, the exine, was extremely thin or absent. These callose remnants were apparent in all microspores at this stage (Figure 4C). The four microspores showed crosswall cohesion bridges that stained with antibodies against unesterified and methyl-esterified pectins in the microspore wall (Figure 4D-H). Following this inter-intine cohesion, additional deposition of sporopollenin with a joint layering of the four microspores further strengthened this connection (Figure 4I).

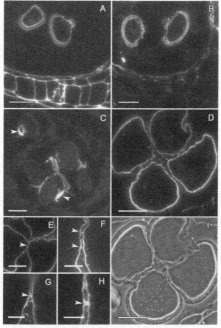

Figure 4. Establishment of pollen cohesion in Annona cherimola.
(A) Microspore walls show methyl-esterified pectins, and also (B) unesterified pectins. (C) As callose is digested, remnants of callose (white arrow) are observed layering the pollen aperture sites. (D-F) Microspores show crosswall cohesion bridges showing the presence of unesterified pectins. (G-H) Details of crosswall cohesion bridges, showing the presence of methyl-esterified pectins. (I) Phase contrast of a mature pollen grain showing internal cohesion and a joint sporopollenin layering. Specific cell components were localized using antibodies against: methyl-esterified pectin (JIM7) (A, G-H), unesterified pectin (JIM5) (B, D, E, F) Callose (C). A-E, I: Bar = 10 μm. F-H: Bar = 3 μm.

Microgametogenesis

As the microspores increased in size, their cytoplasm became vacuolated (Figure 5A) and starch grains were absent (Figure 5B). During this vacuolization, nuclear migration preceded the first mitosis to form bicellular pollen grains. Following the first mitosis, 4-6 days before anthesis, the vacuoles decreased in size (Figure 5C) and starch was again stored (Figure 5D). Young pollen grains had no vacuoles (Figure 5E) and numerous starch grains were present (Figure 5F). The second mitotic division producing the first tricellular pollen grains started some four hours prior to anther dehiscence (Figure 5G) and one day prior to anther dehiscence starch grains began to hydrolyze (Figure 5H). Mitotic divisions were not synchronized within a pollen tetrad and single pollen grains with different numbers of nuclei could be observed in the same tetrad resulting in the coexistence of bicellular and tricellular pollen upon anther dehiscence.

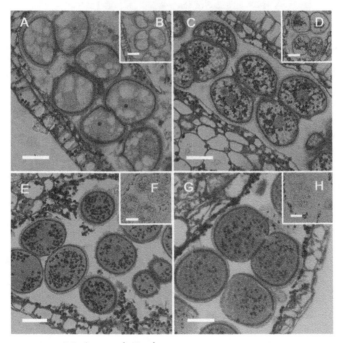

Figure 5. Microgametogenesis in Annona cherimola.
(A) Microspores increase in size as vacuoles appear in the cytoplasm, (B) microspores at this stage do not have starch grains. (C) Microspores following mitosis I; (D) as vacuoles decrease in size, starch grains are stored. (E) Young pollen grains without vacuoles (F), which accumulated starch grains. (G) Close to the time of anther dehiscence, the second mitosis occurs, the tapetum is completely degenerated and (H) starch is digested. Longitudinal sections of anthers stained with PAS and Toluidine blue (A, C, E, G), and with PAS (B, D, F, H) to show starch grains. Bar = 20 μm.

In mature pollen, while intine thickness was similar around the pollen grain, the exine was thinner or absent at the pollen aperture sites where contact points between sibling pollen grains were established (Figure 6A). At these areas unesterified and methyl-esterified pectin bridges were maintained throughout pollen development although these connections seemed to be less strong in mature pollen

(Figure 6B). However, mature pollen tetrads resisted separation during acetolysis, showing the permanence of joint sporopollenin (Figure 6C).

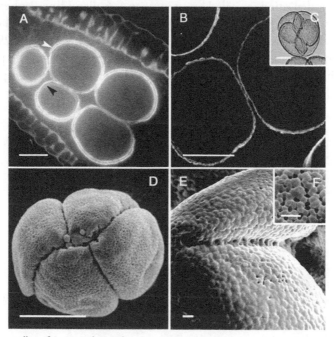

Figure 6. Mature pollen of Annona cherimola.
(A) Intine (black arrow) is similar all around the pollen grain, but exine (white arrow) is thinner in the pollen aperture site. Longitudinal section stained with a 3:1 mixture of Auramine and Calcofluor. (B) Sibling pollen grains have a faint cohesion that showed with JIM 5 antibody the presence of unesterified pectins. (C) Mature pollen tetrad following acetolysis. (D, E, F) Mature pollen observed with scanning electron microscopy (SEM). (D) Mature pollen grains with a globose shape and a radiosymmetric disposition. (E) Exine cohesion helps keeping sibling pollen grains together. (F) Pollen exine shows a tectate perforate appearance. A, B, D: Bar = 20 μm; C: Bar = 10 μm; E, F: Bar = 2 μm.

Scanning electron micrographs revealed that mature pollen had a radiosymetric globose shape, was inaperturated, tectate perforate, and with a diameter of 40 μm (Figure 6D, 6E, 6F). Mature pollen was shed in groups of four sibling pollen grains that stick together having an exine cohesion, clearly visible with high magnification scanning electron microscopy images (Figure 6E).

Tapetum Degeneration

A. cherimola has a secretory tapetum with tapetal-type septa similar to those described in other species of the genus Annona such as A. squamosa [26] and A. glabra [27]. Prior to meiosis, septal initials formed tapetal chambers that host the

PMC (Figure 7A). After meiosis, the tapetum showed a vacuolization and a progressive degeneration as the tapetal chamber enlarged (Figure 7B). The nuclei of the tapetal cells displayed elongated and lobular shapes together with a extremely high chromatin condensation, revealed by an intense 4',6-diamidino-2-phenylindole (DAPI) fluorescence (Figure 7C), typical features of programmed cell death [28], which have also been found in the tapetal nuclei of other species [29]. At the same time, tapetal cells released their cellular contents that coated the pollen grains to form the pollenkit. At anther dehiscence the tapetum was completely degenerated and had disappeared (Figure 7D).

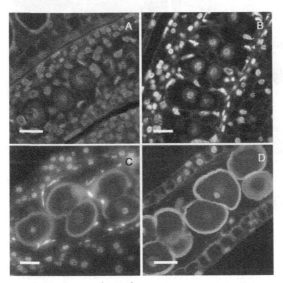

Figure 7. Tapetum degeneration in Annona cherimola.
(A) Pollen mother cells in Prophase I and an active tapetum. (B) Dyad phase in enlarged tapetal chambers. (C) Anther, 4 days before anthesis, showing bicellular pollen and degenerated tapetum, with nuclei displaying elongated shapes and chromatin condensation. (D) Tapetum has disappeared in anthers of flowers at the female stage showing mature pollen. Longitudinal 5 μm resin sections stained with DAPI. Bar = 20 μm.

Discussion

Pollen Development

Pollen in A. cherimola is shed in groups of four, originating from the same meiotic division and, hence, the same tetrad. Pollen development, however, continues beyond tetrad formation and, although held together, pollen grains are fully mature upon anther dehiscence. Meiosis cytokinesis occurred through the formation of ingrowths of callose that are also found in genera of some primitive angiosperms [30,31] including species of the Magnoliid clade as Magnolia tripetala [32] and

Degeneria vitiensis [33] in the Magnoliales, Laurelia novae-zelandiae [34] and Liriodendron tulipifera [35] in the Laurales or Asarum in the Piperales [30] as well as in monocots as Sisyrinchium [36].

Starch accumulated prior to meiosis and the first pollen mitosis and vanished with the onset of these two divisions; this also occurred 6 days before anther dehiscence, preceding the shedding of starchless pollen. The accumulation of starch in PMC has also been reported in other primitive angiosperms, such as Anaxagorea brevipes [37] or Austrobaileya maculata [38], and in other evolutionarily more recent angiosperms [39,40]. Vacuolization also follows a conserved pattern [41,42]. The cytoplasm enlarges through a first vacuolization and, later on, following the first pollen mitosis, small vacuoles appear as starch builds up.

In most species a dehydration process takes place prior to pollen shedding and starch is hydrolyzed to form sucrose that protects pollen against desiccation [43]. Starchless pollen is the most common pollen type in the angiosperms [44], being more frequent in bicellular than in tricellular pollen [45,46]. In A. cherimola, pollen is shed in a highly hydrated stage [47] and this lack of dehydration may explain why the second mitotic division continues in free pollen after pollen shedding, producing a mixed population of bi and tricellular pollen [48]. However, both types of pollen are starchless at anther dehiscence.

Pollen Cohesion

Several reasons could account for the release of pollen in groups of four. In Arabidopsis, failure of different enzymes during the dissolution of the pectic layer that surrounds the PMC wall has been reported in two quartet mutants [10,49]. In our work the immunolocalization of esterified and non-esterified pectins showed that, although they were clearly present in the PMC wall, the pectins disappeared following tetrad formation. A closer examination of the photographs reveals that the PMC wall, which stains for cellulose, remained beyond the tetrad stage. Interestingly quartet mutants of Arabidopsis also show a defect in the degradation of the PMC wall [10]. Cellulase has also been shown to be involved in the breakdown process of the PMC wall [50] and a delay in its action could lead to this phenomenon. However, this failure does not seem permanent, since 25 days later this wall is completely dissolved. Thus, the permanence of the PMC wall appears as a key factor contributing to pollen grouping as pollen tetrads in A. cherimola, similar to the observations in Arabidopsis mutants. A mixture of enzymes is require to break down the complex PMC wall [2], and a failure of one or more of these enzymes could result in a similar final result.

Different events that take place during the retention of this wall may explain pollen adherence once this wall disappears. The confining of pollen within the tapetal chamber keeping the young microspores in close proximity may contribute to this wall maintenance. Surprisingly, the young microspores are apparently separated from their sibling cells allowing some free movement indicated by the strange 180° rotation of the pollen aperture sites. Thus, those aperture sites that originally looked outwards rotate inward to face each other. A similar rotation has been reported previously in other Annonaceae [A. glabra and A. montana [22] and Cymbopetalum [23], and also in species of the Poaceae [51]. This distal-proximal microspore polarity transition in development contrasts with the evolutionary shift from a proximal to a distal aperture that has been long regarded as one of the major evolutionary innovations in seed plants [52]. Proximal apertures predominate in the spores of mosses, lycophytes and ferns while distal apertures are more common in extant seed plants including gymnosperms, cycads and early-divergent angiosperms [52]. In fact, species in the Annonaceae with monad pollen are reported to have distal apertures [see [19] for review]. However, a complete study of 25 Annonaceae genera with species that release aggregated pollen showed proximal apertures [16] and, consequently, the distal-proximal transition observed in pollen development of A. cherimola and other Annonaceae [22,23] could represent a widespread situation in this basal family.

Another reason proposed for this permanent binding of pollen in groups of four could be a failure in the synthesis of the callose layer during microspore separation in the tetrad [8]. However, the results shown in this work in A. cherimola indicate that callose is layered following the standard pattern and vanishes later, after meiosis is completed, similar to the way it occurs in Arabidopsis quartet mutants, in which callose dissolution proceeds normally [53]. However, the use of antibodies against callose showed that callose remains for a while in the area where pollen apertures will form hampering the layering of sporopollenin. Callose remnants in this area have also been reported in other Annonaceae and it has been suggested that these remnants pull the pollen grains to undergo the 180° turning [22,23]. In the formation of the pollen wall, callose dissolution occurs concomitantly with the layering of the exine [54] and the formation of the pollen aperture is related to endoplasmic reticulum blocking the deposition of primexine [3]. The callose remnant at the pollen aperture sites has not been investigated in detail in other species and, given the high conservation of pollen ontogeny in angiosperms, this is a topic worthy of a detailed study. Interestingly, in an Arabidopsis mutant lacking the gene responsible for callose synthesis, pollen develops unusual pore structures [55].

Further binding at the aperture sites could follow this initial adhesion process through the observed joint deposition of pollenkit that has also been reported in other species [4]. Thus, two key processes could contribute to holding together the four pollen grains in A. cherimola, the confined space that permits the delay in the dissolution of the PMC wall and the tapetal chamber and pollen rotation that allows the adhesion of the sticky proximal faces by the formation of small pectin bridges. Later, the join deposition of sporopollenin would further strengthen this initial binding.

Biological Significance of the Pollen Dispersal Unit

A failure or delay in the dissolution of the PMC wall and tapetal chamber appears to be a critical step, resulting in the continued proximity of the four microspores produced by meiosis of a single PMC. However, this phenotype could also result from failure in the different enzymes that dissolve the PMC wall. The distribution of this character, together with the information provided by Arabidopsis mutants, shows that this has occurred independently several times during evolution, suggesting that it must provide some evolutionary advantages [8].

The adaptive advantages derived from aggregated pollen have been reviewed recently [6]. The release of aggregated pollen in insect pollinated species could increase pollination efficiency, since more pollen grains could be transferred in a single pollinator visit and, in this sense, a correlation between pollen tetrads and polyads with a high number of ovules per flower has been shown in a survey of the Annonaceae [17]. The release of aggregated pollen is more advantageous in situations where pollinators are infrequent [6] and in situations of short pollen viability and pollen transport periods. A short pollen viability period has been reported in A. cherimola, [47,56] and a short pollen transport episode is common in several Annonaceae [57] and in other beetle pollinated species of early divergent angiosperm lineages [58].

An additional possible benefit of aggregated pollen is protection against desiccation and entry of pathogens through the thin walls of the pollen aperture sites. Pollen grouped in dyads, tetrads or polyads show a strong proximal reduction of the exine in Annonaceae [59]. A. cherimola pollen is inaperturate and germinates in the proximal face, showing a large area of unprotected intine [47,60]. More evolutionarily recent species present a colpus that, in dehydrated pollen, is just a narrow slit protected by loose pollenkit. Only upon hydration, when the pollen faces a wet surface on the stigma, this slit swells developing a wider colpus through which the pollen tube protrudes [61]. Inaperturate pollen does not have this protection from desiccation and the development of inward facing intines may play a role in protecting pollen against desiccation.

Conclusion

The results obtained in this work support the hypothesis that aggregated pollen could be the result of relatively minor ontogenetic changes beneficial for pollen transfer or/and protection from pollen desiccation. Comparison of the events reported here with those recorded in recent pollen development mutants in Arabidopsis suggests that a simple event along development, the delay in the dissolution of the pollen mother cell wall and tapetal chamber, results in conspicuous morphological changes that lead to the release of pollen in tetrads. A variety of different mutations within the enzymes required to breakdown this wall, may contribute to this common morphology. These changes have occurred and recur in nature and, due to their adaptive advantages for pollen transfer, have been selected during evolution several independent times, representing an example of convergent evolution.

Methods

Plant Material

The research was performed on adult A. cherimola, cv. Campas trees of located in a field cultivar collection at the EE la Mayora CSIC, Málaga, Spain. To study the relationship between flower bud length and developmental stages, tagged flower buds were measured sequentially on the trees. Buds were measured twice a week for 8 weeks from leaf unfolding, when the buds were visible but buried under the leaf petiole until anthesis. A. cherimola, as other members of the Annonaceae, presents protogynous dichogamy [62]. The flower opens in the female stage and remains in this stage until the following day in the afternoon when at a precise time, around 6 pm. under our environmental conditions, it changes to the male stage: the anthers dehisce, the petals open more widely and the stigmas shrivel [48].

Light Microscope Preparations

To follow pollen development, anthers were collected from flower buds of a range of stages, with petals 6, 9, 12, 16, 22, 24 and 30 mm long. Anthers were also collected from flowers one day prior to anthesis and at the female (F) and male (M) stages of mature flowers. The anthers from three flowers of each stage were fixed in glutaraldehyde at 2.5% in 0.03 M phosphate buffer [63], dehydrated in an ethanol series, embedded in Technovit 7100 (Kulzer & Co, Wehrheim, Germany), and sectioned at 2 µm.

Sections were stained with periodic acid-Schiff's reagent (PAS) for insoluble carbohydrates and with PAS/Toluidine Blue for general histological observations [64]. Sections were also stained for cutine and exine with 0.01% auramine in 0.05 M phosphate buffer [65] and for cellulose with 0.007% calcofluor in water [66]. Intine and exine were observed with a 3:1 mixture of 0.01% auramine in water and 0.007% calcofluor in water.

To observe nuclei during pollen development, anthers collected from flowers at the same developmental stages ranging from 9 mm long to anthesis were also fixed in 3:1 (V1/V2) ethanol-acetic acid, embedded as described above, sectioned at 5 μm and stained with a solution of 0.25 mg/ml of 4',6-diamidino-2-phenylindole (DAPI) and 0.1 mg/ml p-phenylenediamine (added to reduce fading) in 0.05 M Tris (pH 7.2) for 1 hr at room temperature in a light-free environment [67]. Preparations were observed under an epifluorescent Leica DM LB2 microscope with 340-380 and LP 425 filters for auramine, calcofluor, and DAPI.

For the study of pollen morphology and pollen size, dehisced anthers were sieved through a 0.26 mm mesh sieve and the pollen was placed in glacial acetic acid for acetolysis. Pollen grains were transferred to a mixture of 9:1 acetic anhydride:concentrated sulphuric acid at 65°C for 10 minutes, then washed with glacial acetic acid and washed again three times with water following a modification of the method by Erdtman (1960) [68].

Scanning Electron Microscopy

Pollen for scanning electron microscopy (SEM) was fresh dried with silica gel and directly attached to SEM stubs using adhesive carbon tabs and observed with a JSM-840 scanning electron microscope (JEOL) operated at 10 kV.

Immunocytochemistry

Immunocytochemistry was performed on Technovit 8100 (Kulzer & Co, Wehrheim, Germany) embedded semithin sections and revealed by fluorochromes, as described previously [69,70]. Anthers from three flowers per developmental stage with petals 6, 9, 12, 16, 22, 24 and 30 mm long and at anthesis were fixed in 4% paraformaldehyde in phosphate buffered saline (PBS) at pH 7.3 overnight at 4°C, dehydrated in an acetone series, embedded in Technovit 8100 (Kulzer), polymerized at 4°C and sectioned at 2 μm. Sections were placed in a drop of water on a slide covered with 3-Aminopropyltrietoxy-silane 2% and dried at room temperature.

Different antibodies were used to localize specific cell components: an anti-RNA mouse monoclonal antibody, D44 [71,72], for total RNA detection;

JIM5 and JIM7 rat monoclonal antibodies (Professor Keith Roberts, John Innes Centre, Norwich, UK) which respectively recognize low and high-methyl-esterified pectins [73] for localization of pectins; and an anti-callose mouse monoclonal antibody (Biosupplies, Parkville, Australia) for callose.

Sections were incubated with PBS for 5 minutes and later with 5% bovine serum albumin (BSA) in PBS for 5 minutes. Then, different sections were incubated for one hour with the primary antibodies: JIM5, JIM7, and anti-RNA undiluted and anti-callose diluted 1/20 in PBS. After three washes in PBS, the sections were incubated for 45 minutes in the dark with the corresponding secondary antibodies (anti-rat, for JIM5 and JIM7, and anti-mouse, for anti-RNA and anti-callose) conjugated with Alexa 488 fluorochrome (Molecular Probes, Eugene, Oregon, USA) and diluted 1/25 in PBS. After three washes in PBS and water, the sections were mounted in Mowiol 4-88 (Polysciences), examined with a Zeiss Axioplan epifluorescent microscope, and photographed with a CCD Digital Leica DFC 350 FX camera.

Authors' Contributions

JL performed most of the experimental analyses, PST had an active contribution to the immunocytochemistry assays, MCR designed and discussed the immunocytochemistry essays, JIH participated in the design of the experiments, MH coordinated the study. All authors contributed to the draft and read and approved the final manuscript.

Acknowledgements

Financial support for this work was provided by the Spanish Ministry of Science and Innovation (Project Grants AGL2004-02290/AGR, AGL2006-13529-C01, AGL2007-60130/AGR, AGL2008-04255 and BFU2008-00203), GIC-Aragón 43, Junta de Andalucía (AGR2742) and the European Union under the INCO-DEV program (Contract 015100). JL. was supported by a grant of the Junta de Andalucía. The authors thank K. Pinney for helpful suggestions to improve the manuscript.

References

1. McCormick S: Control of male gametophyte development. Plant Cell 2004, 16:S142–S153.

2. Scott RJ, Spielman M, Dickinson HG: Stamen development: primordium to pollen. In The molecular biology and biotechnology of flowering. Edited by: Jordan BR. CAB International, Wallingford, UK; 2006:298–331.

3. Blackmore S, Wortley AH, Skvarla JJ, Rowley JR: Pollen wall development in flowering plants. New Phytologist 2007, 174:483–498.

4. Pacini E, Franchi GG: Pollen grain sporoderm and types of dispersal units. Acta Societatis Botanicorum Poloniae 1999, 68:299–305.

5. Walker JW, Doyle JA: Bases of angiosperm phylogeny - palynology. Annals of the Missouri Botanical Garden 1975, 62:664–723.

6. Harder L, Johnson S: Function and evolution of aggregated pollen in angiosperms. International Journal of Plant Sciences 2008, 169:59–78.

7. Pacini E, Franchi GG, Hesse M: The tapetum - its form, function, and possible phylogeny in embryophyta. Plant Systematics and Evolution 1985, 149:155–185.

8. Blackmore S, Crane PR: The systematic implications of pollen and spore ontogeny. In Ontogeny and Systematics. Edited by: Humphries CJ. British Museum (Natural History), London, UK; 1988:83–115.

9. Preuss D, Rhee SY, Davis RW: Tetrad analysis possible in Arabidopsis with mutation of the QUARTET (QRT) genes. Science 1994, 264:1458–1460.

10. Rhee SY, Osborne E, Poindexter PD, Somerville CR: Microspore separation in the quartet 3 mutants of Arabidopsis is impaired by a defect in a developmentally regulated polygalacturonase required for pollen mother cell wall degradation. Plant Physiology 2003, 133:1170–1180.

11. APG II: An update of the Angiosperm Phylogeny Group classification for the orders and families of flowering plants: APG II. Botanical Journal of the Linnean Society 2003, 141:399–436.

12. Soltis DE, Soltis PS, Endress PK, Chase MW: Phylogeny and evolution of angiosperms. Sinauer Associates Incorporated, Sunderland, Massachusetts, USA; 2005.

13. Doyle JA, Le Thomas A: Cladistic analysis and pollen evolution in Annonaceae. Acta Botanica Gallica 1994, 141:149–170.

14. Doyle JA, Le Thomas A: Phylogeny and geographic history of Annonaceae. Geographie Physique et Quaternaire 1997, 51:353–361.

15. Pirie MD, Vargas MP, Botermans M, Bakker FT, Chatrou LW: Ancient paralogy in the cpDNA trnL-F region in Annonaceae: implications for plant molecular systematics. American Journal of Botany 2007, 94:1003–1016.

16. Walker JW: Unique type of angiosperm pollen from the family Annonaceae. Science 1971, 172:565–567.

17. Walker JW: Pollen morphology, phytogeography and phylogeny of the Annonaceae. Contributions of the Gray Herbarium of Harvard University 1971, 202:1–132.

18. Doyle JA, Le Thomas A: Significance of palynology for phylogeny of Annonaceae: experiments with removal of pollen characters. Plant Systematics and Evolution 1997, 206:133–159.

19. Le Thomas A: Ultrastructure characters of the pollen grains of African Annonaceae and their significance for the phylogeny of primitive angiosperms (first part). Pollen et Spores 1980, 22:267–342.

20. Le Thomas A: Ultrastructure characters of the pollen grains of African Annonaceae and their significance for the phylogeny of primitive angiosperms (second part). Pollen et Spores 1981, 23:5–36.

21. Su YC, Saunders RM: Pollen structure, tetrad cohesion and pollen-connecting threads in Pseuduvaria (Annonaceae). Botanical Journal of the Linnean Society 2003, 143:69–78.

22. Tsou CH, Fu YL: Tetrad pollen formation in Annona (Annonaceae): proexine formation and binding mechanism. American Journal of Botany 2002, 89:734–747.

23. Tsou CH, Fu YL: Octad pollen formation in Cymbopetalum (Annonaceae): the binding mechanism. Plant Systematics and Evolution 2007, 263:13–23.

24. Barrett S: Major evolutionary transitions in flowering plant reproduction: an overview. International Journal of Plant Sciences 2008, 169:1–5.

25. Higuchi H, Utsunomiya N: Floral differentiation and development in cherimoya (Annona cherimola Mill.) under warm (30/25 degrees C) and cool (20/15 degrees C) day/night temperatures. Journal of the Japanese Society for Horticultural Science 1999, 68:707–716.

26. Periasamy K, Kandasamy MK: Development of the anther of Annona squamosa L. Annals of Botany 1981, 48:885–893.

27. Tsou CH, David J: Comparative development of aseptate and septate anthers of Annonaceae. American Journal of Botany 2003, 90:832–848.

28. Wu H, Cheung AY: Programmed cell death in plant reproduction. Plant Molecular Biology 2000, 44:267–281.

29. Testillano PS, Corredor E, Solís MT, Chakrabarti N, Raska I, Risueño MC: Changes in nuclear architecture and DNA methylation pattern accompany the

developmental programmed cell death of the tapetum. Proceedings of the 6th. Plant Genomics European Meeting. Tenerife, Spain 2007, 9–10.

30. Furness CA, Rudall PJ, Sampson FB: Evolution of microsporogenesis in angiosperms. International Journal of Plant Sciences 2002, 163:235–260.

31. Nadot S, Furness CA, Sannier J, Penet L, Triki-Teurtroy S, Albert B, Ressayre A: Phylogenetic comparative analysis of microsporogenesis in angiosperms with a focus on monocots. American Journal of Botany 2008, 95:1426–1436.

32. Farr CH: Cell division by furrowing in Magnolia. American Journal of Botany 1918, 5:379–395.

33. Dahl AO, Rowley JR: Pollen of Degeneria vitiensis. Journal of the Arnold Arboretum, Harvard University 1965, 46:308–323.

34. Sampson FB: Cytokinesis in pollen mother cells of angiosperms, with emphasis on Laurelia-novae-zelandiae (Monimiaceae). Cytologia 1969, 34:627–634.

35. Guzzo F, Baldan B, Bracco F, Mariani P: Pollen development in Liriodendron-tulipifera - some unusual features. Canadian Journal of Botany 1994, 72:352–358.

36. Farr CH: Quadripartition by furrowing in Sisyrinchium. Bulletin of the Torrey Botanical Club 1922, 49:51–61.

37. Gabarayeva NI: Pollen wall and tapetum development in Anaxagorea brevipes (Annonaceae) - sporoderm substructure, cytoskeleton, sporopollenin precursor particles, and the endexine problem. Review of Palaeobotany and Palynology 1995, 85:123–152.

38. Zavada M: Pollen wall development of Austrobaileya maculata. Botanical Gazette 1984, 145:11–21.

39. Xiang-Yuan X, Demason DA: Relationship between male and female gametophyte development in rye. American Journal of Botany 1984, 71:1067–1079.

40. Pacini E, Franchi GG: Amylogenesis and amylolysis during pollen grain development. In Sexual reproduction in higher plants. Edited by: Cresti M, Gori P, Pacini E. Springer, Berlin, Heidelberg, New York; 1988:181–186.

41. Maheshwari P: An introduction to the embryology of angiosperms. McGraw-Hill, New York, USA; 1950.

42. Bedinger P: The remarkable biology of pollen. Plant Cell 1992, 4:879–887.

43. Franchi GG, Bellani L, Nepi M, Pacini E: Types of carbohydrate reserves in pollen: localization, systematic distribution and ecophysiological significance. Flora 1996, 191:143–159.

44. Grayum MH: Evolutionary and ecological significance of starch storage in pollen of the Araceae. American Journal of Botany 1985, 72:1565–1577.

45. Brewbaker JL: Distribution and phylogenetic significance of binucleate and trinucleate pollen grains in angiosperms. American Journal of Botany 1967, 54:1069–1083.

46. Baker HG, Baker I: Starch in angiosperm pollen grains and its evolutionary significance. American Journal of Botany 1979, 66:591–600.

47. Lora J, Oteyza MAP, Fuentetaja P, Hormaza JI: Low temperature storage and in vitro germination of cherimoya (Annona cherimola Mill.) pollen. Scientia Horticulturae 2006, 108:91–94.

48. Lora J, Herrero M, Hormaza JI: The coexistence of bicellular and tricellular pollen in Annona cherimola Mill. (Annonaceae): Implications for pollen evolution. American Journal of Botany 2009, 96:802–808.

49. Francis KE, Lam SY, Copenhaver GP: Separation of Arabidopsis pollen tetrads is regulated by QUARTET1, a pectin methylesterase gene. Plant Physiology 2006, 142:1004–1013.

50. Neelam A, Sexton R: Cellulase (endo beta-1,4 glucanase) and cell-wall breakdown during anther development in the sweet pea (Lathyrus odoratus L.): isolation and characterization of partial cDNA clones. Journal of Plant Physiology 1995, 146:622–628.

51. Rowley JR: Formation of the pore in pollen of Poa annua. In Pollen physiology and fertilization. Edited by: Linskens HF. North-Holland Publishing Company, Amsterdam, Netherlands; 1964:59–69.

52. Rudall PJ, Bateman RM: Developmental bases for key innovations in the seed-plant microgametophyte. Trends in Plant Science 2007, 12:317–326.

53. Rhee SY, Somerville CR: Tetrad pollen formation in quartet mutants of Arabidopsis thaliana is associated with persistence of pectic polysaccharides of the pollen mother cell wall. Plant Journal 1998, 15:79–88.

54. Testillano PS, Fadon B, Risueño MC: Ultrastructural localization of the polysaccharidic component during the sporoderm ontogeny of the pollen grain. Review of Paleobotany and Palynology 1995, 85:53–62.

55. Enns LC, Kanaoka MM, Torii KU, Comai L, Okada K, Cleland RE: Two callose synthases, GSL1 and GSL5, play an essential and redundant role in plant and pollen development and in fertility. Plant Molecular Biology 2005, 58:333–349.

56. Rosell P, Herrero M, Sauco VG: Pollen germination of cherimoya (Annona cherimola Mill.). In vivo characterization and optimization of in vitro germination. Scientia Horticulturae 1999, 81:251–265.

57. Ratnayake RMCS, Gunatilleke IAUN, Wijesundara DSA, Saunders RMK: Reproductive biology of two sympatric species of Polyalthia (Annonaceae) in Sri Lanka. I. Pollination by curculionid beetles. International Journal of Plant Sciences 2006, 167:483–493.

58. Davis CC, Endress PK, Baum DA: The evolution of floral gigantism. Current Opinion in Plant Biology 2008, 11:49–57.

59. Le Thomas A, Morawetz W, Waha M: Pollen of paleo- and neotropical Annonaceae: definition of the aperture by morphological and functional characters. In Pollen and Spores: Form and Function. Edited by: Blackmore S, Ferguson IK. Academic Press, London, UK; 1986:375–388.

60. Rosell P, Sauco VG, Herrero M: Pollen germination as affected by pollen age in cherimoya. Scientia Horticulturae 2006, 109:97–100.

61. Heslop-Harrison Y, Heslop-Harrison J: Germination of monocolpate angiosperm pollen: evolution of the actin cytoskeleton and wall during hydration, activation and tube emergence. Annals of Botany 1992, 69:385–394.

62. Gottsberger G: Pollination and evolution in Neotropical Annonaceae. Plant Species Biology 1999, 14:143–152.

63. Sabatini DD, Bensch K, Barrnett RJ: Cytochemistry and electron microscopy - preservation of cellular ultrastructure and enzymatic activity by aldehyde fixation. Journal of Cell Biology 1963, 17:19–58.

64. Feder N, O'Brien TP: Plant microtechnique: some principles and new methods. American Journal of Botany 1968, 55:123–142.

65. Heslop-Harrison Y: Pollen-stigma interaction - pollen-tube penetration in Crocus. Annals of Botany 1977, 41:913–922.

66. Hughes J, McCully ME: Use of an optical brightener in study of plant structure. Stain technology 1975, 50:319–329.

67. Williams JH, Friedman WE, Arnold ML: Developmental selection within the angiosperm style: Using gamete DNA to visualize interspecific pollen competition. Proceedings of the National Academy of Sciences USA 1999, 96:9201–9206.

68. Erdtman G: The acetolysis method. Svensk Botanisk Tidskrift 1960, 54:561–564.

69. Satpute GK, Long H, Seguí-Simarro JM, Risueño MC, Testillano PS: Cell architecture during gametophytic and embryogenic microspore development in Brassica napus L. Acta Physiologia Plantarum 2005, 27:665–674.

70. Solís MT, Pintos B, Prado MJ, Bueno MA, Raska I, Risueño MC, Testillano PS: Early markers of in vitro microspore reprogramming to embryogenesis in olive (Olea europaea L.). Plant Science 2008, 174:597–605.

71. Eilat D, Fischel R: Recurrent utilization of genetic elements in V regions of anti-nucleic acid antibodies from autoimmune mice. Journal of Immunology 1991, 147:361–368.

72. Mena CG, Testillano PS, Gonzalez-Melendi P, Gorab E, Risueño MC: Immunoelectron microscopy detection of RNA in plant nucleoli. Experimental Cell Research 1994, 212:393–408.

73. Knox JP: The use of antibodies to study the architecture and developmental regulation of plant cell walls. International Review of Cytology 1997, 171:79–120.

CITATION

Originally published under the Creative Commons Attribution License. Lora J, Testillano PS, Risueño MC, Hormaza JI, Herrero M. Pollen Development in Annonacherimola mill. (Annonaceae). Implications for the Evolution of Aggregated Pollen. BMC Plant Biology 2009, 9:129   doi:10.1186/1471-2229-9-129.

Distinct Short-Range Ovule Signals Attract or Repel Arabidopsis Thaliana Pollen Tubes in Vitro

Ravishankar Palanivelu and Daphne Preuss

ABSTRACT

Background

Pollen tubes deliver sperm after navigating through flower tissues in response to attractive and repulsive cues. Genetic analyses in maize and Arabidopsis thaliana and cell ablation studies in Torenia fournieri have shown that the female gametophyte (the 7-celled haploid embryo sac within an ovule) and surrounding diploid tissues are essential for guiding pollen tubes to ovules. The variety and inaccessibility of these cells and tissues has made it challenging to characterize the sources of guidance signals and the dynamic responses they elicit in the pollen tubes.

Results

Here we developed an in vitro assay to study pollen tube guidance to excised A. thaliana ovules. Using this assay we discerned the temporal and spatial regulation and species-specificity of late stage guidance signals and character- ized the dynamics of pollen tube responses. We established that unfertilized A. thaliana ovules emit diffusible, developmentally regulated, species-specific attractants, and demonstrated that ovules penetrated by pollen tubes rapidly release diffusible pollen tube repellents.

Conclusion

These results demonstrate that in vitro pollen tube guidance to excised A. thaliana ovules efficiently recapitulates much of in vivo pollen tube behav- iour during the final stages of pollen tube growth. This assay will aid in con- firming the roles of candidate guidance molecules, exploring the phenotypes of A. thaliana pollen tube guidance mutants and characterizing interspecies pollination interactions.

Background

After a pollen grain lands on the surface of the pistil, it absorbs water from the stigma and forms a pollen tube – a long polar process that transports all of the cel- lular contents, including the sperm [1]. Pollen tubes invade the pistil and migrate past several different cell types, growing between the walls of the stigma cells, travelling through the extracellular matrix of the transmitting tissue, and finally arriving at the ovary, where they migrate up the funiculus (a stalk that supports the ovule), and enter the micropyle to deliver the two sperm cells-one fertilizes an egg and other the central cell (Fig. 1a, 1b) [2]. Typically, only one pollen tube enters the ovule through an opening called the micropyle, terminates its journey within a synergid cell, and bursts to release sperm cells-a process defined as pollen tube reception [3].

A combination of genetic and in vitro assays has defined signals that contribute to the early stages of pollen tube guidance. Chemocyanin, a small basic protein from lily stigmas, attracts lily pollen tubes in vitro [4], and in A. thaliana, wild type pollen guidance was abnormal when grown on stigmas over expressing the A. thaliana chemocyanin homolog [5]. Other signals are active in the nutrient-rich extracellular matrix secreted by the female transmitting tissue. A pectin that may promote guidance by mediating adhesion of pollen tubes to this matrix has been identified in lily [6]. Glycoproteins that likely contribute to guidance have also been described: in lily, a lipid transfer protein that contributes to adhesion [6], and in tobacco, two glycoproteins (TTS1 and TTS2) that provide nutritional and guidance cues are known [7,8]. Although potential homologs of these proteins exist in A. thaliana, their role in pollen tube growth is yet to be determined [6].

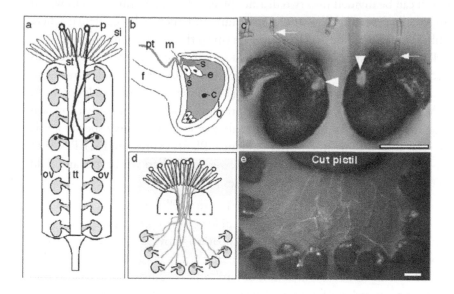

Figure 1. Pollen tube targeting in vitro.

(a) Diagram of a pollinated pistil within an A. thaliana flower. After reaching the stigma (si), pollen (p) extends a tube through the style (st) to reach the transmitting tract (tt) before entering one of the two ovary (ov) chambers to target an ovule. (b) Upon reaching the ovule, the pollen tube (pt, green) either grows up the funiculus (f) or makes a sharp turn towards the micropyle (m) and enters the ovule. Within the ovule, the pollen tube navigates towards the female gametophyte (gray) encased by outer (o) and inner (i) integuments, lyses within one of the two synergid (s) cells that flank an egg cell (e). Upon lysis, one sperm fertilizes the egg cell to form the zygote and the other fuses with the central cell (c) to form the endosperm. The number of pollen tubes drawn is for illustration purposes only and does not reflect the quantity typically observed in an assay. (c) Merged fluorescent and bright field images depicting the final stages of in vitro pollen tube growth. GFP-tagged pollen tubes make a committed turn (arrows) before entering a virgin ovule and lysing (arrow heads). (d) Diagram and (e) merged fluorescent and bright field image of in vitro pollen tube guidance assay. Pollen tubes emerge from the cut portion of the pistil, travel across the agarose medium before entering the excised ovules. Fluorescent green spot within ovules mark successful pollen tube targeting. Scale bars, 100 μm.

After emerging from the transmitting tract, pollen tubes approach the ovule micropyle with remarkable precision. Mutants defective in pollen tube guidance have demonstrated that this process is controlled by a series of molecular signals that involve pollen tubes, ovule tissues, and female gametophytes [1]. The A. thaliana mutants, ino [9] and pop2 [10] point to a role for diploid ovule tissue in pollen tube guidance; these have aberrant interactions between pollen tubes and diploid ovule cells, yet their female gametophytes appear normal, and in the case of pop2, can be fertilized with wild-type pollen [11]. Pollen tubes fail to either reach or enter the micropyle in A. thaliana mutants with nonviable or aberrant female gametophytes yet apparently normal diploid ovule tissue, providing strong support for the role of the haploid germ unit in promoting growth to the micropyle [12,13]. Based on these studies, it was proposed that final stages of pollen tube growth can be divided into two distinct phases: funicular guidance, in which pollen tubes adhere to and grow up the funiculus, and micropylar guidance, where pollen tubes enter the micropyle to deliver sperm to the female gametophyte [13]. Micropylar guidance signals originate at least in part from the two synergid cells contained within the female gametophyte; pollen tubes do not enter ovules in which synergid cells were either laser ablated [14] or defective due to a lesion in A. thaliana MYB98 gene [15]. The maize EA1 protein, which is exclusively expressed in the egg and synergids of unfertilized female gametophytes, may specify a role for these cells in regulating micropylar guidance. Plants expressing EA1 RNAi or antisense constructs produced significantly fewer seeds than wild type, and wild type pollen tubes failed to enter mutant ovules [16].

In vitro assays have been used to characterize intracellular cues such as a Ca2+ gradient at the tip of pollen tubes that is critical for growth. Disrupting this gradient by iontophoretic microinjection or by incubation with Ca2+ channel blockers can change the direction of tube growth [17]. The Ca2+ gradient in pollen tubes is controlled by Rho GTPases; injection of antibodies against these proteins into pollen tubes, or expression of dominant-negative forms of RhoGTPase, causes the tip-focused Ca2+ gradient to diffuse and eliminates tube growth [18], presumably by disrupting F-actin assembly [19]. These pollen tube growth defects can be partially alleviated by adding high concentrations of extracellular Ca2+ [18].

In vitro grown pollen tubes also reorient their growth in response to certain extracellular cues; lily pollen tubes are attracted to chemocyanin [4] and repelled by a point source of nitric oxide [20]. In addition, in vitro grown pearl millet pollen tubes are attracted to ovary extracts [21]. For T. fournieri pollen tube guidance across a simple medium and into the ovule was achieved after pollen tubes were grown through a stigma and style [22]. In this species, the female gametophyte protrudes from the ovule, and pollen tubes enter the micropyle without interacting with funiculus [22]. Thus, the T. fournieri in vitro guidance system serves as a model for the micropylar, but not the funicular guidance phase of pollen tube growth to ovules [23]. Here, we describe an A. thaliana in vitro guidance assay that recapitulates both funicular and micropylar guidance, serving as a model for ovules with encased female gametophytes, an arrangement that is more common among flowering plants. With the sequenced A. thaliana genome, the large collection of mutants affecting reproductive functions, and comparative genomic resources, this assay will greatly facilitate identifying genes that mediate the final phases of pollen tube guidance.

Results

The A. thaliana Stigma and Style Confer Pollen Tube Targeting Competence

Previous studies indicated that pollen tubes germinated in a simple growth medium cannot be guided to the micropyle [14,22]. Consistent with these observations, when such assays are performed for A. thaliana, few ovules are targeted (~3%, Table 1). Hence, we instead removed the upper portion of the pistil (the stigma and style [4,10,14,24]), deposited pollen on the stigma surface and showed that A. thaliana pollen tubes emerged from the style, travelled across an agarose medium to excised ovules and successfully entered the micropyle (Fig. 1a, 1b). To facilitate pollen tube observation, especially after they enter the micropyle and are obscured by the opaque ovule integument cells, we transformed plants with a GFP reporter under the control of the pollen-specific LAT52 promoter [25], and identified GFP-expressing lines with fully functional pollen tubes. Upon reaching the female gametophyte, these tubes burst and release a large spot of GFP (Fig. 1c, 1d, arrowheads), conveniently marking targeted ovules. Pollen tubes that grew within ~100 μm of an unfertilized ovule often made a sharp turn toward an ovule; of the tubes that grew within this range, ~50% successfully entered the micropyle (Table 1). This targeting efficiency is significantly higher than that of tubes germinated on agarose (Table 1). Thus, pollen tubes acquire the ability from pistil tissue to perceive ovule guidance signals, perhaps by absorbing essential nutrients or undergoing critical developmental transitions; a similar phenomenon was reported in T. fournieri [22]. In some cases, pollen grains that germinated on the stigma formed tubes that grew onto the medium, rather than penetrating the pistil and growing through the style. Nonetheless, these tubes successfully targeted excised ovules, suggesting that interaction with the stigma alone is sufficient to confer pollen tube guidance competence (Table 1); targeting efficiency was significantly reduced, however, suggesting that direct contact with female cells, rather than exposure to diffusible factors in the medium, is most important.

Table 1. In vitro ovule targeting efficiency.

Stigma	Style	Ovule	Funiculus	Fraction ovules targeted[a]	%	n[b]
...	–	+	+	4/162	3[*]	6
+	+	+	+	77/145	54	18
+	–	+	+	30/156	19[*]	25
+	+	+	–	61/175	35[*]	21
+	+	+ (fertilized)[c]	+	0/122	0[*]	20
+	+	+ (100°C)[d]	+	0/52	0[*]	8
+	+	+ (25°C)[d]	+	23/57	40	8

[a]Although the same number of ovules were placed in each assay, only those with a pollen tube tip ≤ 100 μm from the micropyle were considered potential targets; [b]number of independent in vitro assays; [c]expanded ovules, indicating embryo growth, were collected 1 day after pollination; [d]ovules immersed in 100°C or 25°C water for 5 min; tissues present (+) or absent (–) after removal by microdissection (rows 1 and 4); assays in which pollen tubes germinated and grew only on the stigma (row 3); [*], significantly different (χ^2, $P < 0.001$) from assays performed with a stigma, style and intact funiculus (row 2).

Characterization of A. thaliana Pollen Tube-Ovule Interactions

As tubes left the style, they dispersed (Fig. 1e and see 1), growing at 2.5 ± 1.0 (s.d) μm/min (n = 20) and up to 3 mm before reaching an ovule (Fig. 2a–f). Near a virgin ovule, however, growth rates decreased (1.2 ± 0.59 (s.d) μm/min (n = 20; Fig. 2a–f) and the tubes often made sharp turns (Fig. 1d, arrows) within 33 ± 20 (s.d) mm of a micropyle (avg. 60 ± 38° (s.d); n = 60). Pollen tube guidance to ovules was abolished when fertilized or heat-treated ovules were used (Table 1), indicating they release a diffusible, heat-labile attractant prior to fertilization. Tube entry into the ovule appears to be not influenced by the number of tubes near a micropyle; targeting was achieved regardless of whether one or multiple tubes were in the vicinity of an ovule. When approaching an ovule, the in vitro grown pollen tubes did not always migrate up a funiculus, Fig. 1c) before entering it. To directly test the role of this tissue, we removed funiculi from ovules, revealing a small but significant decrease in targeting efficiency (Table 1). These results indicate that an interaction between the pollen tube and funiculus is not essential, yet this interaction enhances successful entry into the ovule, perhaps by i) providing a physical support for pollen tubes to reach the micropyle, ii) aiding in the generation and maintenance of a signal gradient, or iii) enhancing the availability of ovule-derived guidance signals.

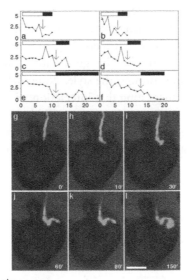

Figure 2. Pollen tube navigation time course.

(a–f) Graphs of position (μm; X axis) versus growth rate (μm/min; Y axis) for six pollen tubes, from the moment they exit the style (white bar), enter the micropyle (arrow) and navigate within the ovule (black bar). (g–l) Merged fluorescent and bright field images of a GFP-tagged pollen tube entering a micropyle (g), navigating past diploid ovule cells (h–j), and pausing upon reaching a synergid (k), and completing lysis (l). Characterstics of pollen tube growth within the ovule shown in G-L were graphed and shown in (e). Time is min ('); scale bar, 50 μm.

After responding to the attractant and entering an ovule, the growth rate of each pollen tube decreased to 0.98 ± 0.34 (s.d) μm/min (n = 19), reaching the female gamtophyte and lysing after a 73 ± 19 (s.d) min delay (n = 19; Fig. 2g–l). While previous work showed that pollen tube growth arrests only after reaching the female gametophyte [9,10], our observations point to additional signals that slow growth within the ovule prior to this arrest. This delay coincides with meandering pollen tube navigation past the integument and nucellus cells (Fig. 2g–k), potentially reflecting guidance by these cells.

Short-Range Guidance Signals from A. thaliana Ovules Are Developmentally Regulated

The data presented above indicate that contact with A. thaliana stigmas and styles enables pollen tubes to respond to diffusible ovule signals. To understand the nature and source of these signals, we examined their activity during ovule development. Previously it was shown that pollen tubes grow randomly or fail to elongate in immature A. thaliana pistils [26]. However, it was impossible to distinguish the contributions of distinct tissues in these experiments. Here, we exploited the modular nature of the in vitro system, varying the age of stigmas, styles, and ovules. While mature flower parts (stage 14 [27] were optimal, the stage of ovule development was critical, with guidance factors completely absent at ~32 hrs (stage 12a) and lower at ~16–24 hrs (stages 12b–c) before flowers mature (Table 2, upper panel). This pattern correlates with synergid development, the suggested source of pollen tube attractants; these cells form after stage 12a [27,28]. Even so, immature ovules promote better pollen tube guidance than heat-treated or fertilized ovules, suggesting that a basal signalling capability is established early and increases as the female gametophyte differentiates. In contrast, the developmental stage of the stigma and style did not significantly alter targeting ($\chi2$; P > 0.1; Table 2, bottom panel), indicating that the signals that confer targeting competence to pollen tubes do not vary over the course of pistil maturation (stages 12a–18) and that they emerge as early as stage 12a.

Short-Range Guidance Signals from Ovules Are Highly Species-Specific

Like many traits that mediate reproduction [29,30], pollen tube guidance signals diverge rapidly – crosses between A. thaliana and its relatives show random or arrested pollen tube growth, even among species separated by <25 million years, MY [13,31]. Because these interspecies crosses utilized intact pistils, it has been impossible to discern the roles of individual tissues; moreover, early blocks in pollen tube migration have often made it difficult to assess interactions at downstream stages, including those near ovules. Here, we examined whether the stigma, style and ovule-derived signalling interactions are shared among A. thaliana relatives separated by ~5, 10, or 20 MY [31] (Table 3). The ability of the stigma and style to promote pollen tube competence was highly conserved ($\chi2$; P > 0.05; Table 3, bottom panel), while the ovule-derived attractant diverged rapidly ($\chi2$; P < 0.01; Table 3, upper panel). For example, A. thaliana pollen tubes inefficiently target ovules from Arabidopsis arenosa (separated by 5 MY from A. thaliana), rarely target Olimarabidopsis pumila ovules (10 MY), and fail to target Capsella rubella or Sysimbrium irio ovules (10 and 20 MY, respectively, Table 3, upper panel). Because C. rubella and S. irio are challenging to transform, it was not possible to test whether ovules from these two species are able to guide self-pollen expressing GFP under our assay conditions. Nonetheless, the ability of A. thaliana pollen to target ovules correlated with phylogenetic separation [30], suggesting that A. thaliana pollen tubes are sufficiently diverged that they fail to recognize attractants from C. rubella and S. irio ovules. Unlike the calcium signals that emanate from synergids [32,33], the proposed source of micropylar guidance signals, our results point to a diffusible, heat-labile ovule-derived signal that is sufficiently complex for rapid divergence – criteria that are most consistent with a protein-based signal. Pollen tubes perceive this signal at a distance of ~100µm from ovules after a 5 hour incubation in the assay. To estimate the molecular weight of this signal, we measured the diffusion rates of fluorescently-labelled dextran molecules under the same conditions in which the in vitro assay was performed, and calculated that the ovule-derived signal could measure up to approximately 85 kD (see methods).

Table 2. Developmental regulation of short-range guidance signals from ovules.

Developmental stage		Fraction ovules targeted[b]	%	n[c]
Stigma, style	Ovule			
14	12a	0/79	0[***]	13
"	12b	15/84	18[**]	7
"	12c	61/335	18[**]	31
"	13	51/159	32	18
"	14	195/398	49	41
"	15	50/120	42	18
"	16	48/119	40	15
"	17	47/104	45	15
"	18	40/103	39	13
12a	14	6/25	24	5
12b	"	14/43	33	8
12c	"	21/55	38	10
13	"	17/64	27	10
14	"	22/46	48	9
15	"	28/66	42	10
16	"	27/61	44	9
17	"	26/64	41	7
18	"	26/63	41	8

[a]Pollen grains were from A. thaliana stage 14 (9); [b, c]as in [a,b]Table 1; [**], significantly different from assays with stigma, style and ovules from stage 14; (χ^2) P < 0.001.

Table 3. Species specificity of short-range guidance signals from ovules.

Species		Fraction ovules targeted[b]	%	n[c]
Stigma, style	Ovule			
A. thaliana	A. thaliana	152/304	50	33
"	A. arenosa	50/161	31*	26
"	O. pumila	5/131	4**	21
"	C. rubella	0/87	0**	12
"	S. irio	0/107	0**	16
A. arenosa	A. thaliana	38/79	48	10
O. pumila	"	72/128	56	17
C. rubella	"	16/48	33	15
S. irio	"	17/36	47	12

[a]Pollen grains were from A. thaliana stage 14 (9); [b,c]as in [a,b]Table 1; *, **, significantly different from stage 14 A. thaliana assays in row 1; (χ^2) $P < 0.05$, $P < 0.001$, respectively.

Targeted A. thaliana Ovules Repel Supernumerary Pollen Tubes in Vitro

Interestingly, while the ovule-derived attractant in the in vitro assay acted to guide multiple pollen tubes toward ovules, only one pollen tube gained access to each micropyle. This is reminiscent of polyspermy blocks in vivo, where only one tube generally migrates up the funiculus and into the ovule [13]. While the mechanisms that prevent multiple tubes from even approaching an ovule are highly efficient, it is nonetheless possible for more than one pollen tube to enter a micropyle. In wild-type maize, heterofertilization results when the egg and central cell are fertilized by different pollen tubes at a frequency of ~1/50 [34] and in A. thaliana, ~1% (wild type) and ~10% (feronia) of ovules are penetrated by multiple pollen tubes [3]. When we performed the in vitro assay with fertilized ovules, many tubes grew within 100 μm, but none entered (Table 1), suggesting that the release of the ovule attractant terminates after fertilization, or alternatively, that a new signal repels additional pollen tubes. To distinguish between these possibilities, we used time-lapse imaging analysis (Fig. 3). While 44% of targeted ovules (n = 143) were approached by additional pollen tubes, in every instance, these tubes did not enter the micropyle. Repelled tubes either stalled near the micropyle or turned sharply away from the targeted ovule (84 ± 42°; n = 61, Fig. 2e–h), a response that was observed as early as 10 min after a successful targeting event. This effect was fairly short-range; only tubes that approached within 27 ± 22 (s.d) μm were repelled. The diffusion rate of a series of dextran molecules through the medium used in this assay allowed us to estimate that a repellent measuring <10kD could diffuse

~27 μm in 10 minutes. Such abrupt turning behaviours were not observed when a single tube approached a virgin ovule. Instead, these tubes changed their growth direction by 60 ± 38° (s.d; n = 60), migrating toward, and not away, from the micropyle.

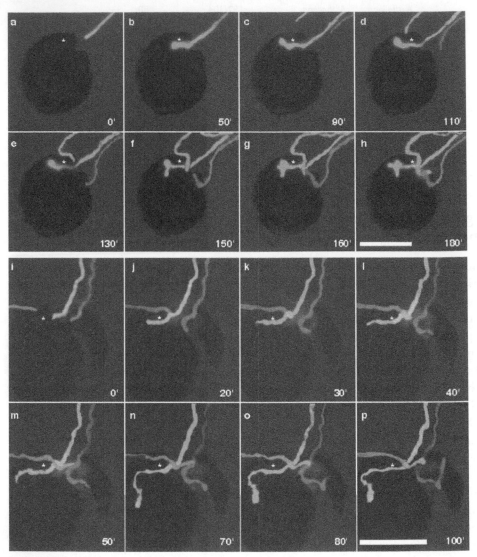

Figure 3. Pollen tubes avoid targeted ovules.

Two examples (a–h and i–p) of pollen tube avoidance by ovules approached by three pollen tubes (false colour: green, the tube that enters the micropyle (*); pink and red, tubes that arrive later and do not enter). Elapsed time in minutes; Scale bars, 100 μm.

In the A. thaliana female sterile feronia and sirene mutants, wild type pollen tubes enter the mutant ovules but fail to cease growth or burst. In addition to this defect, multiple pollen tubes gain access to feronia and sirene mutant ovules [3,35]. Based on these results, it was suggested that repulsion of supernumerary tubes does not initiate until pollen tube reception occurs. Our observations with wild type pollen tubes in this in vitro assay indicate this is not the case – repulsion responses occurred well before tube growth terminated or tubes released their cytoplasm (n = 50, Fig. 3j, 3n). Moreover, while previous work suggested that female gametophyte cells release an inhibitory signal [3,35], our results show that repulsion initiated soon after the pollen tubes entered the micropyle and long before they reached the female gametophyte (Fig. 3). Thus, this work points to a diffusible repulsive signal that is sufficient to override the ovule attractant. This signal may be derived directly from the diploid cells that surround the micropyle, from the female gametophyte, or from the successfully targeted, but unlysed, pollen tubes.

Conclusion

Based on the results described here, we have defined three signaling events that regulate pollen tube guidance in A. thaliana: i) contact-mediated competence conferred by the stigma and style, ii) diffusible ovule-derived attractants and iii) repellents exuded from recently-targeted ovules. The species specificity and diffusion properties of the ovule attractant are consistent with a protein signal, while the abrupt transmission and response to the repellent suggests the activity of a small molecule, a peptide, or post-translational modifications to signals present before fertilization. This investigation also provides a platform to confirm the roles of candidate guidance molecules and to explore the phenotypes of A. thaliana mutants, including those that affect the development of diploid [9,10] and haploid [3,16,35] female tissues or pollen tubes [36]. The ability to characterize interspecies pollination interactions with this assay could lead to improvements in generating novel plant hybrids, a process that often requires in vitro manipulations [37].

Methods

Plant Growth and Material

Pollen was derived from LAT52:GFP transgenic lines (Columbia background). Stigmas, styles, and ovules were from the A. thaliana male sterile mutant, ms1 (CS75, Landsberg background) or from A. arenosa (CS3901), O. pumila

(CS22562), C. rubella (CS22561) and S. irio (CS22653), deposited in the A. thaliana Biological Resources Center, Ohio State University. Female structures are unaffected by ms1, making it a convenient source of virgin pistils without the need for emasculation. No difference was detected between assays performed with materials derived from the Landsberg or Columbia ecotypes (not shown). Seeds were sown in soil and stratified at 4°C for 2 days, and plants were grown under fluorescent light (100 μE) for 16 or 24 hrs/day at 40% humidity. To consistently isolate pistils of varying developmental stages, we correlated the initial day of flowering of our plant population with previously defined floral development stages [26]. First, we confirmed that the youngest open flower is similar to stage 14 [26]. In A. thaliana, flowers continuously arise at the floral apex and are arranged in a spiral, with the younger buds on the inside. This predictable pattern allowed us to select stage 14 flowers as a starting point and identify older flowers (up to stage 18) and younger buds (up to 12a).

In Vitro Pollen Tube Guidance Assay

Growth medium for in vitro manipulations of pollen tubes [10] was determined to be optimal for also growing pollen tubes through a cut pistil. For the in vitro assays described here, pollen growth medium (3 ml) was poured into a 35 mm petri dish (Fisher Scientific, Hampton, USA). This volume of medium was ideal both for pollen tube growth and for microscopically viewing the interactions between pollen tubes and ovules. Excised pistils were pollinated under a dissection microscope (Zeiss Stemi 2000), cut with surgical scissors at the junction between the style and ovary (World Precision Instruments, Sarasota, USA), and placed horizontally on pollen growth medium. Pollen tubes emerged from the pistil ~3 hours after pollination and dispersed along the agarose surface for up to ~3 mm from the pistil.

Unlike previous reports [12,18], ovules were excised dry under a dissection microscope with a 27.5 gauge needle, from pistils that were held horizontally on double-sided tape (Scotch brand, 3M, St. Paul, USA). Excised ovules were immediately placed on the pollen growth medium, ~2 mm from the pistil, a distance that was typically accessible by the emerging pollen tubes. To maximize pollen tube-ovule interactions, 8–10 ovules were placed at the base of a pistil as shown in Fig. 1d. Because pollen tubes tend to disburse and grow randomly after leaving the style, not all ovules, particularly those placed near the cut pistil, are visited by a pollen tube (Fig. 1e).

For time-lapse imaging, ovules were placed with their micropylar end closest to the pistil excision site. Although not essential for targeting, ovules were oriented in this manner to reduce the time elapsed before targeting was achieved.

In vitro assays were typically performed by completely coating stigmas of cut pistils with pollen (>100 grains per stigma); in contrast, for the repulsion assays only 20–30 pollen grains were deposited per stigma, making it possible to clearly observe individual tube behaviour. Based on experiments with limited amounts of pollen, we typically observed 50–80% of the pollen grains produced tubes that emerged from the style.

LAT52:GFP Transgenic Plants

A HindIII fragment encoding GFP expressed from a post-meiotic, pollen-specific LAT52 promoter [25], was cloned into PBI121 (Clonetech, CA) and introduced into A. thaliana (Columbia) plants by Agrobacterium-mediated transformation. Kanamycin-resistant transgenic plants were selected, and a line containing a single transgene insertion, based on segregation of kanamycin resistance and GFP, was chosen for this study. This line had no detectable reproductive defects.

Microscopy

Ovules targeted by pollen tubes in the in vitro assay were counted under a Zeiss fluorescent dissecting stereoscope. For calculating targeting efficiencies, we included only ovule micropyles within 100μm of a growing pollen tube; within this range pollen tubes exhibited responses that are typical for cells undergoing attraction: significant reorientation of growth towards the signal source, followed by a steady advance towards the target. For time-lapse fluorescent microscopy, GFP-labelled pollen tubes were observed using a Zeiss Axiovert 100 fitted with an automated shutter, motorized stage and CCD camera (CoolSNAP fxHQ, Roper Scientific, Inc Tucson, AZ). Images were captured at 10-minute intervals, converted to a TIFF format per the manufacturer's instructions using Slidebook (Intelligent Innovations Imaging, Santa Monica, CA). Pollen tube behaviours (growth rate, angle of turning, distance from micropyle) were measured and images were assembled into movies using ImageJ image analysis software (http://rsb. info.nih.gov/ij/download.html).

Diffusion Rates

To estimate the size of pollen tube signalling molecules, we performed time-lapse imaging of diffusion of a series of fluorescein-conjugated dextrans (Invitrogen, Carlsbad, CA) ranging in molecular weight from 3 to 70 kD on the pollen growth medium used for performing the in vitro pollen tube guidance assay. We dissolved each dextran compound in pollen growth medium, and spotted 2 ul each of 10 ng/ul and 100 ng/ul onto pollen tube guidance assay plates. Time-lapse imaging was performed as described for in vitro pollen tube behaviours except that images were captured once in 30 minutes. The rates of diffusion (3 kD = 5.5μm/min; 10 kD = 2.73 μm/min; 40 kD = 2.43 μm/min and 70 kD = 1.01 μm/min) were measured from these images using ImageJ software. Specifically, the fluorescent intensities along a line drawn from the centre of a dextran spot to the diffused periphery were calculated for an entire time-lapse series. The data was imported into Microsoft Excel and regression analyses were performed. Extrapolating from these values, molecular weights were estimated for attractants that would diffuse 33 μm and repellents that would diffuse 27 μm in 300 minutes (the time required for a typical pollen tube to reach an ovule in the in vitro assay).

Statistical Analysis

To measure the significance of the differences among observed ovule targeting efficiencies, we employed a χ^2 test for consistency in observed frequency distributions with a dichotomous classification and variable sample size [38].

Authors' Contributions

RP carried out experiments described in this study. RP and DP conceived, designed, coordinated this study and drafted the manuscript. Both authors read and approved the final manuscript.

Acknowledgements

We thank M Johnson, A Hall and E Updegraff for helpful suggestions; N Hagemann for Lat52:GFP transgenic plants; MJ Root, S Bond and A Young for technical assistance; J Yang and J Lui for assistance with statistical analysis. Funding was provided by the Department of Energy DE-FG02-96ER20240 and the University of Chicago MRSEC (NSF DMR-0213745).

References

1. Weterings K, Russell SD: Experimental analysis of the fertilization process. Plant Cell 2004, 16 Suppl:S107–18.

2. Lord EM, Russell SD: The mechanisms of pollination and fertilization in plants. Annu Rev Cell Dev Biol 2002, 18:81–105.

3. Huck N, Moore JM, Federer M, Grossniklaus U: The Arabidopsis mutant feronia disrupts the female gametophytic control of pollen tube reception. Development 2003, 130(10):2149–2159.

4. Kim S, Mollet JC, Dong J, Zhang K, Park SY, Lord EM: Chemocyanin, a small basic protein from the lily stigma, induces pollen tube chemotropism. Proc Natl Acad Sci USA 2003, 100(26):16125–16130.

5. Dong J, Kim ST, Lord EM: Plantacyanin plays a role in reproduction in Arabidopsis. Plant Physiol 2005, 138(2):778–789.

6. Mollet JC, Park SY, Nothnagel EA, Lord EM: A lily stylar pectin is necessary for pollen tube adhesion to an in vitro stylar matrix. Plant Cell 2000, 12(9):1737–1750.

7. Cheung AY, Wang H, Wu HM: A floral transmitting tissue-specific glycoprotein attracts pollen tubes and stimulates their growth. Cell 1995, 82(3):383–393.

8. Wu HM, Wong E, Ogdahl J, Cheung AY: A pollen tube growth-promoting arabinogalactan protein from nicotiana alata is similar to the tobacco TTS protein. Plant J 2000, 22(2):165–176.

9. Baker SC, Robinson-Beers K, Villanueva JM, Gaiser JC, Gasser CS: Interactions among genes regulating ovule development in Arabidopsis thaliana. Genetics 1997, 145(4):1109–1124.

10. Palanivelu R, Brass L, Edlund AF, Preuss D: Pollen tube growth and guidance is regulated by POP2, an Arabidopsis gene that controls GABA levels. Cell 2003, 114(1):47–59.

11. Wilhelmi LK, Preuss D: Self-sterility in Arabidopsis due to defective pollen tube guidance. Science 1996, 274(5292):1535–1537.

12. Ray SM, Park SS, Ray A: Pollen tube guidance by the female gametophyte. Development 1997, 124(12):2489–2498.

13. Shimizu KK, Okada K: Attractive and repulsive interactions between female and male gametophytes in Arabidopsis pollen tube guidance. Development 2000, 127(20):4511–4518.

14. Higashiyama T, Yabe S, Sasaki N, Nishimura Y, Miyagishima S, Kuroiwa H, Kuroiwa T: Pollen tube attraction by the synergid cell. Science 2001, 293(5534):1480–1483.

15. Kasahara RD, Portereiko MF, Sandaklie-Nikolova L, Rabiger DS, Drews GN: MYB98 is required for pollen tube guidance and synergid cell differentiation in Arabidopsis. Plant Cell 2005, 17(11):2981–2992.

16. Marton ML, Cordts S, Broadhvest J, Dresselhaus T: Micropylar pollen tube guidance by egg apparatus 1 of maize. Science 2005, 307(5709):573–576.

17. Gilroy S, Trewavas A: Signal processing and transduction in plant cells: the end of the beginning? Nat Rev Mol Cell Biol 2001, 2(4):307–314.

18. Zheng ZL, Yang Z: The Rrop GTPase switch turns on polar growth in pollen. Trends Plant Sci 2000, 5(7):298–303.

19. Gu Y, Fu Y, Dowd P, Li S, Vernoud V, Gilroy S, Yang Z: A Rho family GT-Pase controls actin dynamics and tip growth via two counteracting downstream pathways in pollen tubes. J Cell Biol 2005, 169(1):127–138.

20. Prado AM, Porterfield DM, Feijo JA: Nitric oxide is involved in growth regulation and re-orientation of pollen tubes. Development 2004, 131(11):2707–2714.

21. Reger BJCRPR: Chemotropic responses by pearl millet pollen tubes. Sexual Plant Reproduction 1992, 5:47–56.

22. Higashiyama T, Kuroiwa H, Kawano S, Kuroiwa T: Guidance in vitro of the pollen tube to the naked embryo sac of torenia fournieri. Plant Cell 1998, 10(12):2019–2032.

23. Johnson MA, Lord E: Extracellular Guidance Cues and Intracellular Signaling Pathways that Guide Pollen Tube Growth. In The Pollen Tube. Edited by: Malho R. Heidelberg , Springer-Verlag; 2006:1–20.

24. Cheung AY: Pollen-pistil interactions in compatible pollination. Proc Natl Acad Sci U S A 1995, 92(8):3077–3080.

25. Twell D, Wing R, Yamaguchi J, McCormick S: Isolation and expression of an anther-specific gene from tomato. Mol Gen Genet 1989, 217(2-3):240–245.

26. Kandasamy MK, Nasrallah JB, Nasrallah ME: Pollen Pistil Interactions and Developmental Regulation of Pollen-Tube Growth in Arabidopsis. Development 1994, 120(12):3405–3418.

27. Christensen CA, King EJ, Jordan JR, Drews GN: Megagametogenesis in Arabidopsis wild type and the Gf mutant. Sex Plant Reprod 1997, 10(1):49–64.

28. Christensen CA, Gorsich SW, Brown RH, Jones LG, Brown J, Shaw JM, Drews GN: Mitochondrial GFA2 is required for synergid cell death in Arabidopsis. Plant Cell 2002, 14(9):2215–2232.

29. Swanson WJ, Vacquier VD: The rapid evolution of reproductive proteins. Nat Rev Genet 2002, 3(2):137–144.

30. Ferris PJ, Pavlovic C, Fabry S, Goodenough UW: Rapid evolution of sex-related genes in Chlamydomonas. Proc Natl Acad Sci USA 1997, 94(16):8634–8639.

31. Hall AE, Fiebig A, Preuss D: Beyond the Arabidopsis genome: opportunities for comparative genomics. Plant Physiol 2002, 129(4):1439–1447.

32. Mascarenhas JP, Machlis L: Chemotropic Response of Antirrhinum majus Pollen to Calcium. NATURE 1962, 196(4851):292.

33. Chaubal R, Reger BJ: Calcium in the synergid cells and other regions of pearl-millet ovaries. Sexual Plant Reproduction 1992, 5(1):34–46.

34. Kato A: Heterofertilization exhibited by trifluralin-induced bicellular pollen on diploid and tetraploid maize crosses. Genome 2001, 44(6):1114–1121.

35. Rotman N, Rozier F, Boavida L, Dumas C, Berger F, Faure JE: Female control of male gamete delivery during fertilization in Arabidopsis thaliana. Curr Biol 2003, 13(5):432–436.

36. Johnson MA, von Besser K, Zhou Q, Smith E, Aux G, Patton D, Levin JZ, Preuss D: Arabidopsis hapless mutations define essential gametophytic functions. Genetics 2004, 168(2):971–982.

37. Scholten S, Kranz E: In vitro fertilization and expression of transgenes in gametes and zygotes. Sexual Plant Reproduction 2001, 14:35–40.

38. Lister M: 100 Statistical Tests. New York , SAGE publications; 1993:42.

Copyrights

CITATION

Originally published under the Creative Commons Attribution License. Palanivelu R, Preuss D. Distinct Short-Range Ovule Signals Attract or Repel Arabidopsis Thaliana Pollen Tubes in Vitro. BMC Plant Biology 2006, 6:7 doi:10.1186/1471-2229-6-7.

Index

Printed and bound by CPI Group (UK) Ltd, Croydon, CR0 4YY

23/10/2024

01777695-0008